Introduction to the Analysis of Normed Linear Spaces

AUSTRALIAN MATHEMATICAL SOCIETY LECTURE SERIES

1 Introduction to Linear and Convex Programming, N. CAMERON
2 Manifolds and Mechanics, A. JONES, A GRAY & R. HUTTON
3 Introduction to the Analysis of Metric Spaces, J. R. GILES
4 An Introduction to Mathematical Physiology and Biology, J. MAZUMDAR
5 2-Knots and their Groups, J. HILLMAN
6 The Mathematics of Projectiles in Sport, N. DE MESTRE
7 The Petersen Graph, D. A. HOLTON & J. SHEEHAN
8 Low Rank Representations and Graphs for Sporadic Groups, C. PRAEGER & L. SOICHER
9 Algebraic Groups and Lie Groups, G. LEHRER (ed)
10 Modelling with Differential and Difference Equations, G. FULFORD, P. FORRESTER & A. JONES
11 Geometric Analysis and Lie Theory in Mathematics and Physics, A. L. CAREY & M. K. MURRAY (eds)
12 Foundations of Convex Geometry, W. A. COPPEL
13 Introduction to the Analysis of Normed Linear Spaces, J. R. GILES
14 The Integral: An Easy Approach after Kurzweil and Henstock, L. P. YEE & R. VYBORNY

Australian Mathematical Society Lecture Series. 13

Introduction to the Analysis of Normed Linear Spaces

J. R. Giles

Department of Mathemtics

University of Newcastle, Australia

CAMBRIDGE UNIVERSITY PRESS
Cambridge, New York, Melbourne, Madrid, Cape Town,
Singapore, São Paulo, Delhi, Tokyo, Mexico City

Cambridge University Press
The Edinburgh Building, Cambridge CB2 8RU, UK

Published in the United States of America by Cambridge University Press, New York

www.cambridge.org
Information on this title: www.cambridge.org/9780521653756

First published 2000

A catalogue record for this publication is available from the British Library

ISBN 978-0-521-65375-6 Paperback

for Zeny

wife, mother and
γιαγια

CONTENTS

PREFACE

This text is designed as a basic course in functional analysis for senior undergraduate or beginning postgraduate students. For students completing their final undergraduate year, it is aimed at providing some insight into basic abstract analysis which more than ever, is the contextual language of much modern mathematics. For postgraduate students it is aimed at providing a foundation and stimulus for their further research development.

It is assumed that the student will be familiar with real analysis and have some background in linear algebra and complex analysis. It is also assumed that the student will have studied a course in the analysis of metric spaces such as that given in the author's text

Introduction to the Analysis of Metric Spaces, Cambridge University Press, 1987.

Reference to this text will be made under the abbreviation AMS § __ .

In AMS, most of the example spaces introduced are normed linear spaces and many of the implications of linear structure were explored. For example when closure in metric spaces was discussed it was natural to consider the closure of linear subspaces in normed linear spaces and when continuity was considered it was logical to study the continuity of linear mappings on normed linear spaces. In order to make this text as self-contained as possible, the example spaces are again introduced and the elementary properties of normed linear spaces are treated but in a more sophisticated way. For example, in AMS the properties of finite dimensional normed linear spaces were deduced in a contrived manner from compactness properties of the real numbers, but here the properties are established in a more expeditious manner using topological isomorphism with Euclidean space. Nevertheless, this text may be regarded as a sequel to AMS especially in that we assume all those properties generally associated with the metric topology of the space.

In AMS we were concerned with the particular topological structure of normed linear spaces as metric spaces. The cohesive theme in this text is with the structural properties of normed linear spaces in general, associated especially with dual spaces and the algebra of continuous linear operators on normed linear spaces. The course given here depends fundamentally on the Hahn–Banach Theorem which assumes the Axiom of Choice.

The author follows the approach that a great deal of the analysis of normed linear spaces can be handled with a knowledge of metric space topology and is usefully done so before the student has studied general topology. In fact the wide applicability of the analysis of normed linear spaces would argue for this approach. He would suggest that from an analysis point of view the most useful application of the analysis of general topological spaces is in the analysis of linear topological spaces which, for a proper understanding, requires a knowledge of normed linear space theory.

However, to follow this approach means that some topics have to be omitted, in particular any discussion of weak topologies, which includes the Banach–Alaoglu Theorem and compactness characterisations of reflexivity. But such material is naturally included in a subsequent course on the analysis of linear topological spaces.

Spaces involving the Lebesgue integral are mentioned in the text but the student needs no background knowledge of integration theory for this course and references to the Lebesgue integral can be glossed over without loss of understanding.

A glance at the table of contents shows that the text consists mostly of well established material. However, at this stage of his mathematical education the student needs occasionally to be made aware of the great problem areas in the subject and to glimpse the frontiers where recent advances have been made. It is important that the serious student feel that he is studying a subject which is still on the move and one where he may be able to follow where the current ground is being broken.

So at several points the student is introduced to material which is more recent and is given a guide to literature where major problems have been tackled. For example, in Section 8 the Bishop–Phelps Theorem is proved; this is a result which is constantly used by researchers in the analysis of Banach spaces. When discussing Schauder bases in Section 7 it is of interest to mention the Basis Problem and its solution by Enflo. When treating the representation of compact operators by finite rank operators in Section 15 it is useful to refer to Enflo's contribution. The Invariant Subspace Problem arises naturally in Spectral Theory and it is convenient to give a proof of Lomonosov's Theorem for compact operators in Section 18.

A lecture course using this text can be given in between 25 and 30 lecture hours given an adequate extra tutorial program and provided the students have sufficient background. The material can be tailored to suit such a course length by omitting Section 8 completely and Chapter VI on Spectral Theory.

At the end of each section there is a set of exercises which follows the order of presentation of material in the section.

At the end of the text there is an appendix which includes the set theory results used at various points of the course.

Historical notes are included to show how the subject developed from late nineteenth century beginnings into the analysis of an abstract structure using the axiomatic method. Two recent advances are included to demonstrate the continued vigour of research in the area.

As with AMS, this text has a full index. Rarely are students told that when a mathematician uses a textbook he regularly consults its index. So the index in this text gives references to all the significant places where a particular concept is used.

There are many texts in functional analysis which cover the material presented in this text book in a variety of different ways.

The great standard text for many years was

Angus E. Taylor, *Introduction to functional analysis*, John Wiley, 1958

which was updated by Angus E. Taylor and David C. Lay in 1980. This is a great reference but is generally pitched at more advanced students than our text.

A text more accessible to students of our course is

Erwin Kreysig, *Introductory functional analysis with applications*, John Wiley, 1978.

This text is very popular with engineers and does not rely to any great extent on previous knowledge of general topology.

The text which had a formative influence on the author is

A.L. Brown and A. Page, *Elements of functional analysis*, van Nostrand Reinhold, 1970.

This text is pitched at the same level as ours and does not have a general topology or general integration prerequisite. Our text often gives more detail and introduces recent results.

An expansive text is

George Bachman and Lawrence Narici, *Functional Analysis*, Academic Press, 1966.

But this text covers spectral theory with greater detail and depth.

The text

George F. Simmons, *Introduction to topology and modern analysis*, McGraw–Hill, 1963

is very readable, but it is written with more of a general topology background.

An excellent text, but one also written from a topological point of view is

G.J.O. Jameson, *Topology and normed linear spaces*, Chapman–Hall, 1974.

There are more modern texts such as

Walter Rudin, *Functional Analysis*, McGraw–Hill, 2nd edn, 1991.

This presents a far more sophisticated course than ours, but will be found to be a useful reference.

There are two modern British texts which should be mentioned.

Nicholas Young, *An introduction to Hilbert space*, Cambridge University Press, 1988

concentrates on Hilbert space with applications in mind.

Béla Bollobás, *Linear Analysis*, Cambridge University Press, 1990.

This contains a very fine lecture course and does refer to new results. It moves somewhat faster than our text and with less detail than many students would find comfortable.

I have been influenced by the text

J.R. Ringrose, *Compact non-self-adjoint operators*, van Nostrand Reinhold, 1971

in my approach to spectral theory for compact operators.

The author has given lectures on a course such as this to third year honours students at the University of Newcastle for a period of more than ten years. The lecture notes were first produced in duplicated form but have been modified and expanded considerably in the light of experience, to the form presented here.

Thanks are due to my colleagues and students in the department for their conversations over many years which have had their effect on the final result. My thanks to Philip Charlton for producing the diagrams which appear in the text. I am particularly indebted to Jan Garnsey who has so patiently and competently typed and retyped the final copy from my handwritten manuscript. I am grateful to Brailey Sims for supplying source material for the historical notes.

J.R. Giles
The University of Newcastle
NSW Australia.

I. NORMED LINEAR SPACE STRUCTURE AND EXAMPLES

Everyday Euclidean space which underlies most of our mathematical activity, possesses some remarkable abstract structures which provide a rewarding study in their own right. In particular we could isolate the linear space structure and the metric space structure and the one which is a significantly rich mixture of these two, the normed linear space structure. To study such an abstract structure is profitable for two reasons: the study can be undertaken in a systematic and deductive fashion and there are many diverse mathematical situations which exhibit the same structure to which the developed theory can be applied.

Moreover, the study of normed linear spaces has an intriguing interest which is quite distinct from that of metric spaces. Just as the study of Euclidean space gives rise to matrix theory so the linear structure of normed linear spaces enables us to propagate more normed linear spaces from the continuous linear mappings between them and this is the origin of operator theory.

§1. BASIC PROPERTIES OF NORMED LINEAR SPACES

We begin with a review of the defining structure of normed linear spaces, of the fundamental properties of continuous linear mappings and of the notions of basis in normed linear spaces. Our theory is developed from a knowledge of these and we will use them in our discussion of examples and our subsequent construction of associated normed linear spaces in Chapter II.

1.1 **Definition.** Given a linear space X over $\mathbb{C}$ (or $\mathbb{R}$), a mapping $\|\cdot\|: X \to \mathbb{R}$ is a *norm* for X if it satisfies the following properties:
For all $x \in X$,
(i) $\|x\| \geq 0$,
(ii) $\|x\| = 0$ if and only if $x = 0$,
(iii) $\|\lambda x\| = |\lambda| \, \|x\|$ for all scalar λ,
and for all $x, y \in X$,
(iv) $\|x+y\| \leq \|x\| + \|y\|$.

(The norm is said to assign a "length" to each vector in X.) The pair $(X, \|\cdot\|)$ is called a *normed linear space.* Different norms can be assigned to the same linear space giving rise to different normed linear spaces.

1.2 **Remarks.** The norm generates a special metric on the linear space. Given a normed linear space $(X, \|\cdot\|)$, a function $d: X \times X \to \mathbb{R}$ defined by

$$d(x,y) = \| x-y \|$$

is a metric for X and is called the *metric generated by the norm.*
This metric is an *invariant metric* for X; that is, for any given $z \in X$,

$$d(x+z, y+z) = d(x,y) \qquad \text{for all } x,y \in X.$$

So given $x \in X$ and $r > 0$, the ball

$$B(x; r) \equiv \{y \in X : \| y-x \| < r\} = x + \{z \in X : \| z \| < r\} \equiv x + B(0; r),$$

is the translate by x of the ball $B(0; r)$.
Furthermore, property (iii) implies that given $r > 0$, the ball centred at the origin with radius r,

$$B(0; r) = r\{y \in X : \| y \| < 1\} \equiv r\, B(0; 1),$$

is an r–multiple of the unit ball.
Properties (iii) and (iv) tell us that the norm is a *convex function* on X; that is,
for any $x,y \in X$ we have $\| \lambda x+(1-\lambda)y \| \le \lambda \| x \| + (1-\lambda) \| y \|$ for all $0 \le \lambda \le 1$.
This implies that the ball $B(0; 1)$ is a *convex set* in X; that is,
for any $x,y \in B(0; 1)$ we have $\lambda x + (1-\lambda)y \in B(0; 1)$ for all $0 \le \lambda \le 1$.
But (iii) also tells us that the ball $B(0; 1)$ is *symmetric*; that is,
for any $x \in B(0; 1)$, we have $-x \in B(0; 1)$. □

Since a normed linear space has both algebra and analysis structures, the relationship between these two quite different aspects of the space is a matter of special interest. The fruitfulness of the relationship follows from the way the linear space operations are linked to the norm.

1.3 **Remark**. Given a normed linear space $(X, \|\cdot\|)$, from properties (iii) and (iv) we can derive the important norm inequality

$$\big| \| x \| - \| y \| \big| \le \| x-y \| \quad \text{for all } x,y \in X$$

which implies that the norm is a continuous function on X.
Properties (iii) and (iv) actually relate addition of vectors and multiplication of a vector by a scalar to the norm and imply that these algebraic operations have a continuity property which we call *joint continuity* :
if $x \to x_0$ and $y \to y_0$ then $x + y \to x_0 + y_0$, and if $\lambda \to \lambda_0$ then $\lambda x \to \lambda_0 x_0$.
This can be deduced simply from the inequalities:

$$\| (x+y) - (x_0+y_0) \| \le \| x-x_0 \| + \| y-y_0 \|,$$

and $\| \lambda x-\lambda_0 x_0 \| \le \| \lambda x-\lambda x_0 \| + \| \lambda x_0-\lambda_0 x_0 \| \le |\lambda| \| x-x_0 \| + |\lambda-\lambda_0| \| x_0 \|$. □

A normed linear space is a generalisation of Euclidean space.

1.4 **Example.** *Euclidean n-space and Unitary n-space.* The real (complex) linear space of ordered n-tuples of real numbers $\mathbb{R}^n$ (of complex numbers $\mathbb{C}^n$), where for $x \equiv (\lambda_1, \lambda_2, \ldots, \lambda_n)$ we define the norm

$$\|x\|_2 = \sqrt{\sum_{k=1}^{n} |\lambda_k|^2}$$

in the real case is called *Euclidean n-space* and is denoted by $(\mathbb{R}^n, \|\cdot\|_2)$ and in the complex case is called *Unitary n-space* and is denoted by $(\mathbb{C}^n, \|\cdot\|_2)$. The n-dimensional complex linear space $\mathbb{C}^n$ is a 2n-dimensional real linear space $\mathbb{C}^n(\mathbb{R})$ and it is clear that the norm calculation is not affected by the space being considered as over $\mathbb{R}$ or $\mathbb{C}$. An elegant proof that the norm satisfies the triangle inequality (property (iv)) follows from an exploration of the inner product structure of the space; (see Example 2.2.10 below). □

Other normed linear spaces are formed by taking different norms on the same underlying linear space $\mathbb{R}^n$ (or $\mathbb{C}^n$). The advantage of these norms which at first offend our Euclidean intuition, is that they are often more convenient for computation.

1.5 **Examples.**

(i) $(\mathbb{R}^n, \|\cdot\|_\infty)$, (or $(\mathbb{C}^n, \|\cdot\|_\infty)$).

For $x \equiv (\lambda_1, \lambda_2, \ldots, \lambda_n) \in \mathbb{R}^n$ (or $\mathbb{C}^n$) we define the norm

$$\|x\|_\infty = \max\{ |\lambda_k| : k \in \{1, 2, \ldots, n\}\}.$$

(ii) $(\mathbb{R}^n, \|\cdot\|_1)$, (or $(\mathbb{C}^n, \|\cdot\|_1)$).

For $x \equiv (\lambda_1, \lambda_2, \ldots, \lambda_n) \in \mathbb{R}^n$ (or $\mathbb{C}^n$) we define the norm

$$\|x\|_1 = \sum_{k=1}^{n} |\lambda_k|.$$

In both cases it is much simpler than in the Euclidean case to verify that these norms satisfy the norm properties (i)–(iv). □

Several example spaces can be considered as different forms of a general function space.

1.6 **Examples.** For any nonempty set X, the set $\mathcal{B}(X)$ of bounded real (complex) functions on X is a linear space under pointwise definition of the linear operations and is a normed linear space with norm defined by

$$\|f\|_\infty = \sup\{|f(x)| : x \in X\}.$$

(i) When $X \equiv \{1, 2, \ldots, n\}$ then $\mathcal{B}(X)$ is the linear space $\mathbb{R}^n$(or $\mathbb{C}^n$) and for $x \equiv (\lambda_1, \lambda_2, \ldots, \lambda_n) \in \mathbb{R}^n$ (or $\mathbb{C}^n$),

$$\|x\|_\infty = \max\{|\lambda_k| : k \in \{1, 2, \ldots, n\}\}.$$

(ii) When $X \equiv \mathbb{N}$ the set of natural numbers, then $\mathcal{B}(\mathbb{N})$ is the linear space m (or ℓ_∞) of bounded sequences of scalars and for $x \equiv \{\lambda_1, \lambda_2, \ldots, \lambda_n, \ldots\} \in m$,

$$\| x \|_\infty = \sup \{ |\lambda_k| : k \in \mathbb{N} \}.$$

(iii) When $X \equiv [a,b]$ the bounded closed interval, then $\mathcal{B}[a,b]$ is the linear space of bounded scalar functions on $[a,b]$ and for $f \in \mathcal{B}[a,b]$,

$$\| f \|_\infty = \sup \{ |f(x)| : x \in [a,b] \}. \qquad \square$$

1.7 **Definition.** A normed linear space which is complete as a metric space with its metric generated by the norm, is called a *Banach space.*

Using the completeness of the scalar field we establish the completeness of the general example given in Example 1.6. This also illustrates the general method used to prove completeness in other examples.

1.8 **Theorem.** *For any nonempty set* X, *the normed linear space* $(\mathcal{B}(X), \|\cdot\|_\infty)$ *is a Banach space.*

Proof. Consider a Cauchy sequence $\{f_n\}$ in $(\mathcal{B}(X), \|\cdot\|_\infty)$; then given $\varepsilon > 0$ there exists a $\nu \in \mathbb{N}$ such that

$$\| f_n - f_m \|_\infty < \varepsilon \quad \text{for all } m,n > \nu;$$

that is, $\sup\{ |(f_n - f_m)(x)| : x \in X\} < \varepsilon$ for all $m,n > \nu$.

But then for each $x \in X$,

$$|f_n(x) - f_m(x)| < \varepsilon \quad \text{for all } m,n > \nu;$$

that is, for each $x \in X$, $\{f_n(x)\}$ is a Cauchy sequence of scalars. Since the scalar field is complete, for each $x \in X$ we can define a function f on X by

$$f(x) = \lim_{n\to\infty} f_n(x).$$

We need to show that $f \in \mathcal{B}(X)$.

Since $\{f_n\}$ is Cauchy it is bounded in $(\mathcal{B}(X), \|\cdot\|_\infty)$; that is, there exists a $K > 0$ such that

$$\| f_n \|_\infty < K \quad \text{for all } n \in \mathbb{N},$$

which implies that $|f_n(x)| < K$ for all $x \in X$ and all $n \in \mathbb{N}$.

Therefore, $|f(x)| \leq K$ for all $x \in X$, and so $f \in \mathcal{B}(X)$.

Then we need to show that $\{f_n\}$ actually converges to f in $(\mathcal{B}(X), \|\cdot\|_\infty)$.

We had for each $x \in X$,

$$|f_n(x) - f_m(x)| < \varepsilon \quad \text{for all } m,n > \nu.$$

Fix n and let $m \to \infty$; then we have for each $x \in X$,

$$|f_n(x) - f(x)| \leq \varepsilon \quad \text{for all } n > \nu.$$

So $\| f_n - f \|_\infty = \sup\{ |f_n(x) - f(x)| : x \in X\} \leq \varepsilon$ for all $n > \nu$;

that is, $\{f_n\}$ converges to f in $(\mathcal{B}(X), \|\cdot\|_\infty)$. $\square$

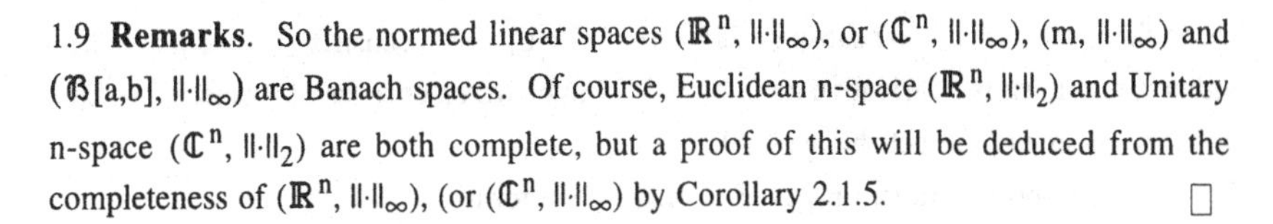
1.9 **Remarks**. So the normed linear spaces $(\mathbb{R}^n, \|\cdot\|_\infty)$, or $(\mathbb{C}^n, \|\cdot\|_\infty)$, $(m, \|\cdot\|_\infty)$ and $(\mathcal{B}[a,b], \|\cdot\|_\infty)$ are Banach spaces. Of course, Euclidean n-space $(\mathbb{R}^n, \|\cdot\|_2)$ and Unitary n-space $(\mathbb{C}^n, \|\cdot\|_2)$ are both complete, but a proof of this will be deduced from the completeness of $(\mathbb{R}^n, \|\cdot\|_\infty)$, (or $(\mathbb{C}^n, \|\cdot\|_\infty)$ by Corollary 2.1.5. □

We will see, as we develop the fundamental theorems in our theory, that completeness plays a strategically important role.

A linear subspace in a normed linear space derives norm structure from the parent space in a natural way.

1.10 **Definition**. Given a normed linear space $(X, \|\cdot\|)$ and a linear subspace Y of X, it is clear that the restriction of the norm $\|\cdot\|$ to Y is also a norm for Y. The restriction is denoted $\|\cdot\|_Y$ and $(Y, \|\cdot\|_Y)$ is a *normed linear subspace* of $(X, \|\cdot\|)$.

The following examples are significant linear subspaces of the example spaces so far introduced.

1.11 **Examples**. In m the linear space of bounded sequences, where for each $x \equiv \{\lambda_1, \lambda_2, \ldots, \lambda_n, \ldots\}$ there exists an $M_x > 0$ such that $|\lambda_n| \le M_x$ for all $n \in \mathbb{N}$,
we have the following linear subspaces.
c the linear subspace of convergent sequences, where for each
$x \equiv \{\lambda_1, \lambda_2, \ldots, \lambda_n, \ldots\}$ there exists a scalar λ such that $\lambda_n \to \lambda$ as $n \to \infty$,
c_0 the linear subspace of sequences which converge to zero,
ℓ_2 the linear subspace of sequences $x \equiv \{\lambda_1, \lambda_2, \ldots, \lambda_n, \ldots\}$ where $\sum |\lambda_n|^2 < \infty$,
ℓ_1 the linear subspace of sequences $x \equiv \{\lambda_1, \lambda_2, \ldots, \lambda_n, \ldots\}$ where $\sum |\lambda_n| < \infty$, and
E_0 the linear subspace of sequences with only a finite number of nonzero entries, where for each $x \equiv \{\lambda_1, \lambda_2, \ldots, \lambda_n, \ldots\}$, $\lambda_n = 0$ for all except a finite number of elements in $\mathbb{N}$.
That ℓ_2 and ℓ_1 are linear subspaces will be verified in Example 2.2.11 and 2.3.6.
Now all of these are normed linear subspaces of $(m, \|\cdot\|_\infty)$. □

1.12 **Examples**. For a metric space (X, d), consider $\mathcal{B}(X, d)$ the linear space of bounded functions on (X, d).
We have the important linear subspace $\mathcal{BC}(X, d)$ of bounded continuous functions on (X, d). When (X, d) is compact, then all continuous functions on (X, d) are bounded so $\mathcal{BC}(X, d) = \mathcal{C}(X, d)$ the linear subspace of continuous functions on (X, d).
When $X \equiv [a,b]$, we could also consider the linear subspace $\mathcal{R}[a,b]$ of $\mathcal{B}[a,b]$ which consists of the Riemann integrable functions on $[a,b]$. Of course, $\mathcal{C}[a,b]$ is a linear subspace of $\mathcal{R}[a,b]$.

We could consider the linear subspace $\mathcal{C}^1[a,b]$ of $\mathcal{C}[a,b]$ which consists of those functions with a continuous first derivative on [a,b] and $\mathcal{C}^\infty[a,b]$ which consists of functions which are infinitely often differentiable on [a,b].
We could also form linear subspaces of $\mathcal{B}(X, d)$ of the form

$$\{f \in \mathcal{B}(X, d) : f(x_0) = 0\} \text{ for a given } x_0 \in X.$$

So we would have $\mathcal{B}_0[0,1]$ the linear subspace $\{f \in \mathcal{B}[0,1] : f(0) = 0\}$.
We would also have $\mathcal{C}_0[0,1]$ the linear subspace $\{f \in \mathcal{C}[0,1] : f(0) = 0\}$.
Again all of these are normed linear subspaces of the appropriate bounded function space with the supremum norm $\|\cdot\|_\infty$. □

Often the simplest way to examine a normed linear space for completeness is to use the following metric space link between completeness and closedness, (see AMS §4).

1.13 **Proposition**. *Consider a metric space* (X, d) *and a subset* Y.
(i) *If* $(Y, d|_Y)$ *is complete then* Y *is a closed subset of* (X, d).
(ii) *If* (X, d) *is complete and* Y *is a closed subset of* (X, d) *then* $(Y, d|_Y)$ *is complete.*

1.14 **Example**. Given a metric space (X, d), the normed linear space $(\mathcal{BC}(X, d), \|\cdot\|_\infty)$ is complete.

Proof. We show that $\mathcal{BC}(X, d)$ is a closed subset of $(\mathcal{B}(X, d), \|\cdot\|_\infty)$ and apply Proposition 1.13.
Given a cluster point f of $\mathcal{BC}(X, d)$ in $(\mathcal{B}(X, d), \|\cdot\|_\infty)$ we show that f is continuous:
Now there exists a sequence $\{f_n\}$ in $\mathcal{BC}(X, d)$ convergent to f ; that is, given $\varepsilon > 0$ there exists a $\nu \in \mathbb{N}$ such that

$$\| f_n - f \|_\infty < \varepsilon \quad \text{when } n > \nu.$$

Consider $x_0 \in X$. Since $f_{\nu+1}$ is continuous at x_0 there exists a $\delta > 0$ such that

$$| f_{\nu+1}(x) - f_{\nu+1}(x_0) | < \varepsilon \quad \text{when } d(x,x_0) < \delta.$$

Therefore,

$$\begin{aligned} | f(x) - f(x_0) | &\leq | f(x) - f_{\nu+1}(x) | + | f_{\nu+1}(x) - f_{\nu+1}(x_0) | + | f_{\nu+1}(x_0) - f(x_0) | \\ &\leq 2 \| f_{\nu+1} - f \|_\infty + | f_{\nu+1}(x) - f_{\nu+1}(x_0) | \\ &< 3\varepsilon \qquad \text{when } d(x,x_0) < \delta; \end{aligned}$$

that is, f is continuous at x_0. □

When (X, d) is compact, the Banach space $\mathcal{C}(X, d)$ has a particularly rich structure, (see AMS §9).

We often encounter situations where we have a linear space with a real function which has all the norm properties except (ii).

1.15 Definition. Given a linear space X over $\mathbb{C}$ (or $\mathbb{R}$), a mapping p: $X \to \mathbb{R}$ is a *seminorm* for X if it satisfies all the norm properties except (ii) and instead of (ii) satisfies

(ii)' $\quad p(x) = 0$ if $x = 0$.

The pair (X,p) is called a *seminormed linear space.*

So a seminorm allows the possibility that there exists some $x \neq 0$ for which $p(x) = 0$.

1.16 Remark. A seminorm p on a linear space X generates a semimetric e on X defined by

$$e(x,y) = p(x-y).$$

As with the norm properties noted in Remarks 1.2, property (iii) implies that the set $\{x \in X : p(x) < 1\}$ has the properties that, given $r > 0$

$$\{y \in X : p(y-x) < r\} = x + r\{z \in X : p(z) < 1\}$$

and is convex and symmetric.

However, if p is not a norm then ker p is a nontrivial linear subspace and is a subset of $\{x \in X : p(x) < 1\}$. □

Given a linear space and a proper linear subspace, another linear space can be generated in a natural way as a quotient space.

1.17 Definition. Given a linear space X and a proper linear subspace M of X, the *quotient space* (or *factor space*) X/M is the linear space of *cosets* $[x] \equiv x + M$, with addition and multiplication by a scalar defined by

$$[x] + [y] \equiv (x + M) + (y + M) = (x + y) + M \equiv [x+y] \quad \text{for all } [x],[y] \in X/M$$

and $\quad \lambda[x] \equiv \lambda(x+M) = \lambda x + M \equiv [\lambda x] \quad$ for all $[x] \in X/M$ and scalar λ.

A seminormed linear space can be transformed into a normed linear space as a quotient space.

1.18 Definition. Given a seminormed linear space (X, p), the *associated normed linear space* is the quotient space X/ker p whose elements are the cosets $[x] \equiv x+\ker p$, with norm

$$\| [x] \| = p(x) \qquad \text{for any } x \in [x].$$

1.19 Example. Consider $\mathcal{R}[a,b]$ the linear space of Riemann integrable functions on [a,b]. Now the function p_1: $\mathcal{R}[a,b] \to \mathbb{R}$ defined by

$$p_1(f) = \int_a^b | f(t) | \, dt$$

is a seminorm on $\mathcal{R}[a,b]$.

On the quotient space $\mathcal{R}[a,b]/\ker p_1$ of cosets $[f] = f+\ker p_1$, we have the associated norm

$$\| [f] \|_1 = p_1(f) \quad \text{for any } f \in [f]. \qquad \square$$

When a quotient space is generated from a normed linear space by a proper closed linear subspace then the quotient space has an associated quotient norm.

1.20 **Definition.** Given a normed linear space $(X, \|\cdot\|)$, and a proper closed linear subspace M of $(X, \|\cdot\|)$, the *quotient norm* $\|\cdot\|$ is defined on the quotient space X/M by

$$\| [x] \| = d(0, x+M) = \inf\{ \| x+m \| : m \in M\}.$$

1.21 **Remark**. It is routine to verify the norm properties (i)–(iv) for the quotient norm. But we note that we need the linear subspace M to be closed to establish that the norm is not just a seminorm; that is, we need to show that $d(0, x+M) = 0$ implies that $x + M = M$ so that $\| [x] \| = 0$ implies that $[x] = 0$. $\square$

It is instructive to see how such a quotient space inherits completeness from the parent space.

1.22 **Theorem**. *Given a Banach space* $(X, \|\cdot\|)$ *and a proper closed linear subspace* M *then the quotient space* $(X/M, \|\cdot\|)$ *is also complete.*

Proof. Consider a Cauchy sequence $\{x_n+M\}$ in $(X/M, \|\cdot\|)$. Then for each $k \in \mathbb{N}$ there exists a $\nu(k) \in \mathbb{N}$ such that

$$\| (x_n + M) - (x_m + M) \| < \frac{1}{2^k} \quad \text{for all } m,n > \nu(k).$$

Consider a subsequence of the form $\{x_{n(k)} + M\}$ where for each $k \in \mathbb{N}$, $n(k) > \nu(k)$.

Then $$\| (x_{n(k)} + M) - (x_{n(k+1)} + M) \| < \frac{1}{2^k}\,.$$

For each $k \in \mathbb{N}$ choose $x_k \in x_{n(k)} + M$ such that

$$\| x_k - x_{k+1} \| < \frac{1}{2^k}\,.$$

The sequence $\{x_k\}$ in $(X, \|\cdot\|)$ has the property that

$$\begin{aligned} \| x_k - x_m \| &\le \| x_k - x_{k+1} \| + \dots + \| x_{m-1} - x_m \| \qquad \text{for } m > k \\ &< \frac{1}{2^{k-1}} \qquad \text{for all } k \in \mathbb{N}, \end{aligned}$$

so $\{x_k\}$ is a Cauchy sequence in $(X, \|\cdot\|)$. But $(X, \|\cdot\|)$ is complete so there exists an $x \in X$ such that $\{x_k\}$ is convergent to x. Then

$$\| (x_{n(k)}+M) - (x+M) \| = \| (x_k+M) - (x+M) \| \le \| x_k - x \|$$

and so $\{x_{n(k)}+M\}$ is convergent to $x + M$ in $(X/M, \|\cdot\|)$.

However, $\{x_{n(k)}+M\}$ is a convergent subsequence of the original Cauchy sequence $\{x_n+M\}$, so $\{x_n+M\}$ is also convergent in $(X/M, \|\cdot\|)$. $\square$

1.23 **Example**. Consider $\mathcal{R}[a,b]$ the linear space of Riemann integrable functions on $[a,b]$ with norm

$$\| f \|_\infty = \sup\{ | f(t) | : t \in [a,b] \}.$$

Now $(\mathcal{R}[a,b], \|\cdot\|_\infty)$ is a Banach space, (see AMS §4).

Furthermore, $\ker p_1$ where

$$p_1(f) = \int_a^b | f(t) | \, dt$$

is a closed linear subspace of $(\mathcal{R}[a,b], \|\cdot\|_\infty)$. On the quotient space $\mathcal{R}[a,b]/\ker p_1$ whose elements are cosets, $[f] = f + \ker p_1$, we have the norm

$$\| [f] \|_\infty = d(0, f+\ker p_1) = \inf\{ \| f+g \|_\infty : g \in \ker p_1 \} .$$

From Theorem 1.22, the quotient space $(\mathcal{R}[a,b]/\ker p_1, \|\cdot\|_\infty)$ is a Banach space. □

1.24 **Continuous linear mappings**

The algebraic study of linear spaces finds its full development in an examination of the homomorphisms or structure preserving mappings between such spaces. As we might expect, these linear mappings are of similar significance in the development of the analysis of normed linear spaces. However, interest in this case is focused on the continuous linear mappings which preserve the topological and norm structure along with the linear structure.

The following characterisation theorem provides an essential tool for discussing these mappings.

1.24.1 **Theorem**. *Given normed linear spaces* $(X, \|\cdot\|)$ *and* $(Y, \|\cdot\|')$, *a linear mapping* $T: X \to Y$

(i) *is continuous if and only if there exists an* $M > 0$ *such that*

$$\| Tx \|' \le M \| x \| \quad \textit{for all } x \in X,$$

(ii) *has a continuous inverse on* $T(X)$ *if and only if there exists an* $m > 0$ *such that*

$$m \| x \| \le \| Tx \|' \quad \textit{for all } x \in X.$$

Proof.

(i) If the condition holds then clearly T is continuous at 0 and the linearity of T implies that T is continuous on X.

Conversely, suppose that the condition does not hold; that is, for each $n \in \mathbb{N}$ there exists $x_n \in X$ such that

$$\| Tx_n \|' > n \| x_n \|.$$

Then
$$\left\| \frac{1}{n} \frac{x_n}{\| x_n \|} \right\| = \frac{1}{n} \to 0 \text{ as } n \to \infty,$$

but
$$\left\| T \left(\frac{1}{n} \frac{x_n}{\| x_n \|} \right) \right\| > 1 \quad \text{for all } n \in \mathbb{N};$$

that is, T is not continuous at 0.

(ii) If there exists an $m > 0$ such that

$$m \| x \| \leq \| Tx \|' \qquad \text{for all } x \in X.$$

then $\ker T = \{0\}$ so T is one-to-one and T^{-1} exists on $T(X)$ and is clearly linear. Writing $x = T^{-1}y$ for $y \in T(X)$ we have

$$m \| T^{-1}y \| \leq \| T(T^{-1}y) \|' = \| y \|'$$

so

$$\| T^{-1} y \| \leq \frac{1}{m} \| y \|' \qquad \text{for all } y \in T(X);$$

that is, from (i) T^{-1} is continuous on $T(X)$.

Conversely, if T^{-1} is a continuous linear mapping on $T(X)$ then, from (i) there exists an $M > 0$ such that

$$\| T^{-1}y \| \leq M \| y \| \qquad \text{for all } y \in T(X).$$

Since $y = Tx$,

$$\| T^{-1}(Tx) \| \leq M \| Tx \|'$$

and so

$$\frac{1}{M} \| x \| \leq \| Tx \|' \qquad \text{for all } x \in X.$$

□

1.24.2 **Remark.** From linearity it follows that a linear mapping T is continuous on a normed linear space $(X, \|\cdot\|)$ if and only if T is continuous at any one point of X. It follows that a linear mapping T is either continuous at every point of X or continuous at no point of X. □

Two of the most commonly occurring linear mappings are as follows.

1.24.3 **Examples.**

Consider the normed linear spaces $(\mathfrak{C}[0,1], \|\cdot\|_\infty)$ and $(\mathfrak{C}^1[0,1], \|\cdot\|_\infty)$.

(i) Consider the linear mapping $I: \mathfrak{C}[0,1] \to \mathfrak{C}^1[0,1]$ defined by

$$I(f)(x) = \int_0^x f(t)\, dt \qquad \text{for all } x \in [0,1].$$

Now

$$| I(f)(x) | \leq | x | \| f \|_\infty \qquad \text{for all } x \in [0,1]$$

and

$$\| I(f) \|_\infty = \max\{| I(f)(x) | : x \in [0,1]\} \leq \| f \|_\infty \text{ for all } f \in \mathfrak{C}[0,1],$$

so I is continuous.

(ii) Consider the linear mapping $D: \mathfrak{C}^1[0,1] \to \mathfrak{C}[0,1]$ defined by

$$D(f)(x) = f'(x) \text{ for all } x \in [0,1].$$

For the sequence $\{f_n\}$ in $\mathfrak{C}^1[0,1]$ where $f_n(x) = \frac{1}{n} \sin n\pi x$ we have $\| f_n \|_\infty = \frac{1}{n} \to 0$ but $\| D(f_n) \|_\infty \doteq 1$ for all $n \in \mathbb{N}$, so D is not continuous. □

1.24.4 **Definitions**. Consider normed linear spaces $(X, \|\cdot\|)$ and $(Y, \|\cdot\|')$ and a linear mapping $T: X \to Y$.

(i) T is said to be a *topological isomorphism* (or a *linear homeomorphism*) if T is also a homeomorphism; that is, T is linear, continuous, invertible and has a continuous inverse on $T(X)$.

$(X, \|\cdot\|)$ and $(Y, \|\cdot\|')$ are said to be *topologically isomorphic* if there exists a topological isomorphism of X onto Y.

(ii) T is said to be an *isometric isomorphism* if

$$\| Tx \|' = \| x \| \quad \text{for all } x \in X.$$

$(X, \|\cdot\|)$ and $(Y, \|\cdot\|')$ are said to be *isometrically isomorphic* if there exists an isometric isomorphism of X onto Y.

A continuous linear mapping with an algebraic inverse is not necessarily a topological isomorphism.

1.24.5 **Example**. Consider Example 1.24.3. The continuous linear mapping

$$I(f)(x) = \int_0^x f(t)\, dt \qquad \text{for all } x \in [0,1]$$

is a one-to-one mapping of $\mathfrak{C}[0,1]$ onto $\mathfrak{C}_0^1[0,1]$ and has an algebraic inverse

$$D(f)(x) = f'(x) \quad \text{for all } x \in [0,1].$$

But D is not continuous so I is not a topological isomorphism. □

Nevertheless, we will see in Section 10, that there are special circumstances under which we can conclude that a continuous linear mapping with an algebraic inverse is a topological isomorphism.

1.24.6 **Remark**. Not every isometry of a normed linear space into a normed linear space is linear even if it maps 0 to 0.

For example, consider the mapping T of a complex normed linear space $(X, \|\cdot\|)$ into itself defined by

$$T(\lambda x) = \bar{\lambda} T(x) \quad \text{for all } x \in X \text{ and scalar } \lambda.$$

However, it was shown by Stefan Mazur and Stanislaus Ulam in 1932 that an isometry of a real normed linear space into a real normed linear space which maps 0 to 0 is linear; (see Stefan Banach, *Théorie des opérations linéaires*, Chelsea reprint 1932 edition, p.166, and Exercise 2.4.13). □

We deduce the following characterisation of topological isomorphisms from Theorem 1.24.1.

1.24.7 **Corollary**. *Given normed linear spaces* $(X, \|\cdot\|)$ *and* $(Y, \|\cdot\|')$, *a linear mapping* $T: X \to Y$ *is a topological isomorphism if and only if there exist* $m, M > 0$ *such that*

$$m \| x \| \le \| Tx \|' \le M \| x \| \qquad \textit{for all } x \in X.$$

1.24.8 **Remarks.**

(i) A topological isomorphism of $(X, \|\cdot\|)$ onto $(Y, \|\cdot\|')$ establishes an identity between the linear and topological structures of $(X, \|\cdot\|)$ and $(Y, \|\cdot\|')$ in that there is a one-to-one correspondence between the points of the spaces which identifies the linear structure and the topological structure.

The relation "being topologically isomorphic to" is an equivalence relation on the set of all normed linear spaces. As we study different properties of normed linear spaces there is interest in determining whether the properties are "linear topological invariants"; that is, if all normed linear spaces which are topologically isomorphic also possess the same property.

(ii) Of course an isometric isomorphism is a topological isomorphism and so has the properties refered to in (i). Clearly an isometric isomorphism is also a metric isometry. An isometric isomorphism of $(X, \|\cdot\|)$ onto $(Y, \|\cdot\|')$ establishes an identity between the normed linear space structures of $(X, \|\cdot\|)$ and $(Y, \|\cdot\|')$. Such mappings are really "normed linear space isomorphisms".

The relation "being isometrically isomorphic to" is an equivalence relation on the set of all normed linear spaces.

We will see in Section 5, how recognition that a normed linear space is isometrically isomorphic to a more familiar normed linear space gives us a firmer grasp of the subject matter. □

It is useful to note that completeness is a linear topological invariant for normed linear spaces.

1.24.9 **Theorem.** *Given topologically isomorphic normed linear spaces* $(X, \|\cdot\|)$ *and* $(Y, \|\cdot\|')$, *if* $(X, \|\cdot\|)$ *is complete then* $(Y, \|\cdot\|')$ *is complete.*

Proof. Suppose T is a topological isomorphism of $(X, \|\cdot\|)$ onto $(Y, \|\cdot\|')$ and that $(X, \|\cdot\|)$ is complete. Then there exist $m, M > 0$ such that

$$m\|x\| \leq \|Tx\|' \leq M\|x\| \quad \text{for all } x \in X.$$

Consider a Cauchy sequence $\{y_n\}$ in $(Y, \|\cdot\|')$ and sequence $\{x_n\}$ in $(X, \|\cdot\|)$ where $y_n = Tx_n$. Then since

$$m\|x_n - x_m\| \leq \|T(x_n - x_m)\|' = \|y_n - y_m\|' \quad \text{for all } m,n \in \mathbb{N},$$

$\{x_n\}$ is a Cauchy sequence in $(X, \|\cdot\|)$. But $(X, \|\cdot\|)$ is complete so there exists an $x \in X$ such that $\{x_n\}$ is convergent to x. Writing $y = Tx$ we have

$$\|y_n - y\|' = \|T(x_n - x)\|' \leq M\|x_n - x\|$$

and so $\{y_n\}$ is convergent to $y \in Y$ and we conclude that $(Y, \|\cdot\|')$ is complete. □

Although the following implication seems somewhat trivial it will be of use in our discussion of the spectrum of a continuous linear operator in Section 17.

1.24.10 **Corollary**. *Consider a topological isomorphism* T *of a Banach space* $(X, \|\cdot\|)$ *into a normed linear space* $(Y, \|\cdot\|')$. *If* T(X) *is dense in* $(Y, \|\cdot\|')$ *then* T(X) = Y.

Proof. From Theorem 1.24.9 we have that $(T(X), \|\cdot\|'_{T(X)})$ is complete. So by Proposition 1.13(i), T(X) is closed in $(Y, \|\cdot\|')$ and therefore T(X) = Y. □

We mentioned that a linear space X can be assigned different norms. There is a natural way of partitioning such norms.

1.24.11 **Definitions**. Consider a linear space X with norms $\|\cdot\|$ and $\|\cdot\|'$. Norm $\|\cdot\|$ is said to be *stronger* than norm $\|\cdot\|'$ and norm $\|\cdot\|'$ is said to be *weaker* than norm $\|\cdot\|$ if the identity mapping from $(X, \|\cdot\|')$ into $(X, \|\cdot\|)$ is continuous.
Norms $\|\cdot\|$ and $\|\cdot\|'$ are said to be *equivalent* if the identity mapping from $(X, \|\cdot\|')$ into $(X, \|\cdot\|)$ is a topological isomorphism.

1.24.12 **Remarks**.
(i) From Theorem 1.24.1 we see that norm $\|\cdot\|$ is stronger than norm $\|\cdot\|'$ if and only if there exists an $M > 0$ such that

$$\|x\|' \leq M\|x\| \qquad \text{for all } x \in X.$$

The topology generated by norm $\|\cdot\|$ is stronger than that generated by norm $\|\cdot\|'$.
(ii) From Corollary 1.24.7 we see that norms $\|\cdot\|$ and $\|\cdot\|'$ are equivalent norms on X if and only if there exist $m, M > 0$ such that

$$m\|x\| \leq \|x\|' \leq M\|x\| \qquad \text{for all } x \in X.$$

Equivalent norms generate the same norm topology on X.
(iii) From Theorem 1.24.9, we see that if a normed linear space $(X, \|\cdot\|)$ is complete then it is complete under any equivalent norm $\|\cdot\|'$.
Again we will see in Section 10 that there is a significant relation between completeness and equivalent norms. □

1.24.13 **Definition**. A linear mapping of a normed linear space $(X, \|\cdot\|)$ into the scalar field of X is called a *linear functional* on X.
A linear mapping of a normed linear space $(X, \|\cdot\|)$ into $(X, \|\cdot\|)$ is called a *linear operator* on X.

In drawing attention to the special properties of continuous linear functionals the following algebraic property is of crucial importance.

1.24.14 **Definitions**. Given a linear space X, a proper linear subspace M is called a linear subspace of *codimension one* or a *hyperplane* if for a given $x_0 \in X \setminus M$, every $x \in X$

can be represented in the form

$$x = \lambda x_0 + y \qquad \text{where } \lambda \text{ is a scalar and } y \in M.$$

It is easily verified that for a linear subspace of codimension one, such a representation holds uniquely.

1.24.15 **Lemma**. *The kernel of a nonzero linear functional* f *on a linear space* X *is a linear subspace of codimension one and a linear subspace* M *of codimension one is the kernel of a nonzero linear functional on* X.

Proof. For $x_0 \in X \setminus \ker f$, $f(x_0) \neq 0$ and for any $x \in X$,

$$x = \frac{f(x)}{f(x_0)} x_0 + y \qquad \text{where } y \in \ker f.$$

For $x_0 \in X \setminus M$, any $x \in X$ can be represented uniquely in the form

$$x = \lambda x_0 + y \quad \text{where } \lambda \text{ is scalar and } y \in M.$$

Define a functional f on X by $f(x) = \lambda$.
Then clearly f is linear and $M = \ker f$. □

1.24.16 **Corollary**. *In a normed linear space* $(X, \|\cdot\|)$, *the kernel of a nonzero linear functional* f *is either closed or dense.*

Proof. Since f is nonzero, if ker f is closed then ker f is not dense.
If ker f is not closed then there exists a cluster point x_0 of ker f where $x_0 \in X \setminus \ker f$. Now from Lemma 1.24.15, $X = \text{sp}\{x_0, \ker f\}$. But $\overline{\ker f}$ is a linear subspace which contains $\text{sp}\{x_0, \ker f\}$. So $X = \overline{\ker f}$. □

We deduce the following characterisation of the continuity of linear functionals.

1.24.17 **Theorem**. *Given a normed linear space* $(X, \|\cdot\|)$, *a linear functional* f *on* X *is continuous if and only if* ker f *is closed.*

Proof. If f is continuous then clearly ker f is closed.
If f is not continuous then for any $r > 0$, f is unbounded on $B(0; r)$. But then since f is linear $f(B(0; r)) = \mathbb{C}$, (or $\mathbb{R}$).
In particular for any $a \in X$ there exists an $x \in B(0; r)$ such that $f(a) = -f(x)$.
Then $a + x \in \ker f \cap (a+B(0; r))$; that is, ker f is dense in X. □

1.24.18 **Remark**. This characterisation of continuity does not extend to linear mappings generally. Consider the identity mapping *id* of $(\mathcal{C}[0,1], \|\cdot\|_1)$ onto $(\mathcal{C}[0,1], \|\cdot\|_\infty)$. Here ker *id* = $\{0\}$ which is closed but *id* is not continuous. □

1.25 Normed linear spaces with a basis

It is useful to see that normed linear spaces can be classified according to what is called their density character.

1.25.1 **Definition.** A metric space (X, d) is said to be *separable* if it contains a countable dense subset.

1.25.2 **Examples.**

(i) Euclidean space $(\mathbb{R}^n, \|\cdot\|_2)$ is separable since it contains the countable set $\mathbb{Q}^n$ as a dense subset. Similarly, Unitary space $(\mathbb{C}^n, \|\cdot\|_2)$ is separable.

(ii) The normed linear space $(c_0, \|\cdot\|_\infty)$ is separable. The linear subspace E_0 is dense in $(c_0, \|\cdot\|_\infty)$ and the subset $E_{0\mathbb{Q}}$ of E_0 consisting of sequences with only rational entries is countable and dense in $(E_0, \|\cdot\|_\infty)$ and so is dense in $(c_0, \|\cdot\|_\infty)$.

(iii) The normed linear space $(\mathcal{C}[a,b], \|\cdot\|_\infty)$ is separable. The Weierstrass' Approximation Theorem, (see AMS §9), tells us that the linear subspace $\mathcal{P}[a,b]$ of polynomials on [a,b] is dense in $(\mathcal{C}[a,b], \|\cdot\|_\infty)$ and the subset $\mathcal{P}_{\mathbb{Q}}[a,b]$ of polynomials with rational coefficients is dense in $(\mathcal{P}[a,b], \|\cdot\|_\infty)$ and so is dense in $(\mathcal{C}[a,b], \|\cdot\|_\infty)$. □

However, there are some significant nonseparable normed linear spaces.

1.25.3 **Example.** The normed linear space $(m, \|\cdot\|_\infty)$ is not separable. Consider the subset E of sequences consisting of digits 0 and 1; this set is uncountable. But also for any $x, y \in E$, $x \neq y$, $\| x-y \|_\infty = 1$. So $\{B(x; \frac{1}{2}) : x \in E\}$ is an uncountable family of disjoint open balls in $(m, \|\cdot\|_\infty)$. For any subset S dense in $(m, \|\cdot\|_\infty)$ we must have that each ball in this family contains at least one point from S. But this implies that S is uncountable. □

In linear space theory the concept of basis is fundamental.

1.25.4 **Definitions.** Given a linear space X, a *Hamel basis* for X is a linearly independent subset of X which spans X. It is clear that a nonempty subset A is a Hamel basis for X if and only if each element x of X has a unique representation as a linear combination of elements of A.

1.25.5 **Remarks.** It follows from Zorn's Lemma, Appendix A.1.7, that

(i) every linear space X has a Hamel basis and

(ii) for any given linear space X any two Hamel bases are numerically equivalent; (see Appendix A.3). □

These remarks enable us to make the following definitions.

1.25.6 **Definitions.** Given a linear space X, the *Hamel dimension* of X is the cardinal number of any Hamel basis for X.
If X has a finite Hamel basis then X is said to be *finite dimensional*, otherwise it is said to be *infinite dimensional*.

1.25.7 **Examples.**

(i) For the linear space E_0, the set $\{e_1, e_2, \ldots, e_n, \ldots\}$ where

$$e_n \equiv \{0, \ldots, 0, \underset{\text{nth place}}{1}, 0, \ldots\}$$

is a Hamel basis for E_0 and the Hamel dimension of E_0 is countably infinite.

(ii) For the linear space c_0, the set $\{e_1, e_2, \ldots, e_n, \ldots\}$given in (i) is not a Hamel basis since its span E_0 is a proper linear subspace of c_0. The set of all sequences $\{\{1, t, t^2, \ldots, t^n, \ldots\} : t \in (0,1)\}$ is a linearly independent subset of c_0 and has cardinality of the continuum. But the cardinality of c_0 is that of the continuum. So a Hamel basis for c_0 has cardinality of the continuum. □

1.25.8 **Remark.** The concepts of Hamel basis and Hamel dimension belong essentially to linear space theory and are not so significant in the analysis of infinite dimensional normed linear spaces because there is no natural relation of a Hamel basis to the topological structure of the space. Hamel basis is mainly of use in drawing parallels between linear space and normed linear space theory and in the construction of counterexamples. □

The basis concept appropriate in the study of infinite dimensional normed linear spaces is the following.

1.25.9 **Definition.** Given a normed linear space $(X, \|\cdot\|)$, a *Schauder basis* for X is a sequence $\{e_n\}$ such that for each element $x \in X$ there exists a unique sequence of scalars $\{\lambda_n\}$ such that $x = \sum_{n=1}^{\infty} \lambda_n e_n$; that is, $\| x - \sum_{k=1}^{n} \lambda_n e_n \| \to 0$ as $n \to \infty$.

For each $n \in \mathbb{N}$, the coefficient λ_n is called the *nth coordinate* of x with respect to the basis $\{e_n\}$.

1.25.10 **Examples.**

(i) In any finite dimensional normed linear space the concepts of Schauder basis and Hamel basis coincide because both in this case are concerned with finite sequences.

(ii) In $(E_0, \|\cdot\|_\infty)$, the sequence $\{e_1, e_2, \ldots, e_n, \ldots\}$ where

$$e_n \equiv \{0, \ldots, 0, \underset{\text{nth place}}{1}, 0, \ldots\}$$

is both a Schauder basis and a Hamel basis.

(iii) In $(c_0, \|\cdot\|_\infty)$, the same sequence $\{e_1, e_2, \ldots, e_n, \ldots\}$ is a Schauder basis since for each $x \equiv \{\lambda_1, \lambda_2, \ldots, \lambda_n, \ldots\} \in c_0$,

$$\| x - \sum_{k=1}^{n} \lambda_k e_k \|_\infty = \sup \{ |\lambda_k| : k > n \} \to 0 \text{ as } n \to \infty$$

and so $x = \sum_{n=1}^{\infty} \lambda_n e_n$.

But this is not a Hamel basis as the Hamel dimension of c_0 is the cardinality of the continuum. □

Confining our definition of Schauder bases to sequences has the following implications.

1.25.11 **Theorem.** *A normed linear space* $(X, \|\cdot\|)$ *with a Schauder basis* $\{e_n\}$ *is separable.*

Proof. We may consider X over the reals and $\| e_n \| = 1$ for all $n \in \mathbb{N}$. The set of all rational linear combinations from $\{e_n\}$ is a countable set.

Given $x = \sum_{n=1}^{\infty} \lambda_k e_k$ and $\varepsilon > 0$ there exists a $\nu \in \mathbb{N}$ such that

$$\| \sum_{k=1}^{n} \lambda_k e_k \| < \varepsilon \qquad \text{for all } n \geq \nu.$$

For each $k \in \{1,2, \ldots, \nu\}$ choose $\alpha_k \in \mathbb{Q}$ such that $|\lambda_k - \alpha_k| < \dfrac{\varepsilon}{\nu}$.

Then $$\| x - \sum_{k=1}^{\nu} \alpha_k e_k \| \leq \| x - \sum_{k=1}^{\nu} \lambda_k e_k \| + \sum_{k=1}^{\nu} |\lambda_k - \alpha_k| < 2\varepsilon.$$

So the set of rational linear combinations from $\{e_n\}$ is dense in $(X, \|\cdot\|)$. □

1.25.12 **Remark.** The following famous problem is referred to in Stefan Banach, *Théorie des opérations linéaires,* Chelsea reprint 1932 edition, p.111.

The basis problem.

Does every separable Banach space have a Schauder basis?

After a great deal of effort by many mathematicians the problem was finally solved in the negative by Per Enflo, *Acta Math.* **130** (1973), 309–317, where he produced a counterexample. □

There are certain natural linear functionals associated with a Schauder basis.

1.25.13 **Definition.** Given a normed linear space $(X, \|\cdot\|)$ with a Schauder basis $\{e_n\}$, for each $k \in \mathbb{N}$ the linear functional f_k on X defined by

$$f_k(e_n) = \delta_{kn}$$

is called the *kth coordinate functional* with respect to the basis $\{e_n\}$.

It is not always true that such coordinate functionals are continuous.

1.25.14 **Example.** Consider the normed linear space $(E_0, \|\cdot\|_\infty)$ with Schauder basis $\{e_1, e_2, \ldots, e_n, \ldots\}$ given in Example 1.25.10(ii). We construct another Schauder basis $\{e_1', e_2', \ldots, e_n', \ldots\}$ as follows.

Write
$$e_1' = e_1$$
and
$$e_n' = e_1 + \frac{1}{n} e_n \qquad \text{for all } n > 1.$$

Consider $x \equiv \{\lambda_1, \lambda_2, \ldots, \lambda_r, 0, \ldots\}$.

Then $x = \left(\lambda_1 - \sum_{k=2}^{r} k\lambda_k\right) e_1' + \sum_{k=2}^{r} k\lambda_k e_k'$ and the coordinates of x are uniquely determined. So $\{e_1', e_2', \ldots, e_n', \ldots\}$ is a Schauder basis for $(E_0, \|\cdot\|_\infty)$.

But further $\{e_n'\}$ is convergent to e_1'. However, for the coordinate functional f_1' corresponding to e_1' we have $f_1'(e_n') = 0$ for $n \geq 2$, but $f_1'(e_1') = 1$ so f_1' is not continuous. □

It is useful to know when such natural functionals are continuous.

1.25.15 **Definition.** Given a normed linear space $(X, \|\cdot\|)$, a Schauder basis $\{e_n\}$ is said to be *monotone* if for each $x = \sum_{k=1}^{\infty} \lambda_k e_k$ the sequence $\{\|\sum_{k=1}^{n} \lambda_k e_k\|\}$ is monotone increasing.

In most classical sequence spaces the obvious Schauder bases are monotone. For monotone bases we have the following satisfying property.

1.25.16 **Theorem.** *Given a normed linear space* $(X, \|\cdot\|)$, *with a monotone Schauder basis* $\{e_n\}$, *the corresponding coordinate functionals are continuous.*

Proof. We may consider $\|e_n\| = 1$ for all $n \in \mathbb{N}$. Then given $n \in \mathbb{N}$,

$$|f_n(x)| = \|f_n(x)e_n\| = \|\sum_{k=1}^{n} \lambda_k e_k - \sum_{k=1}^{n-1} \lambda_k e_k\| \leq \|\sum_{k=1}^{n} \lambda_k e_k\| + \|\sum_{k=1}^{n-1} \lambda_k e_k\|$$
$$\leq 2\|x\| \qquad \text{for all } x \in X,$$

since the basis $\{e_n\}$ is monotone; so f_n is continuous. □

1.26 EXERCISES

1. Consider the linear space m and the linear subspaces E_0, ℓ_1, ℓ_2, c_0, c introduced in Examples 1.11.
   (i) Order these subspaces by set containment.
   (ii) For which of these subspaces do the following real functions define a norm?
      (a) $\|\cdot\|_1$ where for $x \equiv \{\lambda_1, \lambda_2, \ldots, \lambda_n, \ldots\}$,
      $$\|x\|_1 = \sum_{k=1}^{\infty} |\lambda_k|.$$
      (b) $\|\cdot\|_2$ where for $x \equiv \{\lambda_1, \lambda_2, \ldots, \lambda_n, \ldots\}$,
      $$\|x\|_2 = \sqrt{\sum_{k=1}^{\infty} |\lambda_k|^2}.$$
      (c) $\|\cdot\|_c$ where for $x \equiv \{\lambda_1, \lambda_2, \ldots, \lambda_n, \ldots\}$,
      $$\|x\|_c = \sup\left\{\frac{1}{n}\sum_{k=1}^{n} |\lambda_k| : n \in \mathbb{N}\right\}$$
   (iii) Determine whether the following normed linear spaces are closed subspaces of $(m, \|\cdot\|_\infty)$.
      (a) $(E_0, \|\cdot\|_\infty)$ (b) $(\ell_1, \|\cdot\|_\infty)$ (c) $(c_0, \|\cdot\|_\infty)$
   (iv) Determine whether the following normed linear spaces are complete.
      (a) $(c_0, \|\cdot\|_\infty)$ (b) $(\ell_1, \|\cdot\|_\infty)$ (c) $(\ell_1, \|\cdot\|_1)$

2. Consider the linear space $\mathcal{B}[0,1]$ and the linear subspaces $\mathcal{C}[0,1]$, $\mathcal{R}[0,1]$, $\mathcal{C}^1[0,1]$ and $\mathcal{C}_0[0,1]$ introduced in Examples 1.12.
   (i) Order these subspaces by set containment.
   (ii) Determine whether the following normed linear spaces are closed subspaces of $(\mathcal{C}[0,1], \|\cdot\|_\infty)$.
      (a) $\mathcal{C}^1[0,1]$ (b) $\mathcal{C}^\infty[0,1]$.
   (iii) Determine whether the following normed linear spaces are closed subspaces of $(\mathcal{R}[0,1], \|\cdot\|_\infty)$.
      (a) $\mathcal{C}^1[0,1]$ (b) $\mathcal{R}_0[0,1]$.
   (iv) Determine whether $\mathcal{C}^1[0,1]$ is complete with respect to the following norms.
      (a) $\|f\|_\infty = \max\{|f(t)| : t \in [0,1]\}$.
      (b) $$\|f\|_1 = \int_0^1 |f(t)|\,dt.$$
      (c) $\|f\|' = \|f\|_\infty + \|f'\|_\infty$.

3. Given M and N linear subspaces of a normed linear space $(X, \|\cdot\|)$ where $X = M \oplus N$.

(i) Prove that the real function $\|\cdot\|'$ on X where

$$\|\cdot\|' = \| m \| + \| n \|$$

for every $x \in X$ where $x = m + n$ and $m \in M$, $n \in N$, defines a new norm on X.

(ii) Prove that M and N are closed subspaces of $(X, \|\cdot\|')$.

(iii) Prove that if $(X, \|\cdot\|)$ is complete and M and N are closed subspaces of $(X, \|\cdot\|)$ then $(X, \|\cdot\|')$ is complete.

(iv) Prove that if $(X, \|\cdot\|')$ is complete then M and N are closed subspaces of $(X, \|\cdot\|)$.

(v) If $(X, \|\cdot\|')$ is complete is it necessarily true that $(X, \|\cdot\|)$ is complete?

4. Consider a nonempty subset K in a normed linear space $(X, \|\cdot\|)$ and the function

$$d(x, K) \equiv \inf \{\| x-k \| : k \in K\} .$$

(i) Prove that $d(x, K)$ is continuous on $(X, \|\cdot\|)$.

(ii) Prove that K is convex if and only if $d(x, K)$ is a convex function on X.

(iii) Given that K is a closed linear subspace of $(X, \|\cdot\|)$ prove that $d(x, K)$ is a norm on X/K.

5. Consider the linear space $\mathcal{C}[0,1]$ with norms $\|\cdot\|_\infty$ and $\|\cdot\|_1$.

(i) Consider the linear functional p_0 on $\mathcal{C}[0,1]$ where $p_0(f) = f(0)$.

(a) Prove that p_0 is continuous on $(\mathcal{C}[0,1], \|\cdot\|_\infty)$.

(b) Show that p_0 is not continuous on $(\mathcal{C}[0,1], \|\cdot\|_1)$.

(ii) Show that $\|\cdot\|_\infty$ and $\|\cdot\|_1$ are not equivalent norms on $\mathcal{C}[0,1]$.

(iii) Consider the linear subspace $\mathcal{C}_0[0,1]$.

(a) Prove that $\mathcal{C}_0[0,1]$ is closed in $(\mathcal{C}[0,1], \|\cdot\|_\infty)$.

(b) Show that $\mathcal{C}_0[0,1]$ is dense in $(\mathcal{C}[0,1], \|\cdot\|_1)$.

6. (i) For the linear space m, prove that $p: m \to \mathbb{R}$ where, for $x \equiv \{\lambda_1, \lambda_2, \ldots, \lambda_n, \ldots\}$,

$$p(x) = \sum \frac{|\lambda_n|}{2^n}$$

is a norm for m.

(ii) Prove that p is weaker than

(a) the $\|\cdot\|_\infty$-norm for m and

(b) the $\|\cdot\|_1$-norm for ℓ_1.

(iii) Show that neither (ℓ_1, p) nor (m, p) is complete.

7. Consider a linear functional f on a normed linear space $(X, \|\cdot\|)$.

(i) Prove that f is continuous if and only if the seminorm $|f|$ is continuous.

(ii) Give an example to show that a seminorm p on X is not necessarily continuous if ker p is closed.

(iii) Prove that f is discontinuous if and only if ker $|f|$ is dense.

8. (i) Consider the normed linear space $(c, \|\cdot\|_\infty)$ and the functional f on c defined for $x \equiv \{\lambda_1, \lambda_2, \ldots, \lambda_n, \ldots\}$ by

$$f(x) = \sum_{k=1}^{\infty} \frac{\lambda_k}{2^k}.$$

Prove that f is linear and continuous and determine ker f.

(ii) Consider the normed linear spaces $(c_0, \|\cdot\|_\infty)$ and $(\ell_1, \|\cdot\|_1)$ and the mapping $T: c_0 \to \ell_1$ defined for $x \equiv \{\lambda_1, \lambda_2, \ldots, \lambda_n, \ldots\}$ by

$$Tx = \left\{\frac{\lambda_1}{2}, \frac{\lambda_2}{4}, \ldots, \frac{\lambda_n}{2^n}\right\}.$$

(a) Prove that T is linear and continuous and determine ker T.

(b) Determine whether T is a topological isomorphism.

9. Consider the normed linear space $(c, \|\cdot\|_\infty)$ and the functional f on c defined for $x \equiv \{\lambda_1, \lambda_2, \ldots, \lambda_n, \ldots\} \in c$ where $\lambda_n \to \lambda$ by $f(x) = \lambda$.

(i) Prove that f is a continuous linear functional on $(c, \|\cdot\|_\infty)$.

(ii) Consider the mapping $T_1: c \to c_0$ defined by

$$T_1 x = \{\lambda_1 - \lambda, \lambda_2 - \lambda, \ldots, \lambda_n - \lambda, \ldots\}.$$

Prove that T is a continuous linear mapping of $(c, \|\cdot\|_\infty)$ onto $(c_0, \|\cdot\|_\infty)$ but is not one-to-one.

(iii) Consider the mapping $T_2: c \to c_0$ defined by

$$T_2 x = \{\lambda, \lambda_1 - \lambda, \lambda_2 - \lambda, \ldots, \lambda_n - \lambda, \ldots\}.$$

Prove that T is a topological isomorphism of $(c, \|\cdot\|_\infty)$ onto $(c_0, \|\cdot\|_\infty)$.

(iv) Given that $(c_0, \|\cdot\|_\infty)$ is complete deduce that $(c, \|\cdot\|_\infty)$ is complete.

(v) Define an equivalent norm $\|\cdot\|$ on c such that $(c, \|\cdot\|)$ is isomorphically isomorphic to $(c_0, \|\cdot\|)$.

10. (i) Consider the functional p_1 defined on $(c, \|\cdot\|_\infty)$ where, for $x \equiv \{\lambda_1, \lambda_2, \ldots, \lambda_n, \ldots\}$,

$$p_1 = \lim_{n\to\infty} |\lambda_n|.$$

(a) Prove that p_1 is a continuous seminorm on $(c, \|\cdot\|_\infty)$.

(b) Show that the quotient space $(c/\ker p_1, \|\cdot\|)$ is isometrically isomorphic to the scalar field.

(ii) Consider the functional p_2 defined on $(m, \|\cdot\|_\infty)$ where, for $x \equiv \{\lambda_1, \lambda_2, \ldots, \lambda_n, \ldots\}$,

$$p_2(x) = \limsup_{n\to\infty} |\lambda_n|.$$

(a) Prove that p_2 is a continuous seminorm on $(m, \|\cdot\|_\infty)$.

(b) Show that the quotient space $(m/\ker p_2, \|\cdot\|)$ is complete and infinite dimensional.

11. (i) Prove that the linear spaces $\mathcal{C}[a,b]$ and $\mathcal{C}[0,1]$ are isomorphic under the mapping $f \to f^*$ where $f^*(t) = f((1-t)a+tb)$ for $t \in [0,1]$.

(ii) Hence, or otherwise, prove that

(a) $(\mathcal{C}[a,b], \|\cdot\|_\infty)$ and $(\mathcal{C}[0,1], \|\cdot\|_\infty)$ are isometrically isomorphic and

(b) $(\mathcal{C}[a,b], \|\cdot\|_1)$ and $(\mathcal{C}[0,1], \|\cdot\|_1)$ are topologically isomorphic.

12. (i) Prove that the linear spaces,

$Y \equiv \{f \in \mathcal{C}[-\pi,\pi] : f(-\pi) = f(\pi)\}$,

$\mathcal{C}(\Gamma)$ where Γ is the unit circle $\{(\lambda,\mu) \in \mathbb{R}^2 : \lambda^2 + \mu^2 = 1\}$ and

$\mathcal{C}\mathcal{P}(2\pi)$ of continuous periodic functions of period 2π,

are isomorphic under the mapping $f \mapsto f^*$ where

$$f^*(\cos\theta, \sin\theta) = f(\theta) \quad \text{for } \theta \in \mathbb{R}.$$

(ii) Hence, or otherwise, prove that $(Y, \|\cdot\|_\infty)$, $(\mathcal{C}(\Gamma), \|\cdot\|_\infty)$ and $(\mathcal{C}\mathcal{P}(2\pi), \|\cdot\|_\infty)$ are isometrically isomorphic.

(iii) Hence, or otherwise, deduce that

(a) Y is closed in $(\mathcal{C}[-\pi,\pi], \|\cdot\|_\infty)$ and

(b) $\mathcal{C}\mathcal{P}(2\pi)$ is closed in $(\mathcal{B}\mathcal{C}(\mathbb{R}), \|\cdot\|_\infty)$.

13. Prove that norms $\|\cdot\|$ and $\|\cdot\|'$ are equivalent norms for a linear space X if and only if any one of the following conditions holds.

(i) Convergence of sequences is preserved.

(ii) Boundedness of sets is preserved.

(iii) Sequences being Cauchy is preserved.

14. Given a normed linear space $(X, \|\cdot\|)$ and a symmetric convex set K with $0 \in \text{int } K$, the *gauge* p of K is defined by

$$p(x) = \inf\{\lambda > 0 : x \in \lambda K\}.$$

(i) Prove that p is a continuous seminorm on X.

(ii) Prove that if K is bounded then p is an equivalent norm for X.

15. (i) Prove that separability is a linear topological invariant for normed linear spaces.

(ii) Deduce that

(a) $(c, \|\cdot\|_\infty)$ is separable,

(b) $(c, \|\cdot\|_\infty)$ and $(m, \|\cdot\|_\infty)$ are not topologically isomorphic.

16. (i) Prove that the linear subspace E_0 is dense in both $(c_0, \|\cdot\|_\infty)$ and $(\ell_1, \|\cdot\|_1)$.

(ii) Show that $\|\cdot\|_\infty$ and $\|\cdot\|_1$ are not equivalent norms for E_0.

(iii) Deduce that $(\ell_1, \|\cdot\|_1)$ is separable.

17. Consider a separable Banach space $(X, \|\cdot\|)$ with a Schauder basis $\{e_n\}$.

(i) Prove that $\|\cdot\|': X \to \mathbb{R}$ where for $x \equiv \sum_{k=1}^{\infty} \lambda_k e_k$

$$\| x \|' = \sup \{ \| \sum_{k=1}^{n} \lambda_k e_k \| : n \in \mathbb{N} \}$$

is a norm for X and $(X, \|\cdot\|')$ is complete.

(ii) Prove that $\{e_n\}$ is a monotone basis for $(X, \|\cdot\|')$.

18. Consider the normed linear space $(\mathcal{P}[-1,1], \|\cdot\|_\infty)$ of polynomials on $[-1,1]$ with norm

$$\| p(t) \|_\infty = \max \{ | p(t) | : t \in [-1,1] \}.$$

(i) Prove that the sequence $\{e_n\}$ where

$$e_0 = 1 \text{ and } e_n = t^n \text{ for all } n \in \mathbb{N},$$

is a Schauder basis for $\mathcal{P}[-1,1]$ and that the coordinate functionals f_n are given by $f_n(p) = \dfrac{p^{(n)}(0)}{n!}$.

(ii) Prove that the coordinate functional f_0 is continuous but show that for each $n \in \mathbb{N}$, f_n is not continuous.

(iii) Show that $\{e_n\}$ is not a Schauder basis for $(\mathcal{C}[-1,1], \|\cdot\|_\infty)$ although $\mathcal{P}[-1,1]$ is dense in $(\mathcal{C}[-1,1], \|\cdot\|_\infty)$.

§2. CLASSES OF EXAMPLE SPACES

As we said previously, a normed linear space is a generalisation of Euclidean space. However, we could conceivably carry out such a generalisation in several different ways. So we study examples of normed linear spaces by classifying them according to a particular mode of generalisation. We begin by examining the class of examples which preserves finite dimensionality and then study the class which is dimension free but preserves the special inner product structure of Euclidean space.

2.1 Finite dimensional normed linear spaces

Euclidean n-spaces and Unitary n-spaces have the following fundamental topological properties.

2.1.1 Proposition. *An* n*-dimensional Euclidean space (Unitary space)*
(i) *is complete and*
(ii) *has the property that a subset is compact if and only if it is closed and bounded,*
(the Borel–Lebesgue Theorem), (see AMS §3 and §8).

We treat Euclidean n-space as the prototype of all n-dimensional normed linear spaces and the following theorem explains why this is so.

2.1.2 Theorem. *All* n*-dimensional normed linear spaces over* $\mathbb{R}$ *(or* $\mathbb{C}$*) are topologically isomorphic to* $(\mathbb{R}^n, \|\cdot\|_2)$, *(or* $(\mathbb{C}^n, \|\cdot\|_2)$*).*

Proof. Consider an n-dimensional linear space X_n over $\mathbb{R}$ and a basis $\{e_1, e_2, \dots, e_n\}$ for X_n. Now X_n and $\mathbb{R}^n$ are isomorphic under the linear mapping $T: \mathbb{R}^n \to X_n$ defined by

$$T(\lambda_1, \lambda_2, \dots, \lambda_n) = \lambda_1 e_1 + \lambda_2 e_2 + \dots + \lambda_n e_n.$$

We show that T is a topological isomorphism of $(X_n, \|\cdot\|)$ onto $(\mathbb{R}^n, \|\cdot\|_2)$:
Firstly, $T: \mathbb{R}^n \to X_n$ is continuous since

$$\| T(\lambda_1, \lambda_2, \dots, \lambda_n) \| \le \sum_{k=1}^{n} |\lambda_k| \, \| e_k \| \le \sqrt{\sum_{k=1}^{n} \|e_k\|^2} \sqrt{\sum_{k=1}^{n} |\lambda_k|^2}$$

by the Cauchy–Schwarz inequality,

$$= M \, \| (\lambda_1, \lambda_2, \dots, \lambda_n) \|_2 \quad \text{where } M \equiv \sqrt{\sum_{k=1}^{n} \|e_k\|^2}.$$

Secondly, consider the continuous mapping $\|\cdot\| \circ T: \mathbb{R}^n \to \mathbb{R}$ where

$$\|\cdot\| \circ T(\lambda_1, \lambda_2, \dots, \lambda_n) = \| \lambda_1 e_1 + \lambda_2 e_2 + \dots + \lambda_n e_n \|.$$

The unit sphere S(0; 1) being closed and bounded is compact in $(\mathbb{R}^n, \|\cdot\|_2)$, so $\|\cdot\| \circ T$ attains a minimum $m \geq 0$ at some point $(\lambda_1^0, \lambda_2^0, \ldots, \lambda_n^0) \in S(0; 1)$. If $m = 0$ then

$$\|\cdot\| \circ T(\lambda_1^0, \lambda_2^0, \ldots, \lambda_n^0) = \| T(\lambda_1^0, \lambda_2^0, \ldots, \lambda_n^0) \| = 0,$$

but this is impossible since T is one-to-one. Therefore $m > 0$ and we have

$$m \leq \| T(\lambda_1, \lambda_2, \ldots, \lambda_n) \| \text{ for all } (\lambda_1, \lambda_2, \ldots, \lambda_n) \in S(0; 1)$$

and so

$$m \| (\lambda_1, \lambda_2, \ldots, \lambda_n) \|_2 \leq \| T(\lambda_1, \lambda_2, \ldots, \lambda_n) \| \text{ for all } (\lambda_1, \lambda_2, \ldots, \lambda_n) \in \mathbb{R}^n,$$

which implies that T has a continuous inverse on X_n. □

2.1.3 **Remark.** Since "being topologically isomorphic to" is an equivalence relation on the set of all normed linear spaces it follows from Theorem 2.1.2 that all n-dimensional normed linear spaces over the same scalar field are topologically isomorphic. □

Theorem 2.1.2 has many significant corollaries which spell out particular topological properties for finite dimensional normed linear spaces. The first is a particular application of Theorem 2.1.2.

2.1.4 **Corollary.** *All norms for a given finite dimensional linear space are equivalent.*

The next follows directly from Proposition 2.1.1(i) and Theorem 1.24.9.

2.1.5 **Corollary.** *Every finite dimensional normed linear space is complete.*

We can make the following immediate deduction from this corollary using Proposition 1.13(i).

2.1.6 **Corollary.** *Every finite dimensional subspace of a normed linear space is closed.*

Theorem 2.1.2 also enables us to carry over the characterisation of compactness given in the Borel–Lebesgue Theorem to all finite dimensional normed linear spaces.

2.1.7 **Corollary.** *In every finite dimensional normed linear space a subset is compact if and only if it is closed and bounded.*

Proof. Consider a closed and bounded subset A of a finite dimensional normed linear space $(X_n, \|\cdot\|)$ over $\mathbb{R}$. Denote by T the topological isomorphism of $(X_n, \|\cdot\|)$ onto $(\mathbb{R}^n, \|\cdot\|_2)$. Since T is a homeomorphism, T(A) is closed in $(\mathbb{R}^n, \|\cdot\|_2)$ and since T is a topological isomorphism, T(A) is bounded in $(\mathbb{R}^n, \|\cdot\|_2)$. By the Borel–Lebesgue Theorem, T(A) is compact in $(\mathbb{R}^n, \|\cdot\|_2)$. But A is the image of T(A) under the continuous mapping T^{-1} so it is compact in $(X_n, \|\cdot\|)$. □

As we pursue our study of normed linear spaces in general, it is important to note that normed linear spaces possessing the characterisation of compactness given in Corollary 2.1.7 are necessarily finite dimensional.

To show this we need the following technical lemma.

2.1.8 **Riesz Lemma.** *Consider a proper closed linear subspace* M *of a normed linear space* $(X, \|\cdot\|)$. *For each* $0 < \delta < 1$ *there exists an* $x_\delta \in X \setminus M$ *where* $\| x_\delta \| = 1$ *such that* $d(x_\delta, M) \geq \delta$.

Proof. Since M is proper there exists an $x_1 \in X \setminus M$ and since M is closed, $d(x_1, M) \equiv d > 0$. So there exists some $x_0 \in M$ such that $\| x_1 - x_0 \| \leq d/\delta$.

Write $x_\delta \equiv \dfrac{x_1 - x_0}{\| x_1 - x_0 \|}$.

Then for all $x \in M$,

$$\| x_\delta - x \| = \frac{1}{\| x_1 - x_0 \|} \| x_1 - (x_0 + \| x_1 - x_0 \| x) \|$$

and since $x_0 + \| x_1 - x_0 \| x \in M$ we have

$$\| x_\delta - x \| \geq \frac{d}{d/\delta} = \delta.$$

□

2.1.9 **Riesz Theorem.** *A normed linear space is finite dimensional if and only if the closed unit ball is compact.*

Proof. Consider a normed linear space $(X, \|\cdot\|)$ with compact closed unit ball $B[0; 1]$. Now there exists a finite subset $\{x_1, x_2, \ldots, x_n\}$ in $B[0; 1]$ such that

$\bigcup_{k=1}^{n} \{x_k + B(0; \tfrac{1}{2})\} \supseteq B[0; 1]$. Since $M_n \equiv sp\{x_1, x_2, \ldots, x_n\}$ is finite dimensional

it is, by Corollary 2.1.6, a closed linear subspace of $(X, \|\cdot\|)$. Suppose M_n is proper. Then by Riesz Lemma 2.1.8 there exists an $x_{3/4} \in X \setminus M_n$ where $\| x_{3/4} \| = 1$ such that $d(x_{3/4}, M_n) \geq \frac{3}{4}$.

But this contradicts $\bigcup_{k=1}^{n} \{x_k + B(0; \tfrac{1}{2})\} \supseteq B[0; 1]$.

So we conclude that $X = M_n$ and is finite dimensional. Corollary 2.1.7 completes the characterisation. □

Significantly, Theorem 2.1.2 implies that linear mappings on finite dimensional normed linear spaces are automatically continuous.

2.1.10 **Corollary.** *Every linear mapping of a finite dimensional normed linear space into a normed linear space is continuous.*

Proof. Consider an n–dimensional normed linear space $(X_n, \|\cdot\|)$ over $\mathbb{R}$ with a basis $\{e_1, e_2, \ldots, e_n\}$. For any linear mapping T of X_n into a normed linear space $(Y, \|\cdot\|')$ we have, for $x \equiv \lambda_1 e_1 + \lambda_2 e_2 + \ldots + \lambda_n e_n$,

$$Tx = \sum_{k=1}^{n} \lambda_k Te_k.$$

So
$$\| Tx \|' \leq \sum_{k=1}^{n} |\lambda_k| \, \| Te_k \|' \leq K \sqrt{\sum_{k=1}^{n} |\lambda_k|^2}$$

by the Cauchy–Schwarz inequality, where $K \equiv \sqrt{\sum_{k=1}^{n} \|Te_k\|'^2}$.

Since $(X_n, \|\cdot\|)$ is topologically isomorphic to $(\mathbb{R}^n, \|\cdot\|_2)$, there exists an $M > 0$ such that

$$\| (\lambda_1, \lambda_2, \ldots, \lambda_n) \|_2 \leq M \| x \| \quad \text{for all } x \in X$$

so
$$\| Tx \|' \leq KM \| x \| \quad \text{for all } x \in X.$$ □

This corollary gives us useful insight into the nature of convergence in any finite dimensional normed linear space.

2.1.11 **Corollary.** *In a finite dimensional normed linear space* $(X_m, \|\cdot\|)$ *with basis* $\{e_1, e_2, \ldots, e_m\}$, *a sequence* $\{x_n\}$ *is convergent to* x *if and only if each coordinate sequence* $\{\lambda_n^k\}$ *is convergent to* λ^k *for* $k \in \{1, 2, \ldots, m\}$ *where*

$$x_1 \equiv \lambda_1^1 e_1 + \lambda_2^1 e_2 + \ldots + \lambda_m^1 e_m$$
$$x_2 \equiv \lambda_1^2 e_1 + \lambda_2^2 e_2 + \ldots + \lambda_m^2 e_m$$
$$\ldots$$
$$x_n \equiv \lambda_1^n e_1 + \lambda_2^n e_2 + \ldots + \lambda_m^n e_m$$
$$\ldots$$
$$x \equiv \lambda_1 e_1 + \lambda_2 e_2 + \ldots + \lambda_m e_m.$$

Proof. Now
$$\| x_n - x \| \leq \sum_{k=1}^{m} |\lambda_k^n - \lambda_k| \, \| e_k \|$$

so coordinatewise convergence implies convergence.

Conversely, for each $k \in \{1, 2, \ldots, m\}$ the coordinate functional $f_k\colon X_m \to \mathbb{R}$ defined for $x \equiv \lambda_1 e_1 + \lambda_2 e_2 + \ldots + \lambda_m e_m$ by

$$f_k(x) = \lambda_k$$

is linear, so by Corollary 2.1.10, is continuous. □

Again we should note that the property given in Corollary 2.1.10 actually characterises finite dimensionality. To prove this we use the fact that every linear space has a Hamel basis, (see Appendix A.3).

2.1.12 **Theorem.** *A normed linear space* $(X, \|\cdot\|)$ *is finite dimensional if and only if every linear functional on* X *is continuous.*

Proof. Suppose X is infinite dimensional and consider a Hamel basis $\{e_\lambda\}$. We may suppose that $\| e_\lambda \| = 1$ for all λ. Now since X is infinite dimensional $\{e_\lambda\}$ contains a countably infinite subset $\{e_{\lambda_n}\}$. A linear functional is defined on X if we specify its values on a Hamel basis so any linear functional f on X taking values $f(e_{\lambda_n}) = n$ for $n \in \mathbb{N}$, is not continuous on $(X, \|\cdot\|)$. Corollary 2.1.10 completes the characterisation. □

In Corollary 1.24.16 we noticed how the continuity of a linear functional could be determined by its kernel being closed. It is worth noting that a similar characterisation does extend to linear mappings with finite dimensional range.

2.1.13 **Theorem.** *A linear mapping* T *of a normed linear space* $(X, \|\cdot\|)$ *into a finite dimensional normed linear space* $(Y, \|\cdot\|')$ *is continuous if and only if* ker T *is closed.*

Proof. If T is continuous then ker T is closed.

Conversely, suppose that dim T(X) = n and that $\{e_1, \dots, e_n\}$ is a basis for T(X). Then for each $x \in X$

$$Tx = f_1(x)\, e_1 + \dots + f_n(x)\, e_n.$$

Since T is linear, $f_1, \dots, f_n$ are linear functionals on X. For each $k \in \{1, 2, \dots, n\}$ consider $x_k \in X$ such that $Tx_k = e_k$ and $X_n \equiv \operatorname{sp}\{x_1, \dots, x_n\}$.

Now $X = X_n \oplus \ker T$.

Since ker T is a closed linear subspace, d(x, ker T) is a norm for X_n and since X_n is finite dimensional, d(x, ker T) is an equivalent norm for X_n. So there exists a $K > 0$ such that

$$\| x \| \le K\, d(x, \ker T) \quad \text{for all } x \in X_n.$$

Since X_n is finite dimensional f_k is continuous on X_n for each $k \in \{1, 2, \dots, n\}$ so there exists an $M_k > 0$ such that

$$| f_k(x) | \le M_k \| x \| \quad \text{for all } x \in X_n.$$

Now for $z \in X$, $z = x + y$ for $x \in X_n$ and $y \in \ker T$, and

$$| f_k(z) | = | f_k(x) | \le M_k \| x \| \le K M_k\, d(x, \ker T) \le K M_k \| x+y \| = K M_k \| z \|$$

so f_k is continuous on X.

We conclude that T is continuous on X. □

The compactness property which, by the Riesz Theorem 2.1.9 characterises finite dimensional normed linear spaces, enables us to establish the existence of points of best approximation in finite dimensional subspaces of a normed linear space.

2.1.14 **Theorem**. *Given a normed linear space* $(X, \|\cdot\|)$ *and a finite dimensional linear subspace* X_m, *for any* $x_0 \in X \setminus X_m$ *there exists a* $y_0 \in X_m$ *such that*

$$\| x_0 - y_0 \| = d(x_0, X_m).$$

Proof. Writing $d \equiv d(x_0, X_m)$, consider the closed ball $B[x_0; 2d]$.
From Corollary 2.1.6 we have that X_m is closed so $B[x_0; 2d] \cap X_m$ is closed and bounded in $(X_m, \|\cdot\|_{X_m})$. Since X_m is finite dimensional we deduce from Corollary 2.1.7 that that $B[x_0; 2d] \cap X_m$ is compact.
Consider a sequence $\{x_n\}$ in $B[x_0; 2d] \cap X_m$ such that $\| x - x_n \| \to d$ as $n \to \infty$.
Then since $B[x_0; 2d] \cap X_m$ is compact, $\{x_n\}$ has a subsequence $\{x_{n_k}\}$ which is convergent to some $y_0 \in X_m$.

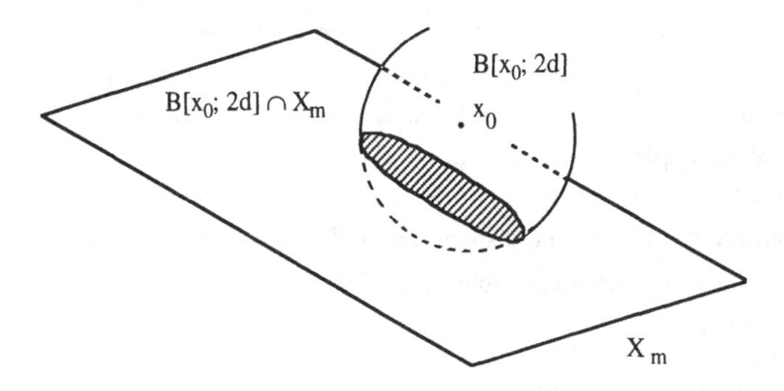

Figure 1. The ball $B[x_0; 2d]$ "scoops" a compact set out of X_m.

Since $\quad d \le \| x_0 - y_0 \| \le \| x_0 - x_{n_k} \| + \| x_{n_k} - y_0 \| \to d \quad$ as $k \to \infty$
we deduce that $\| x_0 - y_0 \| = d$. □

This existence theorem is made somewhat sharper by the following uniqueness theorem.

2.1.15 **Definition**. A normed linear space $(X, \|\cdot\|)$ is said to be *rotund* if for every $x \neq y$, $\| x \| = \| y \| = 1$ we have $\| x+y \| < 2$.

2.1.16 **Theorem**. *Consider a rotund normed linear space* $(X, \|\cdot\|)$ *and a linear subspace* M. *If for* $x_0 \in X \setminus M$ *there exists a* $y_0 \in M$ *such that* $\| x_0 - y_0 \| = d(x_0, M)$ *then* y_0 *is unique.*

Proof. Suppose that there exist y_0 and y_0' in M such that

$$\| x_0 - y_0 \| = \| x_0 - y_0' \| = d(x_0, M).$$

Then $d(x_0, M) = \frac{1}{2}(\| x_0 - y_0 \| + \| x_0 - y_0' \|) \ge \| x_0 - \frac{1}{2}(y_0 + y_0') \| \ge d(x_0, M).$

So $\| x_0 - y_0 \| = \| x_0 - y_0' \| = d(x_0, M)$ and $\| (x_0 - y_0) + (x_0 - y_0') \| = 2d(x_0, M)$.
Since $(X, \|\cdot\|)$ is rotund we deduce that $y_0' = y_0$. □

2.1.17 **Remark.** Theorems 2.1.14 and 2.1.16 guarantee the existence and uniqueness of best approximating points in finite dimensional linear subspaces; the actual calculation of the form of such points is another problem altogether. One of the many advantages of inner product space structure is that it enables the calculation of best approximating points in these spaces. □

2.2. Inner product spaces

The inner product spaces are undoubtedly the most useful class of example spaces. They result from a generalisation of Euclidean space which is dimension free but which retains an algebraic structure enabling fruitful geometrical insight into the nature of the space.

We give our definitions for complex linear spaces but it is to be understood that these include real linear spaces.

2.2.1 **Definition.** Given a linear space X over $\mathbb{C}$, a mapping $(.\,,.): X \times X \to \mathbb{C}$ is an *inner product* on X if it satisfies the following properties:
For all $x,y,z \in X$
(i) $(x+y, z) = (x, z) + (y, z)$
(ii) $(\lambda x, y) = \lambda(x, y)$ for all $\lambda \in \mathbb{C}$
(iii) $(y, x) = \overline{(x, y)}$
(iv) $(x, x) \geq 0$
(v) $(x, x) = 0$ if and only if $x = 0$.
A linear space X with an inner product $(.\,,.)$ is called an *inner product space* (or a *pre-Hilbert space*) and is sometimes denoted formally as a pair $(X, (.\,,.))$.

2.2.2 **Remarks.** Properties (i) and (ii) imply that for any $z \in X$ the mapping
$(.\,, z): X \to C$ is a linear functional on X.
From properties (i), (ii) and (iii) it follows that for all $x,y,z \in X$,

$$(x, y+z) = \overline{(y+z, x)} = \overline{(y, x)} + \overline{(z, x)}$$
$$= (x, y) + (x, z)$$

and

$$(x, \lambda y) = \overline{(\lambda y, x)} = \overline{\lambda}\overline{(y, x)}$$
$$= \overline{\lambda}\,(x, y) \text{ for all } \lambda \in \mathbb{C}.$$

So for any $z \in X$, the mapping $(z\,, .): X \to \mathbb{C}$ is not linear but "conjugate linear".
Any mapping from $X \times X$ into $\mathbb{C}$ which satisfies properties (i), (ii) and (iii) is called an *hermitian form* on X. An hermitian form which satisfies property (iv) is said to be a *positive* hermitian form and a positive hermitian form which satisfies property (v) is said to be a *positive definite* hermitian form. □

From properties (i)–(iv) we derive the following important inequality.

2.2.3 **The Cauchy–Schwarz inequality**. *A positive hermitian form* $(.\,,.)$ *on a linear space* X *satisfies the property*

$$|(x, y)|^2 \leq (x, x)(y, y) \quad \textit{for all } x, y \in X.$$

Proof. If $(x, x) = (y, y) = 0$ then by (iv)

$$(x - (x, y)y, x - (x, y)y) \geq 0$$

simplifying to $-2\,|(x, y)|^2 \geq 0$ which implies that $(x, y) = 0$.
If $(y, y) \neq 0$ then by (iv)

$$\left(x - \frac{(x, y)}{(y, y)}\, y,\; x - \frac{(x, y)}{(y, y)}\, y\right) \geq 0$$

which simplifies to $(x, x)(y, y) - |(x,y)|^2 \geq 0$.
If $(y, y) = 0$ but $(x, x) \neq 0$ we expand

$$\left(y - \frac{(x, y)}{(x, x)}\, x,\; y - \frac{(x, y)}{(x, x)}\, x\right) \geq 0. \qquad \square$$

The following inequality provides the link with normed linear spaces.

2.2.4 **The Minkowski inequality**. *A positive hermitian form* $(.\,,.)$ *on a linear space* X *satisfies the property*

$$\sqrt{(x+y, x+y)} \leq \sqrt{(x, x)} + \sqrt{(y, y)} \qquad \textit{for all } x, y \in X.$$

Proof.

$$\begin{aligned}(x+y, x+y) &= (x, x) + 2\,\mathrm{Re}(x, y) + (y, y)\\ &\leq (x, x) + 2\sqrt{(x, x)(y, y)} + (y, y)\end{aligned}$$

from the Cauchy–Schwarz inequality,

$$\leq \left\{\sqrt{(x, x)} + \sqrt{(y, y)}\right\}^2. \qquad \square$$

2.2.5. **Remarks.**

(i) It follows from the Minkowski inequality that a linear space X with a positive hermitian form $(.\,,.)$ is a seminormed linear space with seminorm $p_2 : X \to \mathbb{R}$ defined by $p_2(x) = \sqrt{(x,x)}$.

A linear space X with an inner product $(.\,,.)$ is a normed linear space with norm $\|\cdot\|_2 : X \to \mathbb{R}$, the *norm generated by the inner product*, where $\|x\|_2 = \sqrt{(x, x)}$.

(ii) In terms of this norm the Cauchy–Schwarz inequality takes the form

$$|(x, y)| \leq \|x\|_2 \|y\|_2 \qquad \text{for all } x, y \in X.$$

This inequality is important in linking the algebraic inner product to the norm topology and implies that the inner product has a continuity property which we call *joint continuity* of the inner product: if $x \to x_0$ and $y \to y_0$ then $(x, y) \to (x_0, y_0)$. This can be deduced simply from the following inequalities.

$$|(x, y) - (x_0, y_0)| \leq |(x, y-y_0)| + |(x-x_0, y_0)| \leq \|x\| \|y-y_0\| + \|x-x_0\| \|y_0\|. \quad \square$$

A linear space X with a positive hermitian form (. , .) is a seminormed linear space. But it can also be transformed in a natural way into an inner product space where the norm generated by the seminorm is the natural norm generated by the seminorm as given in Definition 1.18.

2.2.6 **Theorem.** *Given a linear space* X *with positive hermitian form* (. , .) *generating a seminorm* p_2, *the quotient space* $X/\ker p_2$ *is an inner product space with inner product defined by*

$$([x], [y]) = (x, y) \quad \textit{for any } x \in [x] \textit{ and } y \in [y].$$

Proof. We need to check that the definition of the inner product in $X/\ker p_2$ is well defined.

For any $x, x' \in [x]$ and $y, y' \in [y]$, $p_2(x-x') = 0$ and $p_2(y-y') = 0$. So

$$\begin{aligned} |(x, y) - (x', y')| &\leq |(x-x', y)| + |(x', y-y')| \\ &\leq p_2(x-x')\, p_2(y) + p_2(x')\, p_2(y'-y) \\ &\qquad \text{from the Cauchy–Schwarz inequality,} \\ &= 0. \end{aligned}$$

We need to check that the inner product has property (v).

For any $x \in X$,

$$([x], [x]) = (x, x) = p_2^2(x).$$

So if $([x], [x]) = 0$ then $x \in \ker p_2$, but $\ker p_2$ is the zero element in $X/\ker p_2$. $\square$

An inner product space has the following simple but very significant norm relation.

2.2.7 **The parallelogram law.** *In an inner product space* X,

$$\|x+y\|^2 + \|x-y\|^2 = 2(\|x\|^2 + \|y\|^2) \quad \textit{for all } x, y \in X;$$

(that is, the sum of the squares of the lengths of the diagonals of a parallelogram is equal to the sum of the squares of the lengths of the sides).

Proof.
$$\begin{aligned} \|x+y\|^2 + \|x-y\|^2 &= (x+y, x+y) + (x-y, x-y) \\ &= 2(\|x\|^2 + \|y\|^2) \end{aligned}$$

by straightforward expansion of the inner products. $\square$

2.2.8 **Remark**. P. Jordan and J. von Neumann, *Math. Ann.* **36** (1935), 719–723, proved that if the norm of a normed linear space satisfies the parallelogram law then the norm is generated by an inner product. This is not very difficult to prove, (see Exercise 2.4.9) but it was one of the first characterisations of inner product spaces among normed linear spaces. (A recent collection of such characterisations was compiled by Dan Amir, *Characterisations of inner product spaces*, Birkhäuser, 1986.) □

2.2.9 **Definition**. An inner product space which is complete as a normed linear space is called a *Hilbert space*.

2.2.10 **Example**. Unitary n-space (Euclidean n-space) is an inner product space with inner product $(.\,,.)$ defined on $\mathbb{C}^n$ (or $\mathbb{R}^n$)
for $x \equiv (\lambda_1, \lambda_2, \ldots, \lambda_n)$ and $y \equiv (\mu_1, \mu_2, \ldots, \mu_n)$ by

$$(x, y) = \sum_{k=1}^{n} \lambda_k \bar{\mu}_k.$$

It is quite clear that all the inner product properties (i)–(v) are satisfied. The norm generated by this inner product is the Unitary (Euclidean) norm,

$$\| x \|_2 = \sqrt{\sum_{k=1}^{n} |\lambda_k|^2}.$$

We deduce from the inner product structure that the Cauchy–Schwarz inequality holds and applying this inequality to $(|\lambda_1|, |\lambda_2|, \ldots, |\lambda_n|)$ and $(|\mu_1|, |\mu_2|, \ldots, |\mu_n|) \in \mathbb{R}^n$, we have for $(\lambda_1, \lambda_2, \ldots, \lambda_n)$ and $(\mu_1, \mu_2, \ldots, \mu_n) \in \mathbb{C}^n$ that

$$\begin{aligned} \sum_{k=1}^{n} |\lambda_k \bar{\mu}_k| &= \sum_{k=1}^{n} |\lambda_k| |\mu_k| \\ &\leq \sqrt{\sum_{k=1}^{n} |\lambda_k|^2} \sqrt{\sum_{k=1}^{n} |\mu_k|^2}. \end{aligned}$$

(This derivation of the Cauchy–Schwarz inequality from the inner product structure should be compared with the direct proof, (see AMS §1). □

2.2.11 **Example.** *Hilbert sequence space* ℓ_2 is the linear space whose elements $x \equiv \{\lambda_1, \lambda_2, \ldots, \lambda_n, \ldots\}$ are sequences of scalars such that $\sum |\lambda_n|^2$ is convergent, and with coordinatewise definitions of the linear space operations. This is an inner product space with inner product $(.\,,.)$ defined for $x \equiv \{\lambda_1, \lambda_2, \ldots, \lambda_n, \ldots\}$, $\sum |\lambda_n|^2 < \infty$ and $y \equiv \{\mu_1, \mu_2, \ldots, \mu_n, \ldots\}$, $\sum |\mu_n|^2 < \infty$, by

$$(x, y) = \sum_{k=1}^{\infty} \lambda_k \bar{\mu}_k.$$

Although this is a straightforward generalisation of Example 2.2.10 there are significant points which require checking.

Firstly, we use the Cauchy–Schwarz inequality of Example 2.2.10 to show that an inner product is defined.
For any $n \in \mathbb{N}$,

$$\begin{aligned}\sum_{k=1}^{n} |\lambda_k \bar{\mu}_k| &\le \sqrt{\sum_{k=1}^{n} |\lambda_k|^2}\sqrt{\sum_{k=1}^{n} |\mu_k|^2} \\ &\le \sqrt{\sum_{k=1}^{\infty} |\lambda_k|^2}\sqrt{\sum_{k=1}^{\infty} |\mu_k|^2} < \infty,\end{aligned}$$

so $\sum_{k=1}^{\infty} |\lambda_k \bar{\mu}_k| < \infty$ and $\sum_{k=1}^{\infty} \lambda_k \bar{\mu}_k$ is convergent.
Secondly, we use the Minkowski inequality of Example 2.2.10 to show that ℓ_2 is closed under coordinatewise definition of addition.
For any $n \in \mathbb{N}$,

$$\begin{aligned}\sqrt{\sum_{k=1}^{n} |\lambda_k + \mu_k|^2} &\le \sqrt{\sum_{k=1}^{n} |\lambda_k|^2}\sqrt{\sum_{k=1}^{n} |\mu_k|^2} \\ &\le \sqrt{\sum_{k=1}^{\infty} |\lambda_k|^2}\sqrt{\sum_{k=1}^{\infty} |\mu_k|^2} < \infty,\end{aligned}$$

so $\sum_{k=1}^{\infty} |\lambda_k + \mu_k|^2 < \infty$ and $x + y \in \ell_2$.
The norm generated by this inner product is the Hilbert space norm

$$\| x \|_2 = \sqrt{\sum_{k=1}^{\infty} |\lambda_k|^2}$$

As a prototype Hilbert space we should show that $(\ell_2, \|\cdot\|_2)$ is complete.

Proof of completeness. Consider $\{x_n\}$ a Cauchy sequence in $(\ell_2, \|\cdot\|_2)$ where

$$x_1 \equiv \{\lambda_1^1, \lambda_2^1, \ldots, \lambda_k^1, \ldots\}$$

$$x_2 \equiv \{\lambda_1^2, \lambda_2^2, \ldots, \lambda_k^2, \ldots\}$$

$$\ldots$$

$$x_n \equiv \{\lambda_1^n, \lambda_2^n, \ldots, \lambda_k^n, \ldots\}$$

$$\ldots \qquad ;$$

that is, given $\varepsilon > 0$ there exists a $\nu \in \mathbb{N}$ such that

$$\| x_n - x_m \|_2 = \sqrt{\sum_{k=1}^{\infty} |\lambda_k^n - \lambda_k^m|^2} < \infty \quad \text{for all } m > n > \nu.$$

So for each $k \in \mathbb{N}$,

$$|\lambda_k^n - \lambda_k^m| < \varepsilon \quad \text{for all } m > n > \nu$$

that is, $\{\lambda_k^n\}$ is a Cauchy sequence in $(\mathbb{C}, |\cdot|)$.

But $(\mathbb{C}, |\cdot|)$ is complete, so for each $k \in \mathbb{N}$ we can define

$$\lambda_k = \lim_{n\to\infty} \lambda_k^n$$

and consider the sequence $x \equiv \{\lambda_1, \lambda_2, \ldots, \lambda_n, \ldots\}$.

We need to show that

(i) $x \in \ell_2$, (that is, $\sum_{k=1}^{\infty} |\lambda_k|^2 < \infty$), and

(ii) $\{x_n\}$ converges to x; (that is, $\| x_n - x \|_2 \to 0$ as $n \to \infty$).

Now we have for any $\mu \in \mathbb{N}$ that

$$\sum_{k=1}^{\mu} |\lambda_k^n - \lambda_k^m|^2) < \varepsilon^2 \qquad \text{for all } m > n > \nu$$

so keeping n fixed and increasing m we have

$$\sum_{k=1}^{\mu} |\lambda_k^n - \lambda_k|^2) < \varepsilon^2 \qquad \text{for all } n > \nu.$$

As this holds for all $\mu \in \mathbb{N}$,

$$\sum_{k=1}^{\infty} |\lambda_k^n - \lambda_k|^2) < \varepsilon^2 \qquad \text{for all } n > \nu.$$

From closure under addition we deduce that $x \in \ell_2$ and then this statement implies that $\{x_n\}$ converges to x. □

2.2.12 **Examples**. *Integral inner product spaces.*

(i) $\mathcal{C}[a,b]$ the linear space of all complex (real) continuous functions on the interval [a,b] is an inner product space with inner product (. , .) defined on $\mathcal{C}[a,b]$ by

$$(f, g) = \int_a^b f(t)\,\overline{g(t)}\,dt.$$

Clearly since f is continuous on the bounded closed interval [a,b], f is bounded and

$$\left| \int_a^b f(t)\,\overline{g(t)}\,dt \right| \le \int_a^b |f(t)|\,|g(t)|\,dt \le \|f\|_\infty \|g\|_\infty (b-a),$$

where $\|f\|_\infty \equiv \max\{|f(t)| : a \le t \le b\}$ so the inner product is defined. The norm generated by this inner product is

$$\|f\|_2 = \sqrt{\int_a^b |f(t)|^2\,dt}.$$

However, $(\mathcal{C}[a,b], \|\cdot\|_2)$ is not complete.

Consider the real linear space $\mathcal{C}[0,1]$ and the sequence $\{f_n\}$ where

$$\left.\begin{aligned} f_n(t) &= 1 & 0 \le t \le \tfrac{1}{2} \\ &= 1-(n+1)(t-\tfrac{1}{2}) & \tfrac{1}{2} < t \le \tfrac{1}{2} + \tfrac{1}{n+1} \\ &= 0 & \tfrac{1}{2} + \tfrac{1}{n+1} < t \le 1 \end{aligned}\right\}$$

which is pointwise convergent to f on [0,1] where

$$\left.\begin{aligned} f(t) &= 1 \qquad 0 \le t \le \tfrac{1}{2} \\ &= 0 \qquad \tfrac{1}{2} < t \le 1 \end{aligned}\right\}.$$

Now $\int_0^1 | f_n(t) - f_m(t) |^2 \, dt < \frac{1}{n+1}$ for all m > n so $\{f_n\}$ is a Cauchy sequence in $(\mathcal{C}[0,1], \|\cdot\|_2)$. But $f \notin \mathcal{C}[0,1]$ and there is no continuous function on [0,1] to which $\{f_n\}$ converges with respect to the $\|\cdot\|_2$-norm.

(ii) An attempt to overcome this problem by enlarging the space raises other difficulties. We could consider $\mathcal{R}[a,b]$, the linear space of all Riemann integrable functions on the interval [a,b]. In this case

$$(f, g) = \int_a^b f(t) \, \overline{g(t)} \, dt$$

is a positive Hermitian form but it is not positive definite.

In the real linear space $\mathcal{R}[0,1]$ the function f defined by

$$\left.\begin{aligned} f(t) &= 0 \qquad 0 \le t < 1 \\ &= 1 \qquad t = 1 \end{aligned}\right\}$$

has $\int_0^1 f^2(t) \, dt = 0$.

However, from Theorem 2.2.6 we see that the quotient space $\mathcal{R}[a,b] / \ker p_2$ where $p_2(f) = \sqrt{\int_a^b |f(t)|^2 \, dt}$, can be represented in a natural way as an inner product.

But again the inner product space $(\mathcal{R}[a,b] / \ker p_2, \|\cdot\|_2)$ has the same defect we found in (i); that is, it is not complete.

Consider the interval [0,1] and the following adaptation of the construction of Cantor's Ternary Set, (see AMS §10).

From [0,1] remove $E_1 \equiv (\frac{3}{8}, \frac{5}{8})$, the middle open interval of length $\frac{1}{4}$.

From $[0,1] \setminus E_1$ remove $E_2 \equiv (\frac{5}{32}, \frac{7}{32}) \cup (\frac{25}{32}, \frac{27}{32})$, the union of the middle open intervals of the remaining two intervals each of length $\frac{1}{2} \cdot \frac{1}{8}$.

From $[0,1] \setminus (E_1 \cup E_2)$ remove E_3, the union of the middle open intervals of the remaining four intervals each of length $\frac{1}{4} \cdot \frac{1}{16}$.

In general, from $[0,1] \setminus \bigcup_1^{n-1} E_k$ remove E_n, the union of the middle open intervals of the remaining 2^{n-1} intervals each of length $\frac{1}{2^{n-1}} \cdot \frac{1}{2^{n+1}}$.

Now consider the sequence $\{f_n\}$ of characteristic functions of $\bigcup_1^n E_k$; that is,

$$\left.\begin{aligned} f_n(t) &= 1 \qquad t \in \bigcup_1^n E_k \\ &= 0 \qquad t \in [0,1] \setminus \bigcup_1^n E_k \end{aligned}\right\}$$

Since each f_n has only a finite number of discontinuities, it is Riemann integrable and

$$p_2^2(f_n) = \int_0^1 f_n^2(t)\,dt = \sum_1^n \frac{1}{2^{k+1}} \to \frac{1}{2} \text{ as } n \to \infty$$

and
$$\begin{aligned} p_2^2(f_n - f_m) &= \int_0^1 (f_n(t) - f_m(t))^2\,dt \\ &= \sum_{n+1}^m \frac{1}{2^{k+1}} \quad \text{for } m > n. \end{aligned}$$

So $\{[f_n]\}$ is a Cauchy sequence in $(\mathcal{R}[0,1] / \ker p_2, \|\cdot\|_2)$.

Now $\{f_n\}$ converges pointwise to f the characteristic function of $\bigcup_1^\infty E_k$; that is,

$$\left.\begin{aligned} f_n(t) &= 1 \qquad t \in \bigcup_1^\infty E_k \\ &= 0 \qquad t \in [0,1] \setminus \bigcup_1^\infty E_k \end{aligned}\right\}$$

We recall that a characteristic function on [a,b] is Riemann integrable if and only if given $\varepsilon > 0$ the set of points of discontinuity of the function can be contained in a finite union of intervals of total length $< \varepsilon$. If the set of points of discontinuity of f, which is $[0,1] \setminus \bigcup_1^\infty E_k$, can be covered by a finite union of intervals of total length $\varepsilon < \frac{1}{4}$ then the set $\bigcup_1^\infty E_k$ contains the union of the remaining intervals [0,1] and so $\bigcup_1^\infty E_k$ has total length $> 1 - \varepsilon > \frac{3}{4}$. But the sum of the length of the intervals in $E_n \to \frac{1}{2}$ as $n \to \infty$.

Therefore f is not Riemann integrable on [0,1]. Furthermore, there is no Riemann integrable function f on [0,1] such that $p_2(f_n - f) \to 0$ as $n \to \infty$. We conclude that $\{[f_n]\}$ does not converge in $(\mathcal{R}[a,b] / \ker p_2, \|\cdot\|_2)$.

(iii) The attempt to overcome this defect with integral spaces has led to some very fruitful advances in analysis, in particular to the theory of the Lebesgue integral. The successful strategy is to determine a class of functions which is larger than the class of Riemann integrable functions and then proceed as in (ii). In this way we hope that we have made the space large enough to contain the limits of the Cauchy sequences.

Consider the linear space of complex functions f on [a,b] such that $|f|^2$ is Lebesgue integrable with positive Hermitian form (. , .) defined by

$$(f, g) = \int_{[a,b]} f\bar{g}\,d\mu.$$

The inner product space formed as a quotient by ker p_2 where

$$p_2(f) = \sqrt{\int_{[a,b]} |f|^2 \, d\mu}$$

is complete and is often referred to as *classical Hilbert space* and is denoted $(\mathfrak{L}_2[a,b], \|\cdot\|_2)$. □

There is a theoretically elegant way of representing classical Hilbert space $(\mathfrak{L}_2[a,b], \|\cdot\|_2)$ as the Hilbert space which contains $(\mathfrak{C}[a,b], \|\cdot\|_2)$ as a dense linear subspace. This is helpful conceptually but not so useful in practice. We will discuss this further in Section 7.12.

Perhaps the most important special property of inner product spaces is that they possess a workable notion of orthogonality which is a direct generalisation of orthogonality in Euclidean spaces. This means that in such spaces there is a rich geometrical structure similar to Euclidean space.

We formally define an orthogonality relation by the inner product but the usefulness of the notion depends on its geometrical characterisation in terms of the norm.

2.2.13 **Definition**. Given an inner product space X, for $x,y \in X$ we say that x is *orthogonal* to y if $(x, y) = 0$ and we write $x \perp y$. Given subset M of X we say that x is *orthogonal* to M if x is orthogonal to every element of M and we write $x \perp M$.

2.2.14 **Remarks**. From the defining properties of an inner product we can deduce the following properties about this orthogonality relation.

From property (iii) we see that

x is orthogonal to y if and only if y is orthogonal to x; that is, the orthogonality relation is symmetric.

From property (i) it follows that

if x is orthogonal to a subset M then x is orthogonal to sp M

and with property (iii) that

every element of sp M is orthogonal to x.

From property (ii) we deduce that

0 is orthogonal to every $x \in X$.

From property (v) we have that

0 is the only element orthogonal to itself.

So we deduce from property (v) the result which is so very useful in practice that

if x is orthogonal to every element of the space then $x = 0$; that is, 0 is the only element orthogonal to all the elements of the space. □

We now show that orthogonality defined by the inner product is actually identical to a "least distance" type of orthogonality defined by the norm and important in best approximation problems.

2.2.15 **Theorem**. *In an inner product space* X, *for* x,y ∈ X

$$(x, y) = 0 \text{ if and only if } \| x+\lambda y \| \geq \| x \| \text{ for all scalar } \lambda.$$

Proof. If (x, y) = 0 then

$$\| x+\lambda y \| \| x \| \geq |(x+\lambda y, x)| = | \| x \|^2 + \lambda(y, x) | = \| x \|^2$$

so $\qquad \| x+\lambda y \| \geq \| x \| \qquad$ for all scalar λ.

Conversely, if

$$\| x+\lambda y \| \geq \| x \| \qquad \text{for all scalar } \lambda,$$

then $\qquad \| x+\lambda y \|^2 - \| x \| \|x+\lambda y \| \geq 0$

so Re (x, x+λy) + Re λ(y, x+λy) − | (x, x+λy) | ≥ 0.

Therefore, Re λ(y, x+λy) ≥ 0 for all scalar λ.

For real λ, Re (y, x+λy) ≥ 0 for λ ≥ 0

and ≤ 0 for λ ≤ 0.

But $(y, x+\lambda y) = (y, x) + \lambda \| y \|^2 \to (y, x)$ as $\lambda \to 0$

so we deduce that Re (y, x) = 0.

For λ = iα where α is real, we have

$$\text{Re } \{\alpha(iy, x+\alpha iy)\} \geq 0 \text{ for all real } \alpha$$

so by the same argument we have Re (iy, x) = 0, which implies that Im (y, x) = 0 and we conclude that (x, y) = 0. □

2.2.16. **Remark**. We should explain why the norm condition characterising orthogonality in Theorem 2.2.15 is called "least distance" orthogonality.

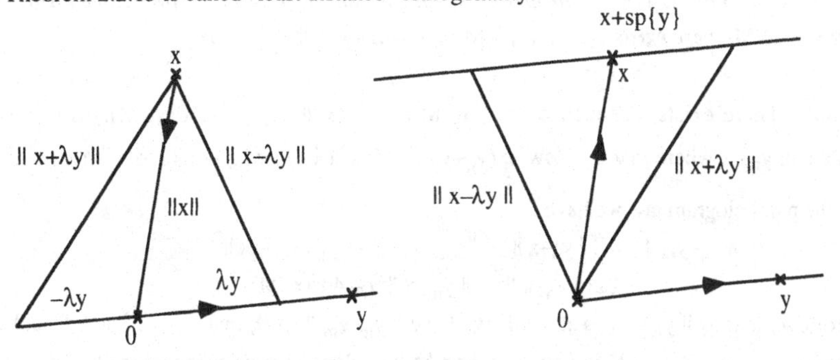

Figure 2. To illustrate "least distance" orthogonality.

Given x,y ∈ X, x,y ≠ 0, consider the two dimensional linear subspace sp{x,y}.

Now $\qquad d(x, sp\{y\}) = \inf\{ \| x+\lambda y \| : \text{scalar } \lambda\} \geq \| x \|$

if and only if x has 0 as a closest point in sp{y}.

Equivalently,

$$d(0, x+sp\{y\}) = \inf\{ \| x+\lambda y \| : \text{scalar } \lambda\} \geq \| x \|$$

if and only if 0 has x as a closest point in $x + sp\{y\}$. □

The following result is an interesting consequence of Theorem 2.2.15 which is not otherwise obvious.

2.2.17 **Corollary**. *In an inner product space* X, *for* $x, y \in X$

$$\| x+\lambda y \| \geq \| x \| \qquad \textit{for all scalar } \lambda$$

if and only if

$$\| y+\lambda x \| \geq \| y \| \qquad \textit{for all scalar } \lambda.$$

2.2.18 **Remark**. Such symmetry of least distance orthogonality does not hold generally for normed linear spaces other than inner product spaces. In fact for normed linear spaces of dimension greater than or equal to three if this property holds then the norm of the space is generated by an inner product, (see Dan Amir, *Characterisations of inner product spaces*, Birkhäuser, 1986, p. 143). □

Given a normed linear space $(X, \|\cdot\|)$, a finite dimensional linear subspace M_n and $x \in X \setminus M_n$, we have seen in Theorem 2.1.14 as a consequence of compactness properties of finite dimensional normed linear spaces, that there exists a closest point to x in M_n. For infinite dimensional linear subspaces generally we do not have such compactness properties and best approximation points do not necessarily exist. However, in Hilbert space the parallelogram law guarantees the existence of best approximation points in all closed linear subspaces.

2.2.19 **Theorem**. *Given any proper closed linear subspace* M *of a Hilbert space* H *and any* $x \in H \setminus M$, *there exists a unique* $y \in M$ *such that* $\| x-y \| = d(x, M)$.

Proof. There exists a sequence $\{y_n\}$ in M such that $\| x-y_n \| \to d(x, M)$. Consider y_n-x and y_m-x where $m \neq n$. Now $\frac{1}{2}(y_n+y_m) \in M$ so $\| \frac{1}{2}(y_n+y_m) - x \| \geq d(x, M)$.

By the parallelogram law we have

$$\begin{aligned} \| y_n-y_m \|^2 &= 2(\| y_n-x \|^2 + \| y_m-x \|^2) - \| y_n+y_m-2x \|^2 \\ &\leq 2(\| y_n-x \|^2 + \| y_m-x \|^2) - 4d^2(x, M). \end{aligned}$$

It follows that as $\| y_n-x \| \to d(x, M)$ we have $\| y_n-y_m \| \to 0$; that is, $\{y_n\}$ is a Cauchy sequence in M. But as H is complete and M is a closed linear subspace of H, $\{y_n\}$ is convergent to some $y \in M$.

Now $\| x-y \| \geq d(x, M)$, but $\| x-y \| \leq \| x-y_n \| + \| y_n - y \| \to d(x, M)$, so we conclude that

$$\| x-y \| = d(x, M).$$

To show that y is unique suppose there is a $z \in M$ such that $\| x-z \| = d(x, M)$. Consider x–y and x–z. Now $\frac{1}{2}(y+z) \in M$ so $\| x - \frac{1}{2}(y+z) \| \geq d(x, M)$.

Again by the parallelogram law we have

$$\| y-z \|^2 = 2(\| x-y \|^2 + \| x-z \|^2) - \| 2x-(y+z) \|^2$$
$$\leq 4d^2(x, M) - 4d^2(x, M) = 0.$$

and we conclude that $z = y$. □

2.2.20 **Remark**. Uniqueness could be deduced directly from the rotundity of the inner product space; (see Exercise 2.4.7 and Theorem 2.1.16). □

Theorem 2.2.19 has the following interpretation in terms of orthogonality.

2.2.21 **Corollary**. *Given any proper closed linear subspace* M *of a Hilbert space* H *there exists a nonzero element* $z \in H \setminus M$ *such that* z *is orthogonal to* M.

Proof. Consider any $x \in H \setminus M$. From Theorem 2.2.19 we have that there exists a unique $y \in M$ such that $\| x-y \| = d(x, M)$. Write $z \equiv x - y$. Then for any $w \in M$ and scalar λ,

$$\| z+ \lambda w \| = \| x- (y-\lambda w) \|$$
$$\geq d(x, M) \quad \text{since } y-\lambda w \in M$$
$$= \| z \|.$$

From Theorem 2.2.15 we deduce that $(z, w) = 0$. But this holds for all $w \in M$ so z is orthogonal to M. □

2.3 The $\mathscr{L}$p spaces ($1 \leq p \leq \infty$)

A significant class of normed linear spaces is determined from a generalisation of the Euclidean norm formula and as such includes the inner product spaces as a special case. We begin with a discussion of these norms in finite dimensional spaces.

2.3.1 **Example**. ℓ_p^n *space* ($1 \leq p \leq \infty$).

The n-dimensional linear space $\mathbb{C}^n$(or $\mathbb{R}^n$) with norm $\|\cdot\|_p$ defined for $x \equiv (\lambda_1, \lambda_2, \ldots, \lambda_n)$ by

$$\| x \|_p = (\sum_{k=1}^{n} | \lambda_k |^p)^{1/p} \qquad 1 \leq p < \infty,$$
$$\| x \|_\infty = \max \{ | \lambda_k | : k \in \{1, 2, \ldots, n\} \}$$

is called an ℓ_p^n *space*. The latter is called the $p = \infty$ case and is included in the class because in $\mathbb{C}^n$(or $\mathbb{R}^n$), for any $x \equiv (\lambda_1, \lambda_2, \ldots, \lambda_n)$

$$\max \{ |\lambda_k | : k \in \{1, 2, \ldots, n\} \} = \lim_{p \to \infty} (\sum_{k=1}^{n} | \lambda_k |^p)^{1/p}, \text{ (see AMS §1).}$$ □

The proof that $\|\cdot\|_p$ obeys the triangle inequality when $p = 1$ or ∞ is straightforward, but when $1 < p < \infty$ the proof depends on the following inequalities.

2.3.2 **Lemma.** *For positive real numbers* α *and* β

$$\alpha\beta \leq \frac{\alpha^p}{p} + \frac{\beta^q}{q}$$

for any given $1 < p < \infty$ *and* $\frac{1}{p} + \frac{1}{q} = 1$.

Proof. This is actually a special case of the G.M.–A.M. inequality which states that for positive real numbers x and y

$$x^{\lambda}y^{1-\lambda} \leq \lambda x + (1-\lambda)y \quad \text{for any given } 0 < \lambda < 1.$$

Now $x^{\lambda} = \exp(\lambda \ln x)$ and $y^{1-\lambda} = \exp((1-\lambda)\ln y)$.

Since the exponential function is convex,

$$x^{\lambda}y^{1-\lambda} = \exp(\lambda \ln x + (1-\lambda)\ln y) \leq \lambda \exp \ln x + (1-\lambda) \exp \ln y = \lambda x + (1-\lambda)y.$$

The required inequality follows by putting

$$x \equiv \alpha^p, \ y \equiv \beta^q \text{ and } \lambda \equiv \frac{1}{p}. \qquad \square$$

2.3.3 **The Hölder inequality.** *In* $\mathbb{C}^n$*(or* $\mathbb{R}^n$*), given* $1 < p < \infty$*, for any* $x \equiv (\lambda_1, \lambda_2, \ldots, \lambda_n)$ *and* $y \equiv (\mu_1, \mu_2, \ldots, \mu_n)$,

$$\left(\sum_{k=1}^{n} |\lambda_k \bar{\mu}_k|\right) \leq \left(\sum_{k=1}^{n} |\lambda_k|^p\right)^{1/p} \left(\sum_{k=1}^{n} |\mu_k|^q\right)^{1/q}$$

where $\frac{1}{p} + \frac{1}{q} = 1$.

Proof. For $x, y \neq 0$, put

$$\alpha \equiv \frac{|\lambda_k|}{\left(\sum_{k=1}^{n} |\lambda_k|^p\right)^{1/p}}, \quad \beta \equiv \frac{|\mu_k|}{\left(\sum_{k=1}^{n} |\mu_k|^q\right)^{1/q}},$$

in the inequality of Lemma 2.3.2 and summing for $k \in \{1,2,\ldots,n\}$ we have

$$\frac{\sum_{k=1}^{n} |\lambda_k \bar{\mu}_k|}{\left(\sum_{k=1}^{n} |\lambda_k|^p\right)^{1/p} \left(\sum_{k=1}^{n} |\mu_k|^q\right)^{1/q}} \leq \frac{1}{p} + \frac{1}{q} = 1. \qquad \square$$

2.3.4 **The Minkowski inequality.** *In* $\mathbb{C}^n$ *(or* $\mathbb{R}^n$*), given* $1 < p < \infty$*, for any* $x \equiv (\lambda_1, \lambda_2, \ldots, \lambda_n)$ *and* $y \equiv (\mu_1, \mu_2, \ldots, \mu_n)$

$$\left(\sum_{k=1}^{n} |\lambda_k + \mu_k|^p\right)^{1/p} \leq \left(\sum_{k=1}^{n} |\lambda_k|^p\right)^{1/p} + \left(\sum_{k=1}^{n} |\mu_k|^p\right)^{1/p}$$

Proof. $\displaystyle\sum_{k=1}^{n} |\lambda_k + \mu_k|^p \leq \sum_{k=1}^{n} |\lambda_k + \mu_k|^{p-1} |\lambda_k| + \sum_{k=1}^{n} |\lambda_k + \mu_k|^{p-1} |\mu_k|.$

Using the Hölder inequality for the sums on the right we have

$$\sum_{k=1}^{n} |\lambda_k+\mu_k|^p \leq \Big(\sum_{k=1}^{n} |\lambda_k+\mu_k|^{(p-1)q}\Big)^{1/q} \Big(\sum_{k=1}^{n} |\lambda_k|^p\Big)^{1/p} + \Big(\sum_{k=1}^{n} |\lambda_k+\mu_k|^{(p-1)q}\Big)^{1/q} \Big(\sum_{k=1}^{n} |\mu_k|^p\Big)^{1/p}$$

But $(p-1)q = p$. When $x + y \neq 0$, dividing by $\big(\sum_{k=1}^{n} |\lambda_k+\mu_k|^p\big)^{1/q}$ we derive the Minkowski inequality. □

We now generalise to consider such norms in infinite dimensional sequence spaces.

2.3.5 **Examples**. *Sequence* ℓ_p *spaces* $(1 \leq p \leq \infty)$

Given $1 < p < \infty$, the set ℓ_p is the linear space of sequences of scalars $x \equiv \{\lambda_1, \lambda_2, \ldots, \lambda_n, \ldots\}$ such that $\sum |\lambda_k|^p$ is convergent, under coordinatewise definition of the linear space operations. The linear space ℓ_p with norm

$$\|x\|_p = \Big(\sum_{k=1}^{\infty} |\lambda_k|^p\Big)^{1/p},$$

is called an ℓp *space.*

The set ℓ_∞ is the linear space of bounded sequences of scalars $x \equiv \{\lambda_1, \lambda_2, \ldots, \lambda_n, \ldots\}$ under coordinatewise definition of the linear space operations. The linear space ℓ_∞ with norm

$$\|x\|_\infty = \sup\{|\lambda_k| : k \in \mathbb{N}\}$$

is called an ℓ_∞ *space* (or m *space*). Again this case is included in the class because for any bounded sequence $x \equiv \{\lambda_1, \lambda_2, \ldots, \lambda_n, \ldots\}$,

$$\sup\{|\lambda_k| : k \in \mathbb{N}\} = \lim_{p\to\infty} \Big(\sum_{k=1}^{\infty} |\lambda_k|^p\Big)^{1/p}, \text{ (see AMS §1).}$$ □

Verification of the particular linear space and norm properties when $p = 1$ or ∞ is straightforward. But when $1 < p < \infty$, particular points we need to verify are that ℓ_p is closed under addition and the norm satisfies the triangle inequality. To prove both we establish an extension of the Minkowski inequality.

2.3.6 **The Minkowski inequality**. *Given* $1 < p < \infty$ *and any* $x \equiv \{\lambda_1, \lambda_2, \ldots, \lambda_n, \ldots\}$ *and* $y \equiv \{\mu_1, \mu_2, \ldots, \mu_n, \ldots\}$ *in* ℓ_p *then* $x + y \in \ell_p$ *and*

$$\Big(\sum_{k=1}^{\infty} |\lambda_k+\mu_k|^p\Big)^{1/p} \leq \Big(\sum_{k=1}^{\infty} |\lambda_k|^p\Big)^{1/p} + \Big(\sum_{k=1}^{\infty} |\mu_k|^p\Big)^{1/p}.$$

Proof. From the finite dimensional Minkowski inequality 2.3.4 we have for any $n \in \mathbb{N}$,

$$\Big(\sum_{k=1}^{n} |\lambda_k+\mu_k|^p\Big)^{1/p} \leq \Big(\sum_{k=1}^{n} |\lambda_k|^p\Big)^{1/p} + \Big(\sum_{k=1}^{n} |\mu_k|^p\Big)^{1/p}$$

$$\leq \Big(\sum_{k=1}^{\infty} |\lambda_k|^p\Big)^{1/p} + \Big(\sum_{k=1}^{\infty} |\mu_k|^p\Big)^{1/p} < \infty \text{ since } x,y \in \ell_p,$$

so $\sum |\lambda_k+\mu_k|^p$ is convergent and $x + y \in \ell_p$ and the general inequality is established. □

Now the normed linear spaces $(\ell_p, \|\cdot\|_p)$ where $1 \le p \le \infty$ are all complete. The proof of completeness follows closely the pattern of proof given for Hilbert sequence space $(\ell_2, \|\cdot\|_2)$ in Example 2.2.11 and is left as an exercise for the reader.

2.3.7 **Examples.** *Integral $\mathcal{L}_p$ spaces* $(1 \le p < \infty)$.

(i) Given $1 \le p < \infty$, $\mathcal{C}[a,b]$ the linear space of all complex (real) continuous functions on the interval [a,b] is a normed linear space with norm

$$\| f \|_p = \left(\int_a^b | f(t) |^p \, dt \right)^{1/p}$$

but the counterexample used in Example 2.2.12(i) shows that $(\mathcal{C}[a,b], \|\cdot\|_p)$ is not complete.

(ii) We may attempt to overcome this problem as in Example 2.2.12(ii) and consider $\mathcal{R}[a,b]$, the linear space of all Riemann integrable functions on the interval [a,b]. Again, given $1 \le p < \infty$, it is easy to see that

$$p_p(f) = \left(\int_a^b | f(t) |^p \, dt \right)^{1/p}$$

is not a norm, but it is a seminorm.

To show that the triangle inequality is satisfied in both cases (i) and (ii), we need to establish the Hölder and Minkowski inequalities for integrals. For $f,g \in \mathcal{R}[a,b]$ where $\int_a^b | f(t) | \, dt$, $\int_a^b | g(t) | \, dt \neq 0$, given $1 \le p < \infty$ and recalling that $\ker p_p = \ker p_1$, (see AMS §1), we put $\alpha \equiv \dfrac{| f(t) |}{\left(\int_a^b | f(t) |^p \, dt \right)^{1/p}}$ and $\beta \equiv \dfrac{| g(t) |}{\left(\int_a^b | g(t) |^q \, dt \right)^{1/q}}$

in the inequality of Lemma 2.3.2 and integrating over [a,b] we obtain the required Hölder inequality. Then following the pattern of proof of the Minkowski inequality 2.3.4 we derive the triangle inequality.

We could form the quotient space $(\mathcal{R}[a,b] / \ker p_p, \|\cdot\|_p)$ as in Example 1.19, but again Example 2.2.12(ii) shows that this space is not complete.

(iii) We successfully overcome this defect as in Example 2.2.12(iii), by the Lebesgue integral. Given $1 \le p < \infty$, consider the linear space of all complex (real) measurable functions f on [a, b] such that $| f |^p$ is Lebesgue integrable with seminorm

$$p_p(f) = \left(\int_{[ab]} | f |^p \, d\mu \right)^{1/p}.$$

The normed linear space formed as a quotient by $\ker p_p$ is complete and is generally referred to as $\mathcal{L}_p$ [a,b] *space*.

For Lebesgue integration of functions on [a,b] we do not confine our attention to bounded functions as we do for Riemann integration. So we should point out that although $\mathcal{R}[a,b] / \ker p_p$ is the same linear space for all $1 \le p < \infty$, the $\mathcal{L}_p[a,b]$ spaces provide different linear spaces for different values of p where $1 \le p < \infty$. □

Again we note that given $1 \le p < \infty$, $(\mathfrak{L}_p[a,b], \|\cdot\|_p)$ can be represented as the Banach space which contains $(\mathfrak{C}[a,b], \|\cdot\|_p)$ as a dense linear subspace. We will discuss this further in Section 7.12.

2.4 EXERCISES

1. Prove that a normed linear space $(X, \|\cdot\|)$ is finite dimensional if and only if
   (i) its unit sphere $S(0; 1)$ is compact,
   (ii) every bounded sequence has a convergent subsequence, or
   (iii) there exists a compact neighbourhood of the origin.

2. (i) Consider a normed linear space $(X, \|\cdot\|)$, and a closed linear subspace M.
   (a) Given $y \notin M$, prove that $M + sp\{y\}$ is closed in $(X, \|\cdot\|)$.
   (b) Given a compact set A in $(X, \|\cdot\|)$ prove that $M + A$ is closed.
   (c) Prove that given any finite dimensional linear subspace N the linear subspace M+N is closed in $(X, \|\cdot\|)$.
   (ii) Is it true that M+N is closed for closed linear subspaces M and N of a normed linear space $(X, \|\cdot\|)$?
   If it is true give a proof, if it is false give a counterexample.
   (Hint: See Exercise 1.26.3.)

3. Consider the linear space $\mathfrak{B}(X)$ of bounded functions on a nonempty set X. Prove that $\mathfrak{B}(X)$ is finite dimensional if and only if X is finite.
   (Hint: Consider $\mathfrak{B}(X)$ with norm $\|f\|_\infty = \sup\{|f(x)| : x \in X\}$.)

4. (i) Consider the $n \times n$ complex matrix (a_{ij}). Prove that the mapping $(\cdot, \cdot): \mathbb{C}^n \times \mathbb{C}^n \to \mathbb{C}$ defined for $x \equiv (\lambda_1, \lambda_2, \ldots, \lambda_n)$ and $y \equiv (\mu_1, \mu_2, \ldots, \mu_n)$ by $(x, y) = \sum_{i,j=1}^{n} a_{ij} \lambda_i \bar{\mu}_j$ is an inner product on $\mathbb{C}^n$ if and only if (a_{ij}) is a positive definite hermitian matrix.
   (ii) Consider $\mathbb{R}^n$ with a norm $\|\cdot\|$. Prove that the norm $\|\cdot\|$ is generated by an inner product if and only if the unit sphere is an ellipsoid.

5. Show that, when $1 \le p \le \infty$, $p \ne 2$, none of the $\|\cdot\|_p$ norms on the ℓ_p-spaces is generated by an inner product.

6. (i) Given a finite dimensional normed linear space $(X_m, \|\cdot\|)$ prove that if for every continuous linear functional f, $\{f(x_n)\}$ is convergent to $f(x)$ then $\{x_n\}$ is convergent to x.

(ii) For an inner product space X, prove that if for every $y \in X$, $\{(x_n, y)\}$ is convergent to (x, y) and $\{\| x_n \|\}$ is convergent to $\| x \|$ then $\{x_n\}$ is convergent to x.

Give an example to show that in an infinite dimensional inner product space, the condition $\{\| x_n \|\}$ being convergent to $\| x \|$ is necessary.

7. (i) Prove that in an inner product space,

$$(x,y) = \| x \| \| y \| \quad \text{for } x,y \neq 0$$

if and only if for some real $\alpha > 0$, $y = \alpha x$.

(ii) Deduce that an inner product space is rotund.

8. (i) Prove the following generalisation of the Hölder inequality 2.3.3.

Given $1 < p < \infty$ *and* $\frac{1}{p} + \frac{1}{q} = 1$ *and any*

$x \equiv \{\lambda_1, \lambda_2, \dots, \lambda_n, \dots\} \in \ell_p$ *and* $y \equiv \{\mu_1, \mu_2, \dots, \mu_n, \dots\} \in \ell_q$

then $\sum \lambda_k \bar{\mu}_k$ *is absolutely convergent and*

$$\sum_{k=1}^{\infty} | \lambda_k \bar{\mu}_k | \leq \Big(\sum_{k=1}^{\infty} | \lambda_k |^p\Big)^{1/p} \Big(\sum_{k=1}^{\infty} | \mu_k |^p\Big)^{1/q}.$$

(ii) Prove that equality holds for $x,y \neq 0$ if and only if for some real $\alpha > 0$, $y = \alpha x$.

(iii) Deduce that for $1 < p < \infty$, the ℓ_p spaces are rotund.

9. Given a normed linear space $(X, \|\cdot\|)$ where the norm satisfies the parallelogram law $\| x+y \|^2 + \| x-y \|^2 = 2(\| x \|^2 + \| y \|^2)$ for all $x, y \in X$, prove that

(i) when X is a real linear space, the mapping $(.\,,\,.): X \times X \to \mathbb{R}$ defined by $(x, y) = \frac{1}{4}(\| x+y \|^2 - \| x-y \|^2)$ is an inner product on X and $\| x \| = \sqrt{(x, x)}$,

(ii) when X is a complex linear space, the mapping $(.\,,\,.): X \times X \to \mathbb{C}$ defined by

$$(x, y) = \tfrac{1}{4}(\| x+y \|^2 - \| x-y \|^2) + \tfrac{1}{4} i (\| x+iy \|^2 - \| x-iy \|^2)$$

is an inner product on X and $\| x \| = \sqrt{(x, x)}$.

(This is the famous Jordan–von Neumann characterisation of inner product spaces referred to in Remark 2.2.8. This expression which defines the inner product by the norm is called the *polarisation formula*.)

10. (i) Prove that in any inner product space, *Apollonius' identity* holds; that is,

$$\| z - \tfrac{1}{2}(x+y) \| = \tfrac{1}{2}\sqrt{2\| z-x \|^2 + 2\| z-y \|^2 - \| x-y \|^2}\ .$$

(This is a formula for the length of a median of a triangle in terms of the lengths of its sides.)

(ii) Prove that in any normed linear space where the norm satisfies Apollonius' identity the norm is generated by an inner product.

11. (i) In any inner product space, prove that $x \perp y$ if and only if

$$\| x+\lambda y \| = \| x-\lambda y \| \text{ for all scalar } \lambda.$$

(ii) Consider an inner product space X.

(a) Prove that if $x \perp y$ then $\| x+y \|^2 = \| x \|^2 + \| y \|^2$

(b) When X is a real linear space, prove that if for $x,y \in X$

$$\| x+y \|^2 = \| x \|^2 + \| y \|^2 \text{ then } x \perp y.$$

When X is a complex linear space, give a counter–example to show that this result does not necessarily hold.

(So when X is real, orthogonality is characterised by *Pythagoras' identity*.)

12. (i) Given any four elements x,y,z,w in a two dimensional inner product space X, prove that

$$\| x-z \| \| y-w \| \le \| x-y \| \| z-w \| + \| y-z \| \| x-w \|$$

(Hint: Consider elements of the Euclidean plane as complex numbers and use the usual inner product for complex numbers.)

(This is *Ptolemy's inequality* which says that in the Euclidean plane, given any quadrilateral the sum of the products of the lengths of opposite sides is greater than or equal to the product of the lengths of the diagonals.)

(ii) Under what conditions on x,y,z,w does equality hold?

(iii) Show that the inequality holds for any four elements x,y,z,w in any inner product space X.

(A normed linear space X which satisfies Ptolemy's inequality has norm generated by an inner product, see Dan Amir, *Characterisations of inner product spaces*, Birkhäuser, 1986, p. 54).

13. A linear operator T on an inner product space X is an *inner product isomorphism* if

$$(Tx, Ty) = (x, y) \quad \text{for all } x,y \in X.$$

(i) Prove that a linear operator T is an isometric isomorphism if and only if it is an inner product isomorphism.

(ii) Prove that if X is finite dimensional then an inner product isomorphism T is onto, but give an example to show that in general an inner product isomorphism T need not be onto.

(iii) Prove that an isometry T from a real inner product space X into a real inner product space Y, where $T(0) = 0$, is an isometric isomorphism.

(iv) Show that an isometry T from a complex inner product space X into a complex inner product space Y, which has $T(0) = 0$, is not necessarily an isometric isomorphism. (See Remark 1.24.6).

14. Consider a Hilbert space H.

(i) Given a closed convex set A in H and $x \in H \setminus A$, prove that there exists a unique element $y \in A$ such that $\| x-y \| = d(x, A)$.

(ii) Given a closed convex set A and a compact set K in H where $A \cap K = \emptyset$, prove that there exist elements $x \in A$ and $y \in K$ such that $\| x-y \| = d(A, K)$.

(iii) Is it true that, given closed convex sets A and B in H, there exist elements $x \in X$ and $y \in B$ such that $\| x-y \| = d(A, B)$? If it is true give a proof, if false give a counterexample.

15. Consider $\mathfrak{C}[-\pi,\pi]$ and the linear subspace M of *odd functions* on $[-\pi,\pi]$; that is $f \in M$ if $f(t) = -f(-t)$ for all $t \in [-\pi,\pi]$.

(i) Prove that the operator F where

$$F(f)(t) = \frac{1}{2}(f(t) + f(-t))$$

is linear and continuous on $(\mathfrak{C}[-\pi,\pi], \|\cdot\|_2)$.

(ii) Deduce that M is a closed linear subspace of $(\mathfrak{C}[-\pi,\pi], \|\cdot\|_2)$.

(iii) Consider the linear subspace N of *even functions* on $[-\pi,\pi]$; that is, $f \in N$ if $f(t) = f(-t)$ for all $t \in [-\pi,\pi]$. Prove that N is a closed linear subspace of $(\mathfrak{C}[-\pi,\pi], \|\cdot\|_2)$ and $\mathfrak{C}[-\pi,\pi] = M \oplus N$ and $M \perp N$.

(iv) Given a normed linear space $(X, \|\cdot\|)$ with closed linear subspaces M and N such that $X = M \oplus N$, is it necessarily true that $(X, \|\cdot\|)$ is complete?

16. Given a subset A in a linear space X, a point $a \in A$ is called a *passing point* of A if there exist points $x,y \in A$, $x \neq y$ and $0 < \lambda < 1$ such that $a = \lambda x + (1-\lambda)y$. A point $a \in A$ which is not a passing point of A is called an *extreme point* of A; that is, an extreme point of A is not in the interior of any line segment in A.

(i) Prove that a one-to-one linear map T from a linear space X into a linear space Y maps passing points of a subset A of X into passing points of T(A) in Y.

(ii) Given a subset A of a normed linear space $(X, \|\cdot\|)$, prove that the extreme points of A lie in the boundary of A.

(iii) Prove that every normed linear space which is isometrically isomorphic to a rotund normed linear space is also rotund.

(iv) Prove that the closed unit ball in $(c_0, \|\cdot\|_\infty)$ has no extreme points.

(v) Show that $(c_0, \|\cdot\|_\infty)$ and $(\ell_1, \|\cdot\|_1)$ are not isometrically isomorphic. (See AMS §4 and §7.)

17. A normed linear space $(X, \|\cdot\|)$ is said to be *uniformly rotund* if given $\varepsilon > 0$, for any $x,y \in X$, $\|x\|, \|y\| \leq 1$ there exists a $\delta(\varepsilon) > 0$ such that $\|x-y\| < \varepsilon$ when $\|\frac{1}{2}(x+y)\| > 1 - \delta$.

(i) Prove that a uniformly rotund normed linear space is rotund.

(ii) Prove that any inner product space is uniformly rotund

(iii) The following inequalities, called *Clarkson's inequalities*, hold in ℓ_p spaces for $2 \leq p < \infty$,

$$\|x+y\|^p + \|x-y\|^p \leq 2^{p-1}(\|x\|^p + \|y\|^p)$$

$$2(\|x\|^p + \|y\|^p)^{q-1} \leq \|x+y\|^q + \|x-y\|^q$$

$$\text{where } \frac{1}{p} + \frac{1}{q} = 1,$$

and for $1 < p \leq 2$ these inequalities hold in reverse.
Use these inequalities to prove that the ℓ_p-spaces where $1 < p < \infty$ are uniformly rotund.

(iv) Prove that in any uniformly rotund normed linear space $(X, \|\cdot\|)$, any sequence $\{x_n\}$ satisfying $\|x_n\| \to 1$, $\|x_n + x_m\| \to 2$ is a Cauchy sequence.

(v) Prove that if M is a proper closed linear subspace of a uniformly rotund Banach space $(X, \|\cdot\|)$, then for any $x \notin M$ there exists a unique $y \in M$ such that $\|x-y\| = d(x, M)$.

18. Consider a proper closed linear subspace M of a Hilbert space H.

(i) Prove that for every $[x] \in H/M$ there exists a unique element $x_0 \in [x]$ such that

$$\|x_0\| = \|[x]\| = \inf\{\|x+m\| : m \in M\}.$$

(ii) Prove that H/M is a Hilbert space with inner product defined by

$$([x], [y]) = (x_0, y_0).$$

19. Consider an incomplete inner product space X with norm $\|\cdot\|$ generated by the inner product, and $(\widetilde{X}, \|\cdot\|)$ the completion of $(X, \|\cdot\|)$; (see AMS §3). Prove that $\widetilde{X}$ is an inner product space with inner product $(.\,,.)$ defined by

$$([x], [y]) = \lim_{n\to\infty} (x_n, y_n)$$

where $\{x_n\}$ and $\{y_n\}$ are Cauchy sequences in X satisfying $\{x_n\} \in [x]$ and $\{y_n\} \in [y]$, and that $([x], [y]) = (x, y)$ for all $x,y \in X$.

§3. ORTHONORMAL SETS IN INNER PRODUCT SPACES

The notion of orthogonality plays an important role in the development of the theory of inner product spaces. We now investigate the special theory developed around the idea of mutually orthogonal sets of elements.

3.1 Definition. Given an inner product space X, an *orthonormal set* of elements $\{e_\alpha\}$ is a set where

$$(e_\alpha, e_\beta) = 0 \qquad \text{for all } \alpha \neq \beta$$

$$\text{and} \qquad \| e_\alpha \| = 1 \qquad \text{for all } \alpha .$$

It is clear that any orthonormal set is linearly independent.

3.2 Examples.

(i) In Hilbert sequence space $(\ell_2, \|\cdot\|_2)$, the set $\{e_n\}$ where

$$e_n \equiv \{0, \ldots, 0, \underset{\text{nth place}}{1}, 0, \ldots\} \qquad \text{for } n \in \mathbb{N}$$

is an orthonormal set.

(ii) In $(\mathcal{C}[-\pi,\pi], \|\cdot\|_2)$ the space of real continuous functions on $[-\pi,\pi]$, with norm

$$\| f \|_2 = \sqrt{\int_{-\pi}^{\pi} f^2(t)dt} \ ,$$

the set of functions

$$\frac{1}{\sqrt{2\pi}}, \frac{1}{\sqrt{\pi}} \cos t, \frac{1}{\sqrt{\pi}} \cos 2t, \ldots, \frac{1}{\sqrt{\pi}} \cos nt, \ldots$$

$$\frac{1}{\sqrt{\pi}} \sin t, \frac{1}{\sqrt{\pi}} \sin 2t, \ldots, \frac{1}{\sqrt{\pi}} \sin nt, \ldots$$

is an orthonormal set. □

Orthonormal sets have great advantages. When we have a linear subspace M of an inner product space X spanned by a sequence $\{x_n\}$ it is often useful to have an orthonormal sequence $\{e_n\}$ which also spans M. The following is an explicit process for generating such an orthonormal sequence.

3.3 Gram–Schmidt orthogonalisation.

Given a sequence $\{x_n\}$ *which is also a linearly independent set in an inner product space* X, *then there exists an orthonormal sequence* $\{e_n\}$ *such that for each* $n \in \mathbb{N}$

$$\text{sp}\{e_k : k \in \{1,2, \ldots, n\}\} = \text{sp}\{x_k : k \in \{1,2, \ldots, n\}\}.$$

Proof and construction. Given $x_1 \neq 0$, we define sequences $\{y_n\}$ and $\{e_n\}$ as follows:

$$y_1 = x_1 \quad , \quad e_1 = \frac{y_1}{\| y_1 \|}$$

$$y_2 = x_2 - (x_2, e_1)\, e_1 \quad , \quad e_2 = \frac{y_2}{\| y_2 \|}$$

$$\dots$$

$$y_n = x_n - \sum_{k=1}^{n-1} (x_n, e_k)\, e_k \quad , \quad e_n = \frac{y_n}{\| y_n \|}$$

$$\dots$$

It is clear from the construction that for any $n \in \mathbb{N}$

$$\mathrm{sp}\{e_k : k \in \{1,2, \dots, n\}\} = \mathrm{sp}\{x_k : k \in \{1,2, \dots, n\}\}\,.$$

For each $n \in \mathbb{N}$, $y_n \neq 0$ since $\{x_1, \dots, x_n\}$ is a linearly independent set. Direct computation shows that $\{e_n\}$ is an orthonormal set. □

3.4 **Example**. In $(\mathcal{C}[-1,1], \|\cdot\|_2)$ the space of real continuous functions on $[-1,1]$ with norm $\| f \|_2 = \sqrt{\int_{-1}^{1} f^2(t)\,dt}$, the sequence of polynomials

$$\{1, t, t^2, \dots, t^n, \dots\}$$

is a linearly independent set. The Gram–Schmidt orthogonalisation of this set yields the orthonormal set

$$\sqrt{\frac{1}{2}},\ \sqrt{\frac{3}{2}}\,t,\ \sqrt{\frac{5}{3}}\left(\frac{3}{2}t^2 - \frac{1}{2}\right),\ \sqrt{\frac{7}{2}}\left(\frac{5}{2}t^3 - \frac{3}{2}t\right)$$

the *Legendre polynomials* defined in general as

$$\sqrt{n+\tfrac{1}{2}}\ P_n(t)$$

where $$P_n(t) = \frac{1}{2^n n!}\frac{d^n}{dt^n}(t^2-1)^n \quad \text{for } n \in \{0, 1, 2, \dots\}.$$ □

3.5 **Definition**. Given an inner product space X with orthonormal set $\{e_\alpha\}$, the set $\{(x, e_\alpha)\}$ is called the set of *orthogonal coefficients* with respect to $\{e_\alpha\}$.

3.6 **Example**. For any continuous function f on $[-\pi,\pi]$, the Fourier coefficients

$$\frac{1}{\sqrt{2\pi}}\int_{-\pi}^{\pi} f(t)\,dt),\quad \frac{1}{\sqrt{\pi}}\int_{-\pi}^{\pi} f(t)\cos nt\,dt),\quad \frac{1}{\sqrt{\pi}}\int_{-\pi}^{\pi} f(t)\sin nt\,dt),\quad \text{for all } n \in \mathbb{N},$$

are orthogonal coefficients with respect to the orthonormal set given in Example 3.2(ii). □

We show the significance of the orthogonal coefficients in representing an element with respect to a finite orthonormal set.

3.7 **Lemma.** *Consider a finite orthonormal set* $\{e_1, e_2, \dots, e_n\}$ *in an inner product space* X. *Then* $\{e_1, e_2, \dots, e_n\}$ *is a basis for* $M_n \equiv \mathrm{sp}\{e_1, e_2, \dots, e_n\}$ *and*
(i) *every* $x \in M_n$ *has unique representation in the form*

$$x = \sum_{k=1}^{n} (x, e_k)\, e_k,$$

(ii) *for every* $x \in X \setminus M_n$, $\sum_{k=1}^{n} (x, e_k)\, e_k$ *is the unique best approximating point to* x *in* M_n.

Proof.
(i) For $x = \alpha_1 e_1 + \alpha_2 e_2 + \dots + \alpha_n e_n$

$$(x, e_k) = \alpha_k \qquad \text{for all } k \in \{1, 2, \dots, n\}.$$

If $x = 0$ then $\alpha_k = 0$ for all $k \in \{1, 2, \dots, n\}$ and so $\{e_1, e_2, \dots, e_n\}$ is linearly independent and is a basis for M_n.
(ii) Now for each $j \in \{1, 2, \dots, n\}$,

$$\Big(x - \sum_{k=1}^{n} (x, e_k)\, e_k, e_j\Big) = (x, e_j) - \sum_{k=1}^{n} (x, e_k)(e_k, e_j) = (x, e_j) - (x, e_j) = 0.$$

So $x - \sum_{k=1}^{n} (x, e_k)\, e_k$ is orthogonal to M_n and by Theorem 2.2.15, $\sum_{k=1}^{n} (x, e_k)\, e_k$ is the unique best approximating point to x in M_n. □

3.8 **Remark.** This lemma has immediate application to the problem of finding points of best approximation in finite dimensional linear subspaces discussed in Theorem 2.1.14. Given a finite dimensional linear subspace X_n in an inner product space X and $x_0 \in X \setminus X_n$, we can compute the unique point in X_n of best approximation to x_0 as follows.
Given a basis $\{x_1, x_2, \dots, x_n\}$ for X_n we generate an orthonormal basis $\{e_1, e_2, \dots, e_n\}$ for X_n by Gram–Schmidt orthogonalisation. Then from Lemma 3.7, $\sum_{k=1}^{n} (x_0, e_k)\, e_k$ is the unique point in X_n of best approximation to x_0. Of course $\sum_{k=1}^{n} (x_0, e_k)\, e_k$ is best approximating with respect to the inner product norm and not necessarily with respect to any other norm on X. □

In moving from a finite to an arbitrarily large orthonormal set we need the following preliminary property.

3.9 **Lemma.** *Given an orthonormal set* $\{e_\alpha\}$ *in an inner product space* X, *for any* $x \in X$, *the set of nonzero orthogonal coefficients of* x *is countable; that is, the set* $S \equiv \{\alpha : (x, e_\alpha) \neq 0\}$ *is countable.*

Proof. Given $\varepsilon > 0$, consider the set

$$S_\varepsilon \equiv \{\alpha : |(x, e_\alpha)| > \varepsilon\}.$$

Suppose that $\{1, 2, \dots, n\} \in S_\varepsilon$. Then from Lemma 3.7(ii),

$$\|x\|^2 \geq \sum_{k=1}^{n} |(x, e_k)|^2 > n\,\varepsilon^2.$$

Therefore
$$n \leq \frac{\|x\|^2}{\varepsilon^2}$$

which implies that S_ε is finite.
But $S = \bigcup\{S_{1/n} : n \in \mathbb{N}\}$, so S is countable. □

We now extend Lemma 3.7(i) for elements in the closed linear space of an orthonormal set in an infinite dimensional inner product space.

3.10 **Theorem.** *Given an orthonormal set* $\{e_\alpha\}$ *in an inner product space* X, *consider* $M \equiv \overline{sp}\{e_\alpha\}$. *For every* $x \in M$, $\Sigma(x, e_\alpha)\, e_\alpha$ *is well defined over all sequential orderings of the set* $S \equiv \{\alpha : (x, e_\alpha) \neq 0\}$ *and* $x = \Sigma\,(x, e_\alpha)\, e_\alpha$.

Proof. Now $x \in M$ can be expressed as an element of $sp\{e_\alpha\}$ or a cluster point of such elements. So given $\varepsilon > 0$ there exists a finite set $\{e_1, e_2, \dots, e_n\}$ in $\{e_\alpha\}$ and a set of scalars $\{\lambda_1, \lambda_2, \dots, \lambda_n\}$ such that

$$\|x - \sum_{k=1}^{n} \lambda_k e_k\| < \varepsilon.$$

But then by Lemma 3.7(ii) applied to $sp\{e_1, e_2, \dots, e_n\}$,

$$\|x - \sum_{k=1}^{n} (x, e_k)e_k\| \leq \|x - \sum_{k=1}^{n} \lambda_k e_k\|.$$

Consider $M_0 \equiv \overline{sp}\{e_\alpha : (x, e_\alpha) \neq 0\}$. Then $\sum_{k=1}^{n} (x, e_k)e_k \in M_0$ and so we deduce that $x \in M_0$.
From Lemma 3.9, the set $S \equiv \{\alpha : (x, e_\alpha) \neq 0\}$ is countable. Consider a sequential ordering $\{\alpha_n\}$ of S. Then given $\varepsilon > 0$ there exists an $n \in \mathbb{N}$ and a set of scalars $\{\mu_1, \mu_2, \dots, \mu_n\}$ such that

$$\|x - \sum_{k=1}^{n} \mu_k e_{\alpha_k}\| < \varepsilon.$$

Again by Lemma 3.7(ii) applied to $sp\,\{e_{\alpha_1}, e_{\alpha_2}, \dots, e_{\alpha_n}\}$

$$\|x - \sum_{k=1}^{n} (x, e_{\alpha_k})\, e_{\alpha_k}\| \leq \|x - \sum_{k=1}^{n} \mu_k e_{\alpha_k}\|$$

which implies that $x - \sum_{k=1}^{\infty} (x, e_{\alpha_k})\, e_{\alpha_k}$ converges to x. Since this holds for any sequential ordering $\{\alpha_n\}$ of S, we have that $\Sigma\,(x, e_\alpha)\, e_\alpha$ is well defined over all sequential orderings of S and we can write

$$x = \Sigma\,(x, e_\alpha)\, e_\alpha. \qquad \square$$

To extend Lemma 3.7(ii) to obtain best approximating elements in the closed linear span of an orthonormal set in an infinite dimensional inner product space, we need completeness of the space.

3.11 **Theorem**. *Given an orthonormal set* $\{e_\alpha\}$ *in a Hilbert space* H, *consider* $M \equiv \overline{sp}\{e_\alpha\}$. *For every* $x \in H \setminus M$, $\Sigma(x, e_\alpha)\, e_\alpha$ *is well defined over all sequential orderings of the set* $S \equiv \{\alpha : (x, e_\alpha) \neq 0\}$ *and* $\Sigma(x, e_\alpha)\, e_\alpha$ *is the unique best approximating point to* x *in* M.

Proof. From Theorem 2.2.19 we have that there exists a $y \in M$, the unique best approximating point to x in M. But then by Theorem 3.10

$$y = \Sigma\,(y, e_\alpha)\, e_\alpha\,.$$

However, from Theorem 2.2.15, x–y is orthogonal to M, so $(x-y, e_\alpha) = 0$ for all e_α; that is, $(y, e_\alpha) = (x, e_\alpha)$ for all α.

So we conclude that

$$y = \Sigma\,(x, e_\alpha)\, e_\alpha\,. \qquad \square$$

Theorems 3.10 and 3.11 enable us to make important existence statements about series representations for elements in an inner product space.

3.12 **Corollary**. *Given an orthonormal set* $\{e_\alpha\}$ *in an inner product space* X, $x \in X$ *has representation in the form* $\Sigma(x, e_\alpha)\, e_\alpha$ *if and only if* $x \in \overline{sp}\,\{e_\alpha\}$. *In any case if* X *is complete then* $\Sigma(x, e_\alpha)\, e_\alpha$ *always defines an element of* X.

To show that an element has such a series representation we need to have methods to determine whether the element belongs to the closed linear span of an orthonormal set. The following fundamental properties of orthogonal coefficients are useful in this regard.

3.13 **Bessel's inequality**.

(i) *Given a finite orthonormal set* $\{e_1, e_2, \ldots, e_n\}$ *in an inner product space* X, *for any* $x \in X$

$$\sum_{k=1}^{n} |(x, e_k)|^2 \leq \|x\|^2.$$

(ii) *Given any orthonormal set* $\{e_\alpha\}$ *in an inner product space* X, *for any* $x \in X$, $\sum |(x, e_\alpha)|^2$ *is well defined over all sequential orderings of the set* $S \equiv \{\alpha : (x, e_\alpha) \neq 0\}$ *and* $\sum |(x, e_\alpha)|^2 \leq \|x\|^2$.

Proof.

(i)
$$0 \leq \|x - \sum_{k=1}^{n} (x, e_k) e_k\|^2 = (x - \sum_{k=1}^{n} (e, e_k) e_k, x - \sum_{j=1}^{n} (x, e_j) e_j)$$
$$= \|x\|^2 - \sum_{k=1}^{n} |(x, e_k)|^2 - \sum_{j=1}^{n} |(x, e_j)|^2 + \sum_{k=1}^{n} \sum_{j=1}^{n} (x, e_k) \overline{(x, e_j)}(e_k, e_j)$$
$$= \|x\|^2 - \sum_{k=1}^{n} |(x, e_k)|^2.$$

(ii) Given $x \in X$, we have from Lemma 3.9 that the set $S \equiv \{\alpha : (x, e_\alpha) \neq 0\}$ is countable, so $\sum |(x, e_\alpha)|^2$ has at most a countable number of nonzero terms. Consider any sequential ordering $\{\alpha_n\}$ of S. Then from (i) we have for any $n \in \mathbb{N}$,
$$\sum_{k=1}^{n} |(x, e_{\alpha_k})|^2 \leq \|x\|^2 \quad \text{so} \quad \sum_{n=1}^{\infty} |(x, e_{\alpha_n})|^2 \leq \|x\|^2.$$
But $\sum |(x, e_{\alpha_n})|^2$ is absolutely convergent so the sum $\sum |(x, e_{\alpha_n})|^2$ is independent of the particular sequential ordering $\{\alpha_n\}$ of S; that is, the sum $\sum |(x, e_\alpha)|^2$ is uniquely defined by $\sum_{n=1}^{\infty} |(x, e_{\alpha_n})|^2$. We conclude that,
$$\sum |(x, e_\alpha)|^2 \leq \|x\|^2.$$
□

3.14 **Definition**. Consider an orthonormal set $\{e_\alpha\}$ in an inner product space X. We say that $x \in X$ satisfies *Parseval's formula* if Bessel's inequality is an equality; that is, if
$$\sum |(x, e_\alpha)|^2 = \|x\|^2.$$
By this we mean that $\sum_{n=1}^{\infty} |(x, e_{\alpha_n})|^2 = \|x\|^2$ for some particular sequential ordering $\{\alpha_n\}$ of the set $S \equiv \{\alpha : (x, e_\alpha) \neq 0\}$. Again the absolute convergence of $\sum |(x, e_{\alpha_n})|^2$ implies that the sum $\sum_{n=1}^{\infty} |(x, e_\alpha)|^2$ is uniquely defined by $\sum_{n=1}^{\infty} |(x, e_{\alpha_n})|^2$.

We now show that an element belongs to the closed linear span of an orthonormal set if and only if it satisfies Parseval's formula.

3.15 **Theorem**. *Given an orthonormal set* $\{e_\alpha\}$ *in an inner product space* X, $x \in \overline{sp}\{x_\alpha\}$ *if and only if* $\sum |(x, e_\alpha)|^2 = \|x\|^2$.

Proof. If $x \in \overline{sp}\{x_\alpha\}$ then by Theorem 3.10, for any given sequential ordering $\{\alpha_n\}$ of the set $S \equiv \{\alpha : (x, e_\alpha) \neq 0\}$,
$$x = \sum_{n=1}^{\infty} (x, e_{\alpha_n}) e_{\alpha_n}.$$

We have then

$$\|x\|^2 = \Big(\sum_{n=1}^{\infty}(x, e_{\alpha_n})e_{\alpha_n}, \sum_{m=1}^{\infty}(x, e_{\alpha_m})e_{\alpha_m}\Big)$$

which by the joint continuity of the inner product implies that

$$\|x\|^2 = \sum_{n=1}^{\infty} |(x, e_{\alpha_n})|^2$$

so we can write

$$\|x\|^2 = \sum |(x, e_\alpha)|^2.$$

Conversely, suppose that Parseval's formula holds. Then for a given particular sequential ordering $\{\alpha_n\}$ of S, given $\varepsilon > 0$ there exists an $\nu \in \mathbb{N}$ such that

$$\|x\|^2 - \sum_{k=1}^{n} |(x, e_{\alpha_k})|^2 < \varepsilon \qquad \text{for all } n > \nu.$$

As in the proof of Bessel's inequality 3.13(i) we have

$$0 \le \|x - \sum_{k=1}^{n}(x, e_{\alpha_k})\, e_{\alpha_k}\|^2 \le \|x\|^2 - \sum_{k=1}^{n} |(x, e_{\alpha_k})|^2 < \varepsilon \quad \text{when } n > \nu.$$

So we conclude that $x \in \overline{sp}\,\{e_\alpha\}$. □

We now examine the particular case of the representation of a Riemann integrable function by its Fourier series.

3.16 **Example**. Consider the real linear space $\mathcal{R}[-\pi,\pi]$ of Riemann integrable functions on $[-\pi,\pi]$. We noted in Example 2.2.12(ii) that

$$(f, g) = \int_{-\pi}^{\pi} fg(t)\, dt$$

is a positive Hermitian form but is not positive definite and

$$p_2(f) = \sqrt{\int_{-\pi}^{\pi} f^2(t)\, dt}$$

is only a seminorm for $\mathcal{R}[-\pi,\pi]$.

If we consider the quotient space $\mathcal{R}[-\pi,\pi]/\ker p_2$ then we have an inner product on this space defined by

$$([f], [g]) = (f, g) \quad \text{for any } f \in [f] \text{ and } g \in [g].$$

We will proceed by dealing with the linear space $\mathcal{R}[-\pi,\pi]$ using our theory for inner product spaces, but keeping in mind that the elements of $\mathcal{R}[-\pi,\pi]$ are only representatives of the equivalence classes of the quotient space $\mathcal{R}[-\pi,\pi]/\ker p_2$ and our results have to be interpreted accordingly.

We show that any Riemann integrable function f on $[-\pi,\pi]$ can be represented by its Fourier series with respect to the orthonormal set

$$\left\{\frac{1}{\sqrt{2\pi}}, \frac{1}{\sqrt{\pi}}\cos nt, \frac{1}{\sqrt{\pi}}\sin nt : n \in \mathbb{N}\right\};$$

that is,

$$f(x) = \frac{1}{2\pi}\int_{-\pi}^{\pi} f(t)dt + \frac{1}{\pi}\sum_{n=1}^{\infty}\left\{\left(\int_{-\pi}^{\pi} f(t)\cos nt\, dt\right)\cos nx + \left(\int_{-\pi}^{\pi} f(t)\sin nt\, dt\right)\sin nx\right\}$$

by which we mean that the Fourier series is convergent to f in the inner product norm; that is, the Fourier series is *mean square* convergent to f.

From the theory of integration, it is clear that if f is Riemann integrable on $[-\pi,\pi]$ then given $\varepsilon > 0$ there exists a step function ϕ on $[-\pi,\pi]$ such that

$$\int_{-\pi}^{\pi} |f-\phi|^2(t)\, dt < \varepsilon .$$

So $\mathcal{R}[-\pi,\pi]$ is contained in the mean square closed linear span of the step functions on $[-\pi,\pi]$.

We show that an elementary step function ϕ defined by

$$\left.\begin{aligned}\phi(t) &= 1 \qquad a \le t \le a+b \text{ where } b \ge 0\\ &= 0 \qquad \text{otherwise}\end{aligned}\right\}$$

satisfies Parseval's formula. It will follow from Theorem 3.15 that the linear space of step functions is contained in the mean square closed linear span of the orthonormal set and so $\mathcal{R}[-\pi,\pi]$ is contained in the mean square closed linear span of the orthonormal set. So we will conclude that any Riemann integrable function f on $[-\pi,\pi]$ has mean square representation by its Fourier series.

Now

$$p_2^2(\phi) = \int_{-\pi}^{\pi} \phi^2(t)\, dt = b.$$

But

$$\sum_{n=1}^{\infty} |(\phi, e_n)|^2 = \frac{1}{2\pi}\left(\int_{-\pi}^{\pi}\phi(t)\, dt\right)^2 + \frac{1}{\pi}\sum_{n=1}^{\infty}\left\{\left(\int_{-\pi}^{\pi}\phi(t)\cos nt\, dt\right)^2 + \left(\int_{-\pi}^{\pi}\phi(t)\sin nt\, dt\right)^2\right\}$$

$$= \frac{b^2}{2\pi} + \frac{1}{\pi}\sum_{n=1}^{\infty}\left\{\left(\int_a^{a+b}\cos nt\, dt\right)^2 + \left(\int_a^{a+b}\sin nt\, dt\right)^2\right\}$$

$$= \frac{b^2}{2\pi} + \frac{1}{\pi}\sum_{n=1}^{\infty}\frac{1}{n^2}\left\{(\sin n(a+b) - \sin na)^2 + (\cos n(a+b) - \cos na)^2\right\}$$

$$= \frac{b^2}{2\pi} + \frac{2}{\pi}\sum_{n=1}^{\infty}\frac{1}{n^2} - \frac{2}{\pi}\sum_{n=1}^{\infty}\frac{\cos nb}{n^2} = \frac{b^2}{2\pi} + \frac{\pi}{3} - \frac{2}{\pi}\sum_{n=1}^{\infty}\frac{\cos nb}{n^2}$$

using the fact that $\sum_{n=1}^{\infty}\frac{1}{n^2} = \frac{\pi^2}{6}$.

Now by Bessel's inequality,

$$\sum_{n=1}^{\infty} |(\phi, e_n)|^2 \leq p_2^2(\phi)$$

so $$b - \left(\frac{b^2}{2\pi} + \frac{\pi}{3} - \frac{2}{\pi}\sum_{n=1}^{\infty} \frac{\cos nb}{n^2}\right) \geq 0.$$

Since $\sum_{n=1}^{\infty} \frac{\cos nt}{n^2}$ is uniformly convergent on $[-\pi,\pi]$ by Weierstrass' M-test, it follows that

$t - \left(\frac{t^2}{2\pi} + \frac{\pi}{3} - \frac{2}{\pi}\sum_{n=1}^{\infty} \frac{\cos nt}{n^2}\right)$ is continuous on $[-\pi,\pi]$, (see AMS §3).

But also from the uniform convergence of $\sum_{n=1}^{\infty} \frac{\cos nt}{n^2}$ it follows, (see AMS §7), that

$$\int_{-\pi}^{\pi} \left\{t - \left(\frac{t^2}{2\pi} + \frac{\pi}{3} - \frac{2}{\pi}\sum_{n=1}^{\infty} \frac{\cos nt}{n^2}\right)\right\} dt = \frac{t^2}{2} - \left(\frac{t^3}{6\pi} + \frac{\pi}{3}t - \frac{2}{\pi}\sum_{n=1}^{\infty} \frac{\sin nt}{n^3}\right)\Bigg]_{-\pi}^{\pi}$$

$$= \pi^2 - \frac{1}{3}\pi^2 - \frac{2}{3}\pi^2 = 0.$$

Since the integrand is continuous and positive we conclude that

$$b = \left(\frac{b^2}{2\pi} + \frac{\pi}{3} - \frac{2}{\pi}\sum_{n=1}^{\infty} \frac{\cos nb}{n^2}\right);$$

that is, ϕ satisfies Parseval's formula

$$\sum_{n=1}^{\infty} |(\phi, e_n)|^2 = p_2^2(\phi).$$

Of course, in classical Fourier analysis interest is centred on uniform and pointwise convergence and not just on mean square convergence. □

3.17 **Definition**. An orthonormal set $\{e_\alpha\}$ in an inner product space X is said to be *maximal* in X if it is not a proper subset of any other orthonormal set in X; that is, if it is maximal in the family of orthonormal subsets of X partially ordered by set inclusion. So $\{e_\alpha\}$ is maximal if and only if there does not exist an $e \in X$, $e \neq 0$ such that $(e_\alpha, e) = 0$ for all α.

3.18 **Examples.**

(i) In Hilbert sequence space $(\ell_2, \|\cdot\|_2)$, consider the orthonormal set $\{e_1, e_2, \dots, e_n, \dots\}$ where

$$e_n \equiv \{0, \dots, 0, \underset{\text{nth place}}{1}, 0, \dots\} \qquad \text{for all } n \in \mathbb{N}.$$

If $e \equiv \{\lambda_1, \lambda_2, \dots, \lambda_n, \dots\} \in \ell_2$ satisfies $(e_n, e) = 0$ for all $n \in \mathbb{N}$ then $\lambda_n = 0$ for all $n \in \mathbb{N}$ and $e = 0$. So $\{e_n\}$ is maximal.

(ii) In $(\mathcal{C}[-\pi,\pi], \|\cdot\|_2)$ the space of real continuous functions on $[-\pi,\pi]$, with norm

$\| f \|_2 = \sqrt{\int_{-\pi}^{\pi} f^2(t)\,dt}$, the orthonormal set of functions $\left\{\frac{\cos nt}{\sqrt{\pi}} : n \in \mathbb{N}\right\}$ is not maximal

since the function $\frac{\sin t}{\sqrt{\pi}}$ satisfies $\int_{-\pi}^{\pi} \cos nt \sin t\, dt = 0$ for all $n \in \mathbb{N}$. □

The following results are similar to those established for Hamel bases in linear spaces. Using the Axiom of Choice in the form of Zorn's Lemma, Appendix A.1, we are able to establish the existence of maximal orthonormal sets.

3.19 **Theorem.** *In any nontrivial inner product space X there exists a maximal orthonormal set. Further, given an orthonormal set in X there exists a maximal orthonormal set containing it as a subset.*

Proof. Given $x \in X$, $x \neq 0$, then $\left\{\frac{x}{\| x \|}\right\}$ is an orthonormal set in X. So X contains an orthonormal set. Consider any orthonormal set $\{e_\alpha\}$ in X and $\mathfrak{F}$ the family of all orthonormal sets containing $\{e_\alpha\}$ as a subset and partially ordered by set inclusion. It is clear that the union of any totally ordered subfamily is an orthonormal set and is an upper bound for the totally ordered subfamily. So by Zorn's Lemma, we deduce that $\mathfrak{F}$ has a maximal orthonormal set. □

The following cardinality result for maximal orthonormal sets is important for the definition of the orthogonal dimension of an inner product space.

3.20 **Theorem.** *In any given inner product space* X, *all maximal orthonormal sets are numerically equivalent.*

Proof. Suppose there exists a finite maximal orthonormal set $\{e_1, e_2, \dots, e_n\}$ and consider $M_n \equiv \overline{sp}\,\{e_1, e_2, \dots, e_n\}$. Suppose that $M_n \neq X$, consider $x \in X \setminus M_n$ and write $x_0 \equiv \sum_{k=1}^{n} (x, e_k)\, e_k$.

Now by Lemma 3.7(ii), we have that $x - x_0 \neq 0$ and $x - x_0$ is orthogonal to M_n.
But this contradicts the maximality of $\{e_1, e_2, \dots, e_n\}$ so $X = M_n$ and X is finite dimensional. As an orthonormal set is linearly independent, $\{e_1, e_2, \dots, e_n\}$ is a basis for X and so every maximal orthonormal set in X has n elements.

Suppose there exist two infinite maximal orthonormal sets $B_1 \equiv \{e_\alpha\}$ and $B_2 \equiv \{f_\beta\}$ in X. For each α, we have from Lemma 3.9 that the set

$$B_2(e_\alpha) \equiv \{\beta : (f_\beta, e_\alpha) \neq 0\}$$

is countable. But also for each β, we have that $(f_\beta, e_\alpha) \neq 0$ for some α, otherwise, $\{e_\alpha\}$ would not be maximal. Therefore $B_2 \subseteq \bigcup_\alpha B_2(e_\alpha)$.

So by Proposition, Appendix A.3.4 there exists a one-to-one mapping of B_2 into B_1. By symmetry of argument and the Schroeder-Bernstein Theorem, Appendix A.2.2 we have our result. □

This result enables us to make the following definition.

3.21 **Definition.** The *orthogonal dimension* of an inner product space X is the cardinal number of a maximal orthonormal set in X.

3.22 **Remark.** For a finite dimensional inner product space a maximal orthonormal set is also a basis for the space.

For an infinite dimensional inner product space, since a maximal orthonormal set is a linearly independent set, the space has infinite Hamel dimension, (see Appendix A.3). However, even for a Hilbert space with countable orthogonal dimension, a Hamel basis has cardinality of the continuum, (see P.R. Halmos, *A Hilbert space problem book*, D. van Nostrand, 1967, p. 170).

3.23 **Definition.** An orthonormal set $\{e_\alpha\}$ in an inner product space X is said to be an *orthonormal basis* for X if for every $x \in X$, we have that $x = \Sigma(x, e_\alpha)\, e_\alpha$; that is, every $x \in X$ can be represented in the inner product norm by its orthogonal expansion with respect to $\{e_\alpha\}$.

3.24 **Remark.** Given an orthonormal set $\{e_\alpha\}$ in an inner product space X, it is clear from Corollary 3.12 that $\{e_\alpha\}$ is an orthonormal basis for X if and only if $X = \overline{\text{sp}}\{e_\alpha\}$. □

3.25 **Example.** In Example 3.16 we showed that every Riemann integrable function on $[-\pi,\pi]$ has mean square representation by its Fourier series. That is, the orthonormal set

$$\left\{\frac{1}{\sqrt{2\pi}}, \frac{1}{\sqrt{\pi}}\cos nt, \frac{1}{\sqrt{\pi}}\sin nt : n \in \mathbb{N}\right\}$$

is an orthonormal basis for this space. □

The following result gives the relation between a maximal orthonormal set and an orthonormal basis.

3.26 **Theorem**. *In an inner product space X, an orthonormal basis is a maximal orthonormal set and if X is complete then a maximal orthonormal set is an orthonormal basis.*

Proof. If an orthonormal set $\{e_\alpha\}$ is not maximal then there exists an $e \in X$, $e \neq 0$ such that $(e_\alpha, e) = 0$ for all α and then $\Sigma(e, e_\alpha)\, e_\alpha = 0$ so $e \neq \Sigma(e, e_\alpha)\, e_\alpha$.
For a complete inner product space X with orthonormal set $\{e_\alpha\}$, we have from Corollary 3.12 that, for any $x \in X$, the sum $\Sigma(x, e_\alpha)\, e_\alpha$ defines an element of X and

$$x - \Sigma(x, e_\alpha)\, e_\alpha \text{ is orthogonal to } e_\beta \text{ for all } \beta.$$

If $\{e_\alpha\}$ is maximal then $x - \Sigma(x, e_\alpha)\, e_\alpha = 0$; that is, $x = \Sigma(x, e_\alpha)\, e_\alpha$.
We conclude that $\{e_\alpha\}$ is an orthonormal basis for X. □

Most of the common inner product spaces are separable and separability is related to countable orthogonal dimension.

3.27 **Theorem**. *An inner product space X is separable if and only if it has a countable orthonormal basis.*

Proof. Suppose that X is separable. Then there exists a countable set $\{x_n\}$ dense in X. First we generate a linearly independent set $\{y_n\}$ such that

$$\overline{sp}\{y_n\} = \overline{sp}(x_n\}= X.$$

Denote by y_1 the element $x_1 \neq 0$ and
by y_2 the element x_{n_1} first in the sequence $\{x_n\}$ such that $y_2 \notin sp\{y_1\}$
by y_k the element x_{n_k} first in the sequence $\{x_n\}$ such that $y_k \notin sp\{y_1, y_2, \ldots, y_{k-1}\}$.
Then for each $k \in \mathbf{N}$,

$$sp\{y_i : i \in \{1, 2, \ldots, k\}\} = sp\{x_i : i \in \{1, 2, \ldots, n_k\}\}$$

and so

$$\overline{sp}\{y_n\} = \overline{sp}(x_n\}= X.$$

We now apply Gram-Schmidt orthogonalisation to the sequence $\{y_n\}$ and obtain an orthonormal sequence $\{e_n\}$ such that for each $n \in \mathbf{N}$,

$$sp\ \{e_k : k \in \{1, 2, \ldots, n\}\} = sp\ \{y_k : k \in \{1, 2, \ldots, n\}\}$$

and

$$\overline{sp}\{y_n\} = \overline{sp}(x_n\}= X.$$

By Remark 3.24, $\{e_n\}$ is an orthonormal basis for X.

Conversely, if $\{e_n\}$ is a countable orthonormal basis for X then $X = \overline{sp}\{e_n\}$. Now the set of all rational linear combinations of elements of $\{e_n\}$ is countable and is clearly dense in X. So we conclude that X is separable. □

3.28 **Remark.** For a separable inner product space X its countable orthonormal basis $\{e_n\}$ is a Schauder basis. To show this we need only verify that an element $x \in X$ has a unique representation in terms of its orthonormal basis. But if $x = \sum_{k=1}^{\infty} \lambda_k e_k$ then since $\{e_n\}$ is an orthonormal set and the inner product is jointly continuous we have that $(x, e_n) = \lambda_n$ for all $n \in \mathbb{N}$. □

We notice from Bessel's inequality 3.13(ii) that in any inner product space X, given an orthonormal sequence $\{e_n\}$, for any $x \in X$, we have that $\sum |(x, e_n)|^2$ is convergent. Completeness is important in specifying those orthogonal sequence expansions which belong to the space.

3.29 **Theorem.** *Consider an orthonormal sequence* $\{e_n\}$ *in a Hilbert space* H *and a sequence of scalars* $\{\lambda_n\}$. *Then* $\sum \lambda_n e_n$ *is convergent if and only if* $\sum |\lambda_n|^2 < \infty$ *and in this case for* $x = \sum \lambda_n e_n$ *we have* $\lambda_n = (x, e_n)$ *for all* $n \in \mathbb{N}$.

Proof. For $m > n$,

$$\| \sum_{k=n}^{m} \lambda_k e_k \|^2 = \sum_{k=n}^{m} |\lambda_k|^2$$

so $\sum \lambda_n e_n$ is Cauchy if and only if $\sum |\lambda_n|^2$ is Cauchy. But the space is complete so $\sum \lambda_n e_n$ is convergent if and only if $\sum |\lambda_n|^2$ is convergent. It follows from the fact that $\{e_n\}$ is an orthonormal sequence and the inner product is jointly continuous that

$$\lambda_n = (x, e_n) \text{ for all } n \in \mathbb{N}.$$

□

This result has application for Fourier analysis in classical Hilbert space $\mathfrak{L}_2[-\pi,\pi]$ of functions f where f^2 is Lebesgue integrable on $[-\pi,\pi]$, factored over $\ker p_2$ where

$$p_2(f) = \sqrt{\int_{[-\pi,\pi]} f^2 \, d\mu}\,.$$

We deduce the following as a corollary.

3.30 **The Riesz-Fischer Theorem.**
Given sequences of scalars $\{\alpha_n\}$ *and* $\{\beta_n\}$ *such that* $\sum |\alpha_n|^2 < \infty$ *and* $\sum |\beta_n|^2 < \infty$, *then the series*

$$\frac{\alpha}{\sqrt{2\pi}} + \frac{1}{\sqrt{\pi}} \sum (\alpha_n \cos nt + \beta \sin nt)$$

converges to some $f \in \mathfrak{L}_2[-\pi,\pi]$ *and* α, α_n *and* β_n *are the Fourier coefficients of* f *with respect to the orthonormal sequence*

$$\left\{\frac{1}{\sqrt{2\pi}}, \frac{\cos nt}{\sqrt{\pi}}, \frac{\sin nt}{\sqrt{\pi}}\right\}.$$

We now seek to classify Hilbert spaces on the basis of orthogonal dimension. This generalises a similar classification of linear spaces by Hamel dimension.

3.31 **Theorem**. *Hilbert spaces over the same scalar field and of the same orthogonal dimension are isometrically isomorphic.*

Proof. Consider Hilbert spaces H_1 and H_2 over the same scalar field with orthonormal bases $\{e_\alpha\}$ and $\{f_\alpha\}$ of the same cardinality.
Consider a one-to-one mapping T of $\{e_\alpha\}$ onto $\{f_\alpha\}$. Given $x \in H$, we have that

$$x = \Sigma(e, e_\alpha)\, e_\alpha.$$

Consider the extension of T as a mapping of H_1 and H_2 defined by

$$Tx = \Sigma(e, e_\alpha)\, T(e_\alpha).$$

We show that such an extension is well defined.
By Lemma 3.9, $\Sigma(e, e_\alpha)$ has only a countable number of nonzero terms so $\Sigma(e, e_\alpha)\, T(e_\alpha)$ has only a countable number of nonzero terms.
By Bessel's inequality 3.13(ii), $\Sigma\, |(e, e_\alpha)|^2 < \infty$
so by Theorem 3.29, $\Sigma(e, e_\alpha)\, T(e_\alpha)$ defines an element in H_2.
Now for each $x \in H$, $\|Tx\|^2 = \Sigma\, |(e, e_\alpha)|^2 = \|x\|^2$ by Theorem 3.15, so T is norm preserving.
But from the linearity and continuity properties of the inner product we deduce that T is linear. So T is an isometric isomorphism of H_1 into H_2.
For $y \in H_2$, $\quad \Sigma\, |(y, f_\alpha)|^2 \|y\|^2 < \infty$
by Theorem 3.15, so

$$x = \Sigma(y, f_\alpha)\, T^{-1}(f_\alpha) \in H_1$$

and

$$Tx = \Sigma(y, f_\alpha)\, f_\alpha = y,$$

which shows that T is onto. □

This general result has the following implications.

3.32 **Corollary**.

(i) *Every real (complex)* n*-dimensional inner product space is isometrically isomorphic to* $(\mathbb{R}^n, \|\cdot\|_2)$ $((\mathbb{C}^n, \|\cdot\|_2))$.

(ii) *Every real (complex) separable Hilbert space is isometrically isomorphic to real (complex) Hilbert sequence space* $(\ell_2, \|\cdot\|_2)$.

3.33 EXERCISES

1. Consider the linear space $\mathcal{C}[-\pi,\pi]$ of real continuous functions on $[-\pi,\pi]$ with norm

$$\| f \|_2 = \sqrt{\int_{-\pi}^{\pi} f^2(t)\,dt}$$

and the linear subspace M spanned by $\{\sin t, \sin 2t\}$.

(i) For the function f_0 on $[-\pi,\pi]$ where $f_0(t) = t$, find the unique best approximating function g_0 in M.

(ii) Show that g_0 is also best approximating with respect to the norm

$$\| f \|_\infty = \max\{ | f(t) | : t \in [-\pi,\pi] \}$$

but is not unique.

2. Consider the sequence of polynomials,

$$\{1, t, t^2, \ldots, t^n, \ldots\} \text{ on } [-1,1].$$

Prove that the orthonormal sequence of polynomials which results from Gram-Schmidt orthogonalisation of this sequence in the space $(\mathcal{C}[-1,1], \|\cdot\|_2)$ has the form

$$\{e_0, e_1, e_2, \ldots, e_n, \ldots\}$$

where $sp\{1, t, \ldots, t^n\} = sp\{e_0, e_1, \ldots, e_n\}$ for $n \in \{0, 1, 2, 3, \ldots\}$ and

$$e_n(t) = \sqrt{2n + \frac{1}{2}}\, P_n(t) \text{ using the notation } P_n(t) = \frac{1}{2^n n!}\frac{d^n}{dt^n}(t^2-1)^n.$$

Determine the form of the first three Legendre polynomials, P_0, P_1 and P_2.

3. Consider the linear space $\mathcal{C}[-1,1]$ of continuous real functions on $[-1,1]$ with norm

$$\| f \|_2 = \sqrt{\int_{-1}^{1} f^2(t)\,dt}$$

and the linear subspace M spanned by the Legendre polynomials

$$\left\{ \sqrt{\tfrac{1}{2}}, \sqrt{\tfrac{3}{2}}\,t, \sqrt{\tfrac{5}{8}}\,(3t^2-1) \right\}.$$

(i) For the function f_0 on M where $f_0(t) = | t |$ find the unique best approximating function g_0 in M.

(ii) Is g_0 also best approximating with respect to the norm

$$\| f \|_\infty = \max\{ | f(t) | : t \in [-1,1] \}$$

and if so is it unique?

4. Consider the linear space $\mathcal{C}[-\pi,\pi]$ of real continuous functions on $[-\pi,\pi]$ with norm

$$\|f\|_2 = \sqrt{\int_{-\pi}^{\pi} f^2(t)\,dt}\ .$$

(i) Prove that the linear subspace M of even functions is a closed linear subspace of $(\mathcal{C}[-\pi,\pi], \|\cdot\|_2)$.

(ii) Prove that the set $\{\frac{1}{\sqrt{2\pi}}, \frac{1}{\sqrt{\pi}}\cos nt : n \in \mathbb{N}\}$ is an orthonormal basis for M.

(iii) For the function f_0 on $[-\pi,\pi]$ where $f_0(t) = e^t$ find the unique best approximating element g_0 in M.

(iv) Although $(\mathcal{C}[-\pi,\pi], \|\cdot\|_2)$ is not complete verify that the best even function g_0 approximation to f_0 found by computation in (iii) is actually in $\mathcal{C}[-\pi,\pi]$.

5. (i) The *Chebyshev polynomials* $\{T_n\}$ on $[-1,1]$ are defined as follows:

$$T_0(t) = 1, \quad T_1(t) = t \quad \text{and}$$
$$T_{n+1}(t) = 2t\,T_n(t) - T_{n-1}(t) \quad \text{for all } n \in \mathbb{N}.$$

Prove that the best approximation to $f_0(t) = t^3$ from $\text{sp}\{1, t, t^2\}$ in $(\mathcal{C}[-1,1], \|\cdot\|_\infty)$ is the function

$$g_0(t) = t^3 - \frac{1}{4}\,T_3(t)\ .$$

(ii) Find $a, b, c \in \mathbb{R}$ to minimise the integral

$$\int_{-1}^{1} |\,t^3 - a - bt - ct^2\,|\,dt\ .$$

6. (i) Given an inner product space X with an orthonormal sequence $\{e_n\}$, prove that for any $x \in X$, $\lim_{n\to\infty} (x, e_n) = 0$.

(ii) Consider the linear space $\mathcal{C}[-\pi,\pi]$ of real continuous functions on $[-\pi,\pi]$ with inner product

$$(f, g) = \int_{-\pi}^{\pi} fg(t)\,dt.$$

Deduce for any $f \in \mathcal{C}[-\pi,\pi]$,

$$\lim_{n\to\infty} \int_{-\pi}^{\pi} f(t)\sin nt\,dt = 0 \quad \text{and} \quad \lim_{n\to\infty} \int_{-\pi}^{\pi} f(t)\cos nt\,dt = 0$$

(This is a special case of the *Riemann–Lebesgue Lemma*).

(iii) Does this result extend to the linear space $\mathcal{R}[-\pi,\pi]$ of Riemann integrable functions on $[-\pi,\pi]$ with positive hermitian form

$$(f, g) = \int_{-\pi}^{\pi} fg(t)\,dt\ ?$$

(iv) For $f_0 \in \mathcal{C}[-\pi,\pi]$ we are given that

$$\int_{-\pi}^{\pi} f_0(t) \sin nt \, dt = 0 \qquad \text{for all } n \in \mathbb{N}.$$

Prove that f_0 is an even function.

7. In Hilbert sequence space $(\ell_2, \|\cdot\|_2)$ consider the linear subspace $E \equiv sp\{x_0, E_0\}$ where $x_0 \equiv \{1, \frac{1}{2}, \ldots, \frac{1}{n}, \ldots\}$ and the orthonormal set $\{e_{2n}\}$ where $e_{2n} \equiv \{0, \ldots, 0, \underset{2n\text{ th place}}{1}, 0, \ldots\}$. Show that the orthogonal series for x, $\Sigma(x_0, e_{2n})e_{2n}$ is not convergent in E. Explain why this is so.

8. Consider the real linear space $\mathcal{R}[-\pi,\pi]$ with semi–norm

$$p_2(f) = \sqrt{\int_{-\pi}^{\pi} f^2(t)\,dt} \; .$$

Prove that $\mathcal{C}[-\pi,\pi]$ is mean square dense in $\mathcal{R}[-\pi,\pi] / \ker p_2$.

9. (i) Prove that if an inner product space has a countable Hamel basis then it has a countable orthonormal basis.

(ii) prove that if an inner produce space has a Schauder basis then it has a countable orthonormal basis.

10. Consider an orthonormal sequence $\{e_n\}$ in a Hilbert space H. Prove that $\{e_n\}$ is maximal if and only if

$$(x, y) = \sum_{k=1}^{\infty} (x, e_n)\overline{(y, e_n)} \quad \text{for all } x, y \in H.$$

11. Consider an inner product space X with an orthonormal set $\{e_\alpha\}$.

(i) Prove that if Parseval's relation holds for each $x \in X$ then $\{e_\alpha\}$ is maximal.

(ii) Prove that if X is complete then the converse holds.

12. Consider the linear space of real functions f on $\mathbb{R}$ where for each f the set $\{x \in \mathbb{R} : f(x) \neq 0\}$ is countable and $\Sigma\, |f(x)|^2 < \infty$, with inner product

$$(f, g) = \Sigma\, f(x)\, g(x) \,.$$

Prove that this space is a nonseparable Hilbert space.

II. SPACES OF CONTINUOUS LINEAR MAPPINGS

The richness of linear space structure enables us to construct new families of normed linear spaces from old through the linear mappings between them. Generating and relating spaces in this way is an activity which distinguishes the analysis of normed linear spaces from that of metric spaces. But further it is this process which is the basis of linear operator theory with its important areas of application.

§4. NORMING MAPPINGS AND FORMING DUALS AND OPERATOR ALGEBRAS

The natural association of a norm with the linear space of continuous linear mappings between normed linear spaces is the key factor enabling us to consider new normed linear spaces. The newly formed spaces inherit properties from the original spaces. Of particular interest are dual spaces and operator algebras which have a structural richness not available for general normed linear spaces.

4.1 **Definitions.**

(i) Given linear spaces X and Y over the same scalar field, the set $\mathfrak{L}(X,Y)$ of linear mappings of X into Y is a linear space under pointwise definition of addition and multiplication by a scalar; that is, for $T,S \in \mathfrak{L}(X,Y)$,

$$T+S: X \to Y \text{ is defined by } (T+S)(x) = Tx + Sx$$

and $$\alpha T: X \to Y \text{ is defined by } (\alpha T)(x) = \alpha Tx.$$

There is no difficulty in verifying that $T+S$ and $\alpha T \in \mathfrak{L}(X,Y)$ and that the linear space properties hold.

(ii) Given normed linear spaces $(X, \|\cdot\|)$ and $(Y, \|\cdot\|')$ over the same scalar field, the set $\mathfrak{B}(X,Y)$ of continuous linear mappings of X into Y is a linear subspace of $\mathfrak{L}(X,Y)$. Closure under the linear operations follows from the closure of continuity under these operations.

These spaces are in fact generalisations of the finite dimensional case.

4.2 **Example.** When $X = \mathbb{R}^n$ (or $\mathbb{C}^n$) and $Y = \mathbb{R}^m$ (or $\mathbb{C}^m$) then $\mathfrak{B}(X,Y)$ is isomorphic to the linear space $M_{m\times n}$ of $m \times n$ matrices with entries from $\mathbb{R}$ (or $\mathbb{C}$).

Proof. From Corollary 2.1.10 it follows that in this case $\mathfrak{L}(X,Y) = \mathfrak{B}(X,Y)$.
Given a basis $\{e_1,e_2,\ldots,e_n\}$ of X and a basis $\{f_1,f_2,\ldots,f_m\}$ of Y and a linear mapping $T: X \to Y$ then for each $k \in \{1,2,\ldots,n\}$

$$Te_k = \sum_{j=1}^{m} \alpha_{jk} f_j$$

and $[\alpha_{jk}]$ is a uniquely determined $m \times n$ matrix of scalars.
Since each $x \in X$ has unique representation

$$x = \sum_{k=1}^{n} \lambda_k e_k$$

and T is linear

$$Tx = \sum_{k=1}^{n} \lambda_k Te_k = \sum_{j=1}^{m} \sum_{k=1}^{n} (\alpha_{jk} \lambda_k) f_j .$$

Conversely, given an $m \times n$ matrix $[\alpha_{jk}]$ of scalars, the mapping $T: X \to Y$ defined by

$$Tx = \sum_{j=1}^{m} \sum_{k=1}^{n} (\alpha_{jk} \lambda_k) f_j$$

is clearly linear.
Moreover, it is easily verified that this one-to-one correspondence between $\mathfrak{B}(X,Y)$ and $M_{m\times n}$, the linear space of $m \times n$ matrices of corresponding scalars, is also a linear mapping. □

However, $\mathfrak{B}(X,Y)$ derives norm structure from the norms on X and Y.

4.3 **Theorem.** *Given normed linear spaces* $(X, \|\cdot\|)$ *and* $(Y, \|\cdot\|')$, $\mathfrak{B}(X,Y)$ *is a normed linear space with norm*

$$\|T\| = \sup\{ \|Tx\|' : \|x\| \le 1\}.$$

Proof. Since T is continuous there exists an $M > 0$ such that

$$\|Tx\|' \le M\|x\| \qquad \text{for all } x \in X$$

so

$$\|Tx\|' \le M \qquad \text{for all } \|x\| \le 1$$

and therefore the norm $\|\cdot\|$ on $\mathfrak{B}(X,Y)$ is well defined. We check those norm properties which do not follow directly from the definition.
When $\|T\| = 0$ then $\|Tx\|' = 0$ for all $\|x\| \le 1$ and so $\|Tx\|' = 0$ for all $x \in X$. From the norm properties of $\|\cdot\|'$ we see that $Tx = 0$ for all $x \in X$ and so $T = 0$.
For the triangle inequality we have for $S,T \in \mathfrak{B}(X,Y)$

$$\begin{aligned}\| S+T \| &= \sup\{ \| (S+T)x \|' : \| x \| \le 1 \} \\ &\le \sup\{ \| Sx \|' + \| Tx \|' : \| x \| \le 1 \} \\ &\le \sup\{ \| Sx \|' : \| x \| \le 1 \} + \sup\{ \| Tx \|' : \| x \| \le 1 \} \\ &= \| S \| + \| T \|.\end{aligned}$$

□

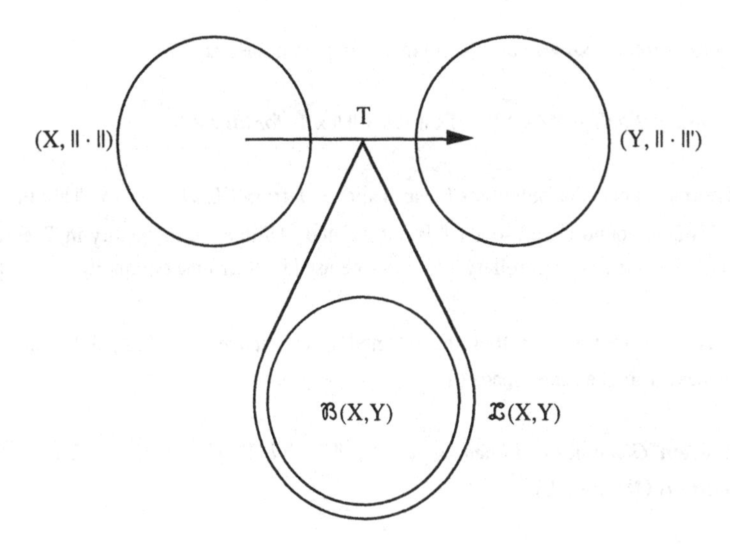

Figure 3. The generation of $\mathfrak{L}(X,Y)$ and $\mathfrak{B}(X,Y)$ from $(X, \|\cdot\|)$ and $(Y, \|\cdot\|')$.

There is an alternative expression for the norm of a continuous linear mapping which is useful.

4.4 **Corollary**. *For* $T \in \mathfrak{B}(X,Y)$,

$$\| T \| = \inf\{ M : \| Tx \|' \le M \| x \| \textit{ for all } x \in X \}.$$

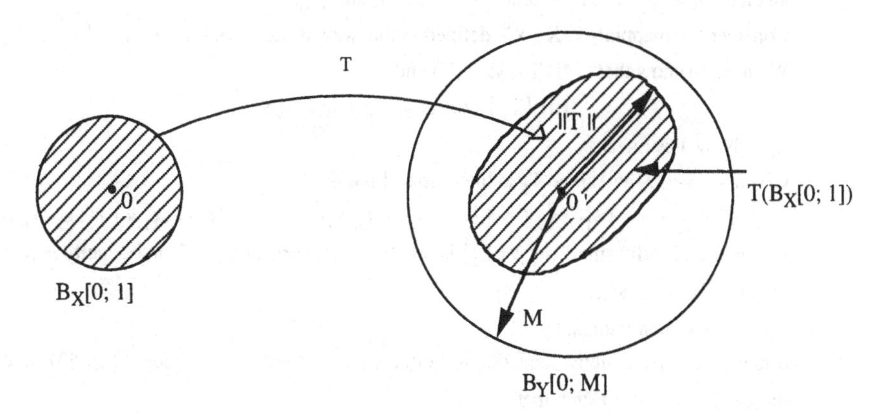

Figure 4. Visualisation of the alternative expression for $\| T \|$.

Proof. Since $\| T \| = \sup\{ \| Tx \|' : \| x \| \leq 1\}$

then $\| T \| \geq \| Tx \|'$ for all $\| x \| \leq 1$

so $\| T \| \geq \inf\{M : \| Tx \|' \leq M \text{ for all } \| x \| \leq 1\}$.

But clearly, $\inf\{M : \| Tx \|' \leq M \text{ for all } \| x \| \leq 1\} \geq \sup\{ \| Tx \|' : \| x \| \leq 1\} = \| T \|$. □

From Corollary 4.4 we have the following important inequality.

4.5 **Corollary.** *For* $T \in \mathcal{B}(X,Y)$, $\| Tx \|' \leq \| T \| \| x \|$ *for all* $x \in X$.

4.6 **Remark.** For a discontinuous linear mapping T from $(X, \|\cdot\|)$ into $(Y, \|\cdot\|')$, the set $T(B[0; 1])$ is unbounded and so $\| T \|$ is not defined. Unlike the inequality in Theorem 1.24.1(i), the inequality in Corollary 4.5 cannot be used to determine continuity. □

It is important to see that the normed linear space $(\mathcal{B}(X,Y), \|\cdot\|)$ inherits completeness from the range space $(Y, \|\cdot\|')$.

4.7 **Theorem.** *Given normed linear spaces* $(X, \|\cdot\|)$ *and* $(Y, \|\cdot\|')$, *if* $(Y, \|\cdot\|')$ *is complete then so also is* $(\mathcal{B}(X,Y), \|\cdot\|)$.

Proof. Consider a Cauchy sequence $\{T_n\}$ in $(\mathcal{B}(X,Y), \|\cdot\|)$; that is, given $\varepsilon > 0$ there exists a $\nu \in \mathbb{N}$ such that

$$\| T_n - T_m \| < \varepsilon \quad \text{for all } m,n > \nu.$$

So for any given $x \in X$,

$$\begin{aligned} \| T_n x - T_m x \|' &\leq \| T_n - T_m \| \| x \| \\ &< \varepsilon \| x \| \quad \text{for all } m,n > \nu ; \end{aligned}$$

that is, $\{T_n x\}$ is a Cauchy sequence in $(Y, \|\cdot\|')$. Since $(Y, \|\cdot\|')$ is complete there exists an element in Y, which we denote by Tx, such that $\{T_n x\}$ converges to Tx.

Consider the mapping $T: X \to Y$ defined in this way as the pointwise limit of $\{T_n\}$.

We need to show that (i) $T \in \mathcal{B}(X,Y)$ and

(ii) $\{T_n\}$ converges to T in $(\mathcal{B}(X,Y), \|\cdot\|)$.

(i) Now T is linear:

Given $x_1, x_2 \in X$ and since T_n is linear for all $n \in \mathbb{N}$,

$$\| T(x_2+x_2) - Tx_1 - Tx_2 \|' \leq \| T(x_1+x_2) - T_n(x_1+x_2) \|' + \| Tx_1 - T_n x_1 \|' + \| Tx_2 - T_n x_2 \|'$$

which is arbitrarily small since $\{T_n\}$ is pointwise convergent to T. Homogeneity is proved by a similar argument.

But T is also continuous:

Since $\{T_n\}$ is a Cauchy sequence in $(\mathcal{B}(X,Y), \|\cdot\|)$ it is bounded, (see AMS §3); that is, there exists an $M > 0$ such that

$$\| T_n \| \leq M \quad \text{for all } n \in \mathbb{N}.$$

Therefore, given $x \in X$

$$\| Tx \|' \leq \| Tx - T_n x \|' + \| T_n x \|' \leq M \| x \|$$

since $\{T_n\}$ is pointwise convergent to T.

We conclude that $T \in \mathfrak{B}(X,Y)$.

(ii) We had, for all $x \in X$

$$\| T_n x - T_m x \|' < \varepsilon \| x \| \qquad \text{for all } m,n > \nu .$$

So keeping n fixed and increasing m we have for all $\| x \| \leq 1$,

$$\| T_n x - Tx \|' \leq \varepsilon \quad \text{for all } n > \nu.$$

Therefore, $\| T_n - T \|' \leq \varepsilon \quad$ for all $n > \nu$;

that is, $\{T_n\}$ converges to T in $(\mathfrak{B}(X,Y), \|\cdot\|)$. □

We will find the following extension property of continuous linear mappings particularly useful.

4.8 **Theorem**. *Consider a normed linear space* $(X, \|\cdot\|)$ *and a Banach space* $(Y, \|\cdot\|')$ *and* T *a continuous linear mapping from a dense linear subspace* A *of* $(X, \|\cdot\|)$ *into* $(Y, \|\cdot\|')$. *There exists a uniquely determined continuous linear extension* $\tilde{T}: X \to Y$ *and* $\| \tilde{T} \| = \| T \|_A$.

Proof. To define $\tilde{T}$ consider $x \in X \setminus A$ and a sequence $\{a_n\}$ in A such that $\{a_n\}$ converges to x in $(X, \|\cdot\|)$. Then $\{a_n\}$ is a Cauchy sequence in A and

$$\| Ta_n - Ta_m \|' = \| T(a_n - a_m) \|' \leq \| T \| \| a_n - a_m \|'$$

so that $\{Ta_n\}$ is a Cauchy sequence in $(Y, \|\cdot\|')$. Since $(Y, \|\cdot\|')$ is complete there exists a $y \in Y$ such that $\{Ta_n\}$ is convergent to y in $(Y, \|\cdot\|')$. We define $\tilde{T}x = y$.

To show that this definition is independent of the choice of sequence $\{a_n\}$ convergent to x we note that for any other sequence $\{a_n'\}$ convergent to x we have

$$\| Ta_n' - Ta_n \| \leq \| T \| \| a_n' - a_m \|$$

so $\{Ta_n'\}$ converges to the same limit as $\{Ta_n\}$ in $(Y, \|\cdot\|')$.

Clearly the mapping $\tilde{T}: X \to Y$ defined in this way is an extension of T to X.

To prove linearity of $\tilde{T}$ we consider sequences $\{a_n\}$ convergent to x_1 and $\{a_n'\}$ convergent to x_2 in $(X, \|\cdot\|')$. Then

$$\| \tilde{T}(x_1+x_2) - \tilde{T}x_1 - \tilde{T}x_2 \|' \leq \| \tilde{T}(x_1+x_2) - T(a_n + a_n') \|' + \| Ta_n - \tilde{T}x_1 \|' + \| Ta_n' - \tilde{T}x_2 \|'.$$

So $\tilde{T}$ is additive and homogeneity follows by a similar argument.

Since A is a subspace of $(X, \|\cdot\|)$ we have

$$\| \tilde{T} \| > \| T \|_A .$$

But for sequence $\{a_n\}$ convergent to x in $(X, \|\cdot\|)$ we have

$$\| \tilde{T}x \|' < \| \tilde{T}x - Ta_n \|' + \| Ta_n \|' \leq \| \tilde{T}x - Ta_n \|' + \| T \|_A \| a_n \|'$$

so $\qquad \|\tilde{T}x\|' \leq \|T\|_A \|x\| \qquad$ for all $x \in X$

and $\qquad \|\tilde{T}\| \leq \|T\|_A$.

Therefore, $\qquad \|\tilde{T}\| = \|T\|_A$.

The uniqueness of $\tilde{T}$ follows from the fact that A is dense in $(X, \|\cdot\|)$. □

This theorem has the following consequences.

4.9 **Corollary**. *Consider a normed linear space* $(X, \|\cdot\|)$, *a Banach space* $(Y, \|\cdot\|')$ *and a linear subspace* A *dense in* $(X, \|\cdot\|)$. *Then* $(\mathfrak{B}(A,Y), \|\cdot\|)$ *and* $(\mathfrak{B}(X,Y), \|\cdot\|)$ *are isometrically isomorphic.*

There are special cases of continuous linear mappings which are important to identify and discuss. Such cases are consequences of the presence of linear structure and will be recognised as a development of those treated in a purely linear algebra study.

4.10 Dual spaces

Given normed linear spaces $(X, \|\cdot\|)$ and $(Y, \|\cdot\|')$ our first special case arises when we take the range space Y as the scalar field of the domain space X. We then have the normed linear space of continuous linear functionals on $(X, \|\cdot\|)$.

4.10.1 Definitions.

(i) Given a linear space X over $\mathbb{C}$ (or $\mathbb{R}$), the *algebraic dual* (or *algebraic conjugate*) space is the linear space $\mathfrak{L}(X,\mathbb{C})$ (or $\mathfrak{L}(X,\mathbb{R})$) usually denoted by $X^\#$.

(ii) Given a normed linear space $(X, \|\cdot\|)$ over $\mathbb{C}$ (or $\mathbb{R}$), the *dual* (or *conjugate*) space is the normed linear space $(\mathfrak{B}(X,\mathbb{C}), \|\cdot\|)$ $\big($or $(\mathfrak{B}(X,\mathbb{R}), \|\cdot\|)\big)$ usually denoted by $(X, \|\cdot\|)^*$. When we are thinking of the dual as a linear space we denote it by X^* and as a normed linear space with its norm, by $(X^*, \|\cdot\|)$. The norm on X^* is given by

$$\|f\| = \sup\{|f(x)| : \|x\| \leq 1\}$$

and from Corollary 4.5 we have the inequality

$$|f(x)| \leq \|f\| \|x\| \qquad \text{for all } x \in X.$$

4.10.2 **Remark**. Given a normed linear space $(X, \|\cdot\|)$, it is obvious that X^* is a linear subspace of $X^\#$. However, it follows from Theorem 2.1.12 that $X^* = X^\#$ if and only if X is finite dimensional so when X is infinite dimensional X^* is always a proper linear subspace of $X^\#$. □

Since the scalar field is complete, Theorem 4.7 provides the following important property of dual spaces.

4.10.3 **Corollary**. *Whether a normed linear space* $(X, \|\cdot\|)$ *is complete or not, its dual space* $(X^*, \|\cdot\|)$ *is always complete.*

Furthermore, Theorem 4.8 gives us this additional information.

4.10.4 **Corollary**. *Consider a normed linear space* $(X, \|\cdot\|)$ *and a linear subspace* A *dense in* $(X, \|\cdot\|)$. *Then* $(A, \|\cdot\|_A)^*$ *is isometrically isomorphic to* $(X, \|\cdot\|)^*$.

We should recall firstly, the precise algebraic situation for the finite dimensional case.

4.10.5 **Theorem**. *For an* n*-dimensional linear space* X_n *over* $\mathbb{C}$ (or $\mathbb{R}$) *with basis* $\{e_1, e_2, \dots, e_n\}$, *the algebraic dual* $X_n^{\#}$ *is also an* n*-dimensional linear space with basis* $\{f_1, f_2, \dots, f_n\}$ *where for* $k \in \{1,2, \dots, n\}$

$$f_k(e)_j \left.\begin{aligned} &= 1 \quad \textit{when} \quad j = k \\ &= 0 \qquad\qquad j \neq k \end{aligned}\right\} .$$

Proof. Since each $x \in X_n$ has a unique representation in terms of the basis $\{e_1,e_2, \dots, e_n\}$, X_n is isomorphic to $\mathbb{C}^n$ (or $\mathbb{R}^n$) under the mapping

$$x \mapsto (\lambda_1, \lambda_2, \dots, \lambda_n) \text{ where } x \equiv \sum_{k=1}^{n} \lambda_k e_k .$$

For each $k \in \{1, 2, \dots, n\}$ consider the linear functional f_k on X_n defined for $x \equiv \sum_{k=1}^{n} \lambda_k e_k$ by $f_k(x) = \lambda_k$.

Consider the set $\{f_1, f_2, \dots, f_n\}$ in $X_n^{\#}$.

For any linear functional f on X_n,

$$f(x) = \sum_{k=1}^{n} \lambda_k f(e_k) = \sum_{k=1}^{n} f(e_k)\, f_k(x) ;$$

that is,

$$f = \sum_{k=1}^{n} f(e_k)\, f_k.$$

So $\{f_1, f_2, \dots, f_n\}$ spans $X_n^{\#}$.

If $\sum_{k=1}^{n} \alpha_k f_k = 0$ then $\sum_{k=1}^{n} \alpha_k f_k(x) = 0$ for all $x \in X_n$ so for $\sum_{k=1}^{n} \bar{\alpha}_k e_k$ we have $\sum_{k=1}^{n} |\alpha_k|^2 = 0$ which implies that $\alpha_k = 0$ for all $k \in \{1,2, \dots, n\}$; that is, $\{f_1, f_2, \dots, f_n\}$ is linearly independent.

We conclude that $\{f_1, f_2, \dots, f_n\}$ is a basis for $X_n^{\#}$. It is called the basis of $X_n^{\#}$ *dual* to the basis $\{e_1, e_2, \dots, e_n\}$ for X_n. □

One of the simplest examples concerns continuous linear functionals generated by an inner product in an inner product space.

4.10.6 **Example.** In an inner product space X we noted in Remarks 2.2.2 that given $z \in X$, the functional f_z defined on X by

$$f_z(x) = (x, z) \qquad \text{for all } x \in X$$

is linear.

But the Cauchy–Schwarz inequality gives us that

$$|f_z(x)| = |(x, z)| \leq \|z\| \|x\| \quad \text{for all } x \in X$$

which implies that f_z is continuous on X.

Now clearly,

$$\|f_z\| = \inf\{M : |f_z(x)| \leq M \|x\| \text{ for all } x \in X\} \leq \|z\|.$$

But $$f_z\left(\frac{z}{\|z\|}\right) = \left(\frac{z}{\|z\|}, z\right) = \|z\|$$

so $$\|f_z\| = \sup\{|f_z(x)| : \|x\| \leq 1\} = \|z\|.$$ □

A continuous linear functional need not attain its norm on the closed unit ball. We will see later that whether it does or not has some significance.

4.10.7 **Example.** Consider the Banach space $(c_0, \|\cdot\|_\infty)$. Given $y \equiv \{\mu_1, \mu_2, \ldots, \mu_n, \ldots\} \in \ell_1 \setminus \{0\}$, consider the linear functional f_y defined on c_0 by

$$f_y(x) = \sum \lambda_n \bar{\mu}_n \quad \text{for } x \equiv \{\lambda_1, \lambda_2, \ldots, \lambda_n, \ldots\} \in c_0.$$

Now

$$\begin{aligned} |f_y(x)| &\leq \sup\{|\lambda_n| : n \in \mathbf{N}\}(\sum |\mu_n|) \\ &\leq \|y\|_1 \|x\|_\infty \quad \text{for all } x \in c_0 \end{aligned}$$

so f_y is continuous and $\|f_y\| \leq \|y\|_1$.

Now consider the sequence $\{x_n\}$ where

$$\begin{aligned} x_1 &\equiv \{\operatorname{sgn} \mu_1, 0, \ldots \} \\ x_2 &\equiv \{\operatorname{sgn} \mu_1, \operatorname{sgn} \mu_2, 0, \ldots \} \\ &\ldots \\ x_n &\equiv \{\operatorname{sgn} \mu_1, \ldots, \operatorname{sgn} \mu_n, 0, \ldots\} \\ &\ldots \end{aligned}$$

Then $x_n \in c_0$ and $\|x_n\|_\infty \leq 1$ for all $n \in \mathbf{N}$ and

$$f_y(x_n) = \sum_{k=1}^{n} |\mu_k| \to \|y\|_1 \quad \text{as } n \to \infty.$$

So $\|f_y\| = \|y\|_1$, but there is no element $x_0 \in c_0$, $\|x_0\|_\infty = 1$ such that $f_y(x_0) = \|f_y\| = \|y\|_1$. □

It is instructive to consider the following geometrical interpretation of the norm of a nonzero continuous linear functional on a normed linear space. This geometrical interpretation depends on a duality which exists between the linear functionals and certain linear subspaces.

4.10.8 **Definition**. Given a linear subspace M of codimension 1 and $x_0 \notin M$, the set $x_0 + M$, the *translate* of M through x_0 is called a *hyperplane* (more strictly an *affine hyperplane*) *parallel* to M.

We make the following deduction from Lemma 1.24.15.

4.10.9 **Corollary**. *Every hyperplane* M *is of the form*

$$x_0 + \ker f = \{x \in X : f(x) = f(x_0)\}$$

where f *is a nonzero linear functional and* $x_0 \in M$. *Also for any scalar* λ *and nonzero linear functional* f *the set* $\{x \in X : f(x) = \lambda\}$ *is a hyperplane.*

Proof. We prove the second part. For $x_0 \notin \ker f$ we have

$$f\left(\frac{\lambda x_0}{f(x_0)}\right) = \lambda$$

so
$$\{x \in X : f(x) = \lambda\} = \frac{\lambda x_0}{f(x_0)} + \ker f.$$ □

4.10.10 **Remark**. Note that the correspondence between nonzero linear functionals and linear subspaces of codimension one is not one-to-one because $\ker \lambda f = \ker f$ for a nonzero linear functional f and any $\lambda \neq 0$. However, there is a one-to-one correspondence between non–zero linear functionals f and hyperplanes of the form

$$M_f \equiv \{x \in X : f(x) = 1\}.$$ □

The norm of a nonzero continuous linear functional has a geometrical meaning in terms of the distance from the origin of its corresponding hyperplane.

4.10.11 **Theorem**. *Given a normed linear space* $(X, \|\cdot\|)$, *for a nonzero continuous linear functional* f *on* X,

$$\|f\| = \frac{1}{d(0, M_f)}.$$

Proof. Since f is a continuous linear functional,

$$|f(x)| \leq \|f\| \|x\| \quad \text{for all } x \in X.$$

So for all $x \in M_f$,
$$\|x\| \geq \frac{1}{\|f\|}.$$

Therefore
$$d(0, M_f) \geq \frac{1}{\|f\|}.$$

But also, since f is continuous, from Theorem 1.24.17 we have that ker f is closed, and so M_f is also closed. As $0 \notin M_f = \overline{M}_f$ then $d(0, M_f) > 0$. Given $0 < \varepsilon < d(0, M_f)$, consider $x_0 \in M_f$ such that

$$\|x_0\| < d(0, M_f) + \varepsilon.$$

Consider $x \in X \setminus \ker f$. Then $f(x) \neq 0$ and $\frac{x}{f(x)} \in M_f$ and

$$\| x_0 \| < d(0, M_f) + \varepsilon \leq \frac{\| x \|}{| f(x) |} + \varepsilon .$$

Then $$| f(x) | < \frac{\| x \|}{\| x_0 \| - \varepsilon} < \frac{\| x \|}{d(0, M_f) - \varepsilon}$$

Therefore, $$| f(x) | \leq \frac{1}{d(0, M_f)} \| x \| \quad \text{for all } x \in X,$$

so $$\| f \| \leq \frac{1}{d(0, M_f)} .$$

We conclude that

$$\| f \| = \frac{1}{d(0, M_f)} . \qquad \square$$

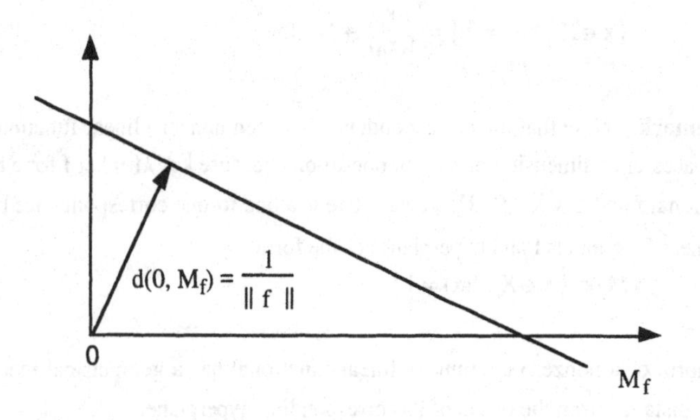

Figure 5. The distance of the hyperplane M_f from 0.

The following example illustrates how Theorem 4.10.11 generalises the Euclidean situation.

4.10.12 **Example.** In Euclidean space $(\mathbb{R}^3, \|\cdot\|_2)$ consider the hyperplane M given by

$$a\lambda + b\mu + c\nu = 1$$

which is $\{ x \in \mathbb{R}^3 : f_z(x) = (x,z) = 1 \}$ where $x \equiv (\lambda,\mu,\nu)$ and $z \equiv (a,b,c)$.

Now $d(0, M) = \frac{1}{\sqrt{a^2+b^2+c^2}}$ which is $\frac{1}{\| z \|} = \frac{1}{\| f_z \|}$ using Example 4.10.6. $\square$

Still exploring the link between a nonzero continuous linear functional f and the hyperplane M_f, it is worth noticing how the attainment of the norm of f on the unit sphere is related to the existence of closest points in M_f.

4.10.13 **Corollary**. *Given a nonzero continuous linear functional* f *on a normed linear space* $(X, \|\cdot\|)$, *for* $x_0 \in X \setminus \ker f$

$$\frac{\| x_0 \|}{f(x_0)} = d(0, M_f) \text{ if and only if } f\left(\frac{x_0}{\| x_0 \|}\right) = \| f \|.$$

4.10.14 **Remark**. We pursue further the relation between non–zero continuous linear functionals and their corresponding hyperplanes. Consider a continuous linear functional f on a normed linear space $(X, \|\cdot\|)$, where $\| f \| = 1$ and there exists an $x_0 \in X$, $\| x_0 \| = 1$ such that $f(x_0) = 1$. We say that the hyperplane M_f is a *tangent hyperplane* to the closed unit ball at x_0. By this we mean that $x_0 \in M_f$ and $| f(x) | \leq 1$ for all $x \in X$ where $\| x \| \leq 1$. Notice that $d(0, M_f) = 1 = \| x_0 \|$.

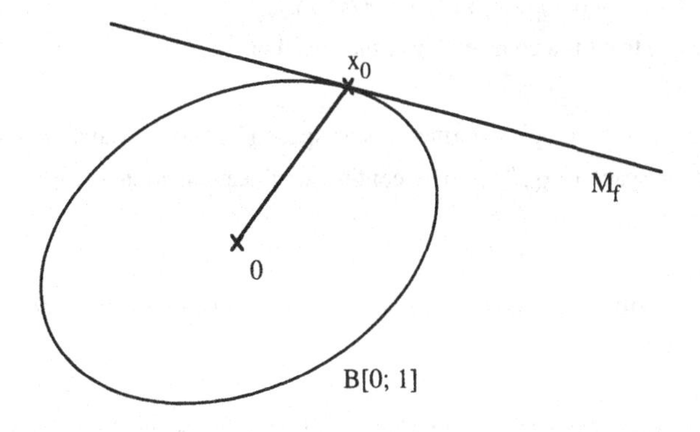

Figure 6. M_f is a tangent hyperplane to the closed unit ball at x_0 .

We have been dealing with linear spaces over $\mathbb{C}$ or $\mathbb{R}$. Now a linear space X over $\mathbb{C}$ can also be regarded as a linear space over $\mathbb{R}$; we will denote this associated linear space by $X_{\mathbb{R}}$. There is in fact a one-to-one correspondence between $X^{\#}$ and $(X_{\mathbb{R}})^{\#}$.

4.10.15 **Theorem**. *Given a complex linear functional* f *on a complex linear space* X, *then* Re f *is a real linear functional on the associated real linear space* $X_{\mathbb{R}}$.
Given a real linear functional $f_{\mathbb{R}}$ *on* $X_{\mathbb{R}}$, *then a complex linear functional* f *on* X *is defined by*

$$f(x) = f_{\mathbb{R}}(x) - i\, f_{\mathbb{R}}(ix).$$

Proof. Given f on X it is clear that Re f is real linear functional on $X_{\mathbb{R}}$.
However, for each $x \in X$,

$$f(x) = \text{Re}\, f(x) + i\, \text{Im}\, f(x).$$

But $f(ix) = i\,f(x)$ so $\operatorname{Re} f(ix) + i \operatorname{Im} f(ix) = \operatorname{Re} i\,f(x) + i \operatorname{Im} i\,f(x) = -\operatorname{Im} f(x) + i \operatorname{Re} f(x)$.
Equating real parts we have

$$\operatorname{Im} f(x) = -\operatorname{Re} f(ix)$$

so we have the following decomposition of f in terms of Re f,

$$f(x) = \operatorname{Re} f(x) - i \operatorname{Re} f(ix) \qquad \text{for all } x \in X.$$

This suggests that, given $f_{\mathbb{R}}$ on $X_{\mathbb{R}}$ we can define f on X by

$$f(x) = f_{\mathbb{R}}(x) - i\,f_{\mathbb{R}}(ix).$$

Clearly f is additive and homogeneous for real scalars. We need to check homogeneity for complex scalars. It is sufficient to consider $\alpha = i\beta$ where β is real. Then

$$\begin{aligned} f(\alpha x) &= f_{\mathbb{R}}(\alpha x) - i\,f_{\mathbb{R}}(i\alpha x) \\ &= \beta(f_{\mathbb{R}}(ix) + i\,f_{\mathbb{R}}(x)) \\ &= \alpha(f_{\mathbb{R}}(x) - i\,f_{\mathbb{R}}(ix)) = \alpha\, f(x). \end{aligned}$$

So we conclude that f is a complex linear functional on X. □

For any given complex normed linear space $(X, \|\cdot\|)$ we have an associated real normed linear space $(X_{\mathbb{R}}, \|\cdot\|)$. The continuous linear functionals on these spaces are closely related.

4.10.16 **Theorem**. *There is a one-to-one norm preserving correspondence between* X^* *and* $(X_{\mathbb{R}})^*$.

Proof. It is clear from the formula relating f on X to $f_{\mathbb{R}}$ on $X_{\mathbb{R}}$ given in Theorem 4.10.15, that f is continuous on $(X, \|\cdot\|)$ if and only if $f_{\mathbb{R}}$ is continuous on $(X_{\mathbb{R}}, \|\cdot\|)$.
We show that this correspondence is norm preserving.

Clearly, $\quad |f(x)| \ge |\operatorname{Re} f(x)| \quad$ for all $x \in \mathbb{R}$

so $\quad \|f\| \ge \|f_{\mathbb{R}}\|$.

However, given $\varepsilon > 0$ there exists an $x_0 \in X$, $\|x_0\| \le 1$ such that

$$|f(x_0)| > \|f\| - \varepsilon.$$

Now if $f(x_0) = e^{i\theta} |f(x_0)|$ then $f(e^{-i\theta}x_0) = |f(x_0)| = |f_{\mathbb{R}}(e^{-i\theta}x_0)|$
so $|f_{\mathbb{R}}(e^{-i\theta}x_0)| > \|f\| - \varepsilon$.
We conclude that $\|f_{\mathbb{R}}\| = \|f\|$. □

4.10.17 **Remark**. We should discuss the relation between hyperplanes in both spaces X and $X_{\mathbb{R}}$ because the order relation associated with $\mathbb{R}$ implies that for $\lambda \in \mathbb{R}$ the real hyperplane $\{x \in X_{\mathbb{R}}: f_{\mathbb{R}}(x) = \lambda\}$ separates the space into easily definable half spaces

$$\{x \in X_{\mathbb{R}}: f_{\mathbb{R}}(x) \ge \lambda\} \text{ and } \{x \in X_{\mathbb{R}}: f_{\mathbb{R}}(x) \le \lambda\}.$$

Now although a complex linear subspace of X is a real linear subspace of $X_{\mathbb{R}}$, a real linear subspace of $X_{\mathbb{R}}$ is not necessarily a complex linear subspace of X. Further, for a nonzero complex linear functional f on X, ker f is a real linear subspace of codimension 2 in $X_{\mathbb{R}}$. From the formula relating f on X to $f_{\mathbb{R}}$ on $X_{\mathbb{R}}$ given in Theorem 4.10.15 we have that

$$(\ker f)_{\mathbb{R}} = \ker f_{\mathbb{R}} \cap \ker f_{\mathbb{R}}(i\,.)$$

This not only implies that $(\ker f)_{\mathbb{R}}$ is contained in $\ker f_{\mathbb{R}}$ but also tells us that $(\ker f)_{\mathbb{R}}$ is uniquely determined by $f_{\mathbb{R}}$. Therefore for complex linear functionals on a complex linear space X, there is a one-to-one correspondence between $(\ker f)_{\mathbb{R}}$ and $\ker f_{\mathbb{R}}$ in $X_{\mathbb{R}}$ although $(\ker f)_{\mathbb{R}}$ is of codimension 1 in $\ker f_{\mathbb{R}}$.

Given a complex normed linear space $(X, \|\cdot\|)$ and a continuous linear functional f on X,

$$M_f \equiv \{x \in X_{\mathbb{R}} : f(x) = 1\}$$

is contained as a closed real affine hyperplane of the corresponding closed real hyperplane

$$M_{f_{\mathbb{R}}} \equiv \{x \in X_{\mathbb{R}} : f_{\mathbb{R}}(x) = 1\},$$

but both are at the same distance $\frac{1}{\|f\|} = \frac{1}{\|f_{\mathbb{R}}\|}$ from the origin. □

4.11 Operator algebras

Given normed linear spaces $(X, \|\cdot\|)$ and $(Y, \|\cdot\|')$ our other special case arises when we take the range space $(Y, \|\cdot\|')$ to be the same as the domain space $(X,\|\cdot\|)$. We then have the normed linear space of continuous linear operators on $(X, \|\cdot\|)$.

But such operator spaces actually have extra algebraic structure.

4.11.1 **Definitions**. Given an algebra A over $\mathbb{C}$ (or $\mathbb{R}$), a norm $\|\cdot\|$ on A is said to be an *algebra norm* if it satisfies the additional norm property:

(v) For all $x,y \in A$, $\|xy\| \le \|x\|\,\|y\|$, (the *submultiplicative inequality*).

The pair $(A, \|\cdot\|)$ is called a *normed algebra.* A normed algebra which is complete as a normed linear space is called a *complete normed algebra* (or a *Banach algebra*).
Different norms can be assigned to the same algebra A giving rise to different normed algebras. Equivalent norms for A as a linear space are not necessarily equivalent algebra norms; that is, they do not necessarily preserve the submultiplicative inequality. If A has a multiplicative identity e and $\|e\| = 1$, then $(A, \|\cdot\|)$ is said to be a *unital normed algebra.*

4.11.2 **Remark**. Property (v) actually links multiplication to the norm and implies that multiplication is jointly continuous: if $x \to x_0$ and $y \to y_0$ then $xy \to x_0y_0$. This can be deduced simply from the inequality

$$\|xy - x_0y_0\| \le \|x(y-y_0)\| + \|(x-x_0)y_0\| \le \|x\|\,\|y-y_0\| + \|x-x_0\|\,\|y_0\|. \quad □$$

4.11.3 **Theorem**. *Given a normed linear space* $(X, \|\cdot\|)$ *over* $\mathbb{C}$ (or $\mathbb{R}$), *the space of linear operators, usually denoted by* $\mathfrak{L}(X)$, *and the space of continuous linear operators, usually denoted by* $\mathfrak{B}(X)$, *are noncommutative algebras under multiplication defined by composition.*
Further, $(\mathfrak{B}(X), \|\cdot\|)$ *is a unital normed algebra and is complete if* $(X, \|\cdot\|)$ *is complete.*

Proof. Given $T,S \in \mathfrak{L}(X)$ we define TS on X by

$$TS = T\circ S$$

where $\qquad T\circ S(x) = T(S(x))$ for all $x \in X$.

Given $T,S \in \mathfrak{B}(X)$, it is clear that TS is also continuous and so $\mathfrak{B}(X)$ is a subalgebra of $\mathfrak{L}(X)$.

Now for $T,S \in \mathfrak{B}(X)$ and all $x \in X$

$$\begin{aligned} \| TS(x) \| &= \| T(S(x)) \| \\ &\le \| T \| \| Sx \| && \text{since } T \text{ is continuous} \\ &\le \| T \| \| S \| \| x \| && \text{since } S \text{ is continuous.} \end{aligned}$$

So

$$\begin{aligned} \| TS \| &= \sup \{ \| TS(x) \| : \| x \| \le 1 \} \\ &\le \| T \| \| S \| . \end{aligned}$$

The identity operator I is the multiplicative identity and clearly $\| I \| = 1$.
The completeness of $(\mathfrak{B}(X), \|\cdot\|)$ follows from Theorem 4.7. □

It is worth noting that an operator algebra is a generalisation of the linear space M_n of $n \times n$ matrices.

4.11.4 **Example**. When $X = \mathbb{R}^n$ (or $\mathbb{C}^n$) then $\mathfrak{B}(X)$ is algebra isomorphic to the linear space M_n of $n \times n$ matrices with entries from $\mathbb{R}$ (or $\mathbb{C}$).

Proof. Here we are dealing with a special case of Example 4.2. We need only verify that given a basis $\{e_1, e_2, \ldots, e_n\}$ of X then the linear mapping $T \to [\alpha_{ij}]$ of $\mathfrak{B}(X)$ into M_n, preserves multiplication.
Now M_n is a noncommutative algebra under matrix multiplication,

$$[\alpha_{ij}]\,[\beta_{ij}] = \Big[\sum_{j=1}^{n} \alpha_{ij}\,\beta_{jk}\Big].$$

If $T \mapsto [\alpha_{ij}]$ and $S \mapsto [\beta_{ij}]$ then

$$Te_k = \sum_{i=1}^{n} \alpha_{ik}\, e_i$$

and

$$S(Te_k) = \sum_{j=1}^{n} \beta_{jk} \sum_{i=1}^{n} \alpha_{ij}\, e_i = \sum_{i=1}^{n}\sum_{j=1}^{n} \alpha_{ij}\,\beta_{jk}\, e_i$$

so $S \circ T \mapsto \Big[\sum_{j=1}^{n} \alpha_{ij}\,\beta_{jk}\Big]$. □

4.11.5 **Remark**. The operator algebra $(\mathfrak{B}(X), \|\cdot\|)$ can be regarded as the prototype of noncommutative unital normed algebras. We will notice that the operator algebra $(\mathfrak{B}(H), \|\cdot\|)$ where H is a Hilbert space is of particular importance in itself and is a prototype of a special class of noncommutative unital normed algebras. □

Many of the properties of the operator algebra $(\mathfrak{B}(X), \|\cdot\|)$ are derived directly from its being a unital normed algebra. So we will develop some elementary normed algebra theory. There are considerable advantages in doing so, in that we are made aware of the essential structure from which the properties are derived, (they mostly do not depend on the underlying space $(X, \|\cdot\|)$), and the theory applies in a wider setting than that of operator algebras.

4.11.6 **Definition**. Given an algebra A with identity e, an element $x \in A$ is said to have a *multiplicative inverse* if there exists an element $x^{-1} \in A$ such that

$$xx^{-1} = x^{-1}x = e.$$

It is easy to see that if an element $x \in A$ has a multiplicative inverse then it is unique.
An element in A which has a multiplicative inverse is said to be *regular*, an element which does not is said to be *singular*.
Clearly the set of regular elements of A form a group under multiplication.

Linear algebra considerations provide the following characterisation for operator algebras.

4.11.7 **Theorem**. *Given a normed linear space* $(X, \|\cdot\|)$, *an element* $T \in \mathfrak{L}(X)$ *is regular in* $\mathfrak{L}(X)$ *if and only if* T *is one-to-one and onto and an element* $T \in \mathfrak{B}(X)$ *is regular in* $\mathfrak{B}(X)$ *if and only if* T *is one-to-one, onto and* T^{-1} *is continuous.*

This generalises the finite dimensional situation.

4.11.8 **Example**. A linear operator T on a finite dimensional linear space X_n is regular if and only if T is one-to-one. Given a basis $\{e_1,e_2, \ldots, e_n\}$ for X_n and the linear operator T on X_n has matrix representation $[\alpha_{ij}]$ with respect to this basis, then T is regular if and only if $\det[\alpha_{ij}] \neq 0$. □

It is clear that in any algebra with identity, the identity is a regular element. We show that in any unital Banach algebra, elements sufficiently close to the identity are also regular.

4.11.9 **Definition**. Given an algebra A and $x \in A$, we write $x^1=x$ and define inductively the positive integral powers of x by, $x^n = x(x^{n-1})$. Clearly,

$$x^n.x^m = x^m.x^n = x^{m+n} \qquad \text{for all } m,n \in \mathbb{N}.$$

Given a normed algebra $(A, \|\cdot\|)$ and $x \in A$, it follows by induction from the submultiplicative inequality that

$$\| x^n \| \leq \| x \|^n \quad \text{for all } n \in \mathbf{N}.$$

4.11.10 **Theorem.** *Given a unital Banach algebra* $(A, \|\cdot\|)$, *if* $x \in A$ *is such that* $\limsup_{n\to\infty} \| x^n \|^{1/n} < 1$ *then* $\sum x^n$ *converges and* $e-x$ *is regular and* $(e-x)^{-1} = e + \sum_{n=1}^{\infty} x^n$.

Proof. Choose real α such that

$$\limsup_{n\to\infty} \| x^n \|^{1/n} < \alpha < 1.$$

Then there exists a $\nu \in \mathbf{N}$ such that

$$\| x^n \| < \alpha^n \qquad \text{for all } n > \nu.$$

Therefore the series $\sum x^n$ is absolutely convergent and since $(A, \|\cdot\|)$ is complete, $\sum x^n$ is convergent; (see AMS §3). Write $s_n \equiv e + \sum_{k=1}^{n} x^k$ and $s \equiv e + \sum_{k=1}^{\infty} x^k$. A simple calculation shows that $(e-x)s_n = s_n(e-x) = e-x^{n+1}$.

Now $s_n \to s$ and $\| x^n \| \to 0$ as $n \to \infty$ so by the joint continuity of multiplication we have $(e-x)s = s(e-x) = e$; that is, $e-x$ is regular and

$$(e-x)^{-1} = e + \sum_{k=1}^{\infty} x^n. \qquad \square$$

4.11.11 **Corollary.** *In a unital Banach algebra, any* $x \in A$ *such that* $\| e-x \| < 1$ *is regular.*

Proof. If $\| e-x \| < 1$ then from the submultiplicative inequality we have

$$\| (e-x)^n \|^{1/n} < 1 \quad \text{for all } n \in \mathbf{N}$$

and so from Theorem 4.11.10, $x = e - (e-x)$ is regular. $\square$

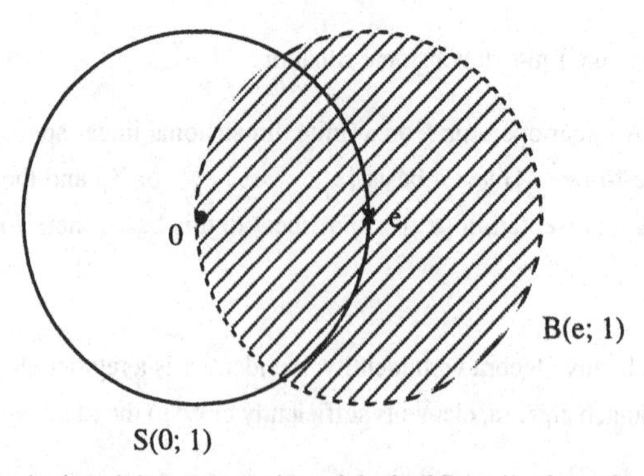

Figure 7. A pictorial representation of the ball B(e; 1) of regular elements in any unital Banach algebra.

As an aside, it is of interest to note that the following limit property holds simply as a consequence of the submultiplicative property of the norm.

4.11.12 **Proposition.** *In a normed algebra* $(A, \|\cdot\|)$, *for any* $x \in A$, $\lim_{n\to\infty} \|x^n\|^{1/n}$ *always exists and is equal to* $\inf\{\|x^n\|^{1/n} : n \in \mathbb{N}\}$.

Proof. Write $\nu \equiv \inf\{\|x^n\|^{1/n} : n \in \mathbb{N}\}$. Given $\varepsilon > 0$ choose $m \in \mathbb{N}$ such that

$$\|x^m\|^{1/m} < \nu + \varepsilon.$$

For any $n \in \mathbb{N}$ write $n = p_n m + q_n$ where p_n and q_n are integers and $p_n \geq 0$ and $0 \leq q_n \leq m - 1$. Then by the submultiplicative property, for any $n \in \mathbb{N}$

$$\begin{aligned}\nu \leq \|x^n\|^{1/n} &= \|x^{mp_n} x^{q_n}\|^{1/n} \\ &\leq \|x^m\|^{p_n/n} \|x\|^{q_n/n} \\ &< (\nu + \varepsilon)^{mp_n/n} \|x\|^{q_n/n}.\end{aligned}$$

Since $\frac{q_n}{n} \to 0$ as $n \to \infty$ we have $\frac{mp_n}{n} \to 1$ as $n \to \infty$.

Therefore $\lim_{n\to\infty} \|x^n\|^{1/n}$ exists and is equal to ν. □

4.11.13 **Remark.** Given a unital Banach algebra $(A, \|\cdot\|)$ and an element $a \in A$, Theorem 4.11.10 shows that $e - \lambda x$ is regular for scalar λ if $\limsup_{n\to\infty} \|(\lambda x)^n\|^{1/n} < 1$; that is, for all scalars λ such that

$$|\lambda| < 1 / \limsup_{n\to\infty} \|x^n\|^{1/n}.$$

For that range of scalars λ,

$$(e-\lambda x)^{-1} = e + \sum_{n=1}^{\infty} \lambda^n x^n.$$

This series is called the *Neumann series* for x. □

We now apply our theory to the solution of an integral equation.

4.11.14 **Example.** *Volterra integral equations*

Consider the integral equation of the form

$$f(x) = g(x) + \lambda \int_a^x k(x,t)\, f(t)\, dt$$

where g is a given continuous function on $[a,b]$,

f is the required solution function on $[a,b]$,

k the kernel of the equation is a given continuous function on the triangular region

$$\{(x,t) : a \leq t \leq x,\ a \leq x \leq b\},$$

and λ is a scalar.

The equation has a unique solution function $f \in \mathfrak{C}[a,b]$ for each given $g \in \mathfrak{C}[a,b]$ and parameter λ.

Proof. The Volterra operator K on $\mathfrak{C}[a,b]$ defined by

$$(K(f))(x) = \int_a^x k(x,t)\, f(t)$$

is a continuous linear operator on $(\mathfrak{C}[a,b], \|\cdot\|_\infty)$, (see AMS §7). In terms of the Volterra operator, the integral equation can be written

$$(I - \lambda K)(f) = g.$$

From Remark 4.11.13 we see that $I - \lambda K$ is regular for all

$$|\lambda| < 1 / \limsup_{n\to\infty} \|K^n\|^{1/n}$$

and the solution function will be given by

$$f = \left(I + \sum_{n=1}^{\infty} \lambda K^n \right)(g) ;$$

that is, we have a series solution for the Volterra integral equation.
In fact, we show that

$$\limsup_{n\to\infty} \|K^n\|^{1/n} = 0$$

and so we have such a solution for all scalars λ.
Now

$$\begin{aligned} |K(f)(x)| &\le (x-a) \sup\{|k(x,t)|\,|f(t)| : a \le t \le x\} \\ &\le M \|f\|_\infty (x-a) \end{aligned}$$

where $M \equiv \sup\{|k(x,t)| : a \le t \le x,\ a \le x \le b\}$.
We prove by induction that

$$|K^n(f)(x)| \le M^n \|f\|_\infty \frac{(x-a)^n}{n!} \quad \text{for all } a \le x \le b :$$

The previous statement is case $n = 1$. Suppose that the statement is true for some $n \in \mathbb{N}$. Then

$$|K^{n+1}(f)(x)| = |K(K^n f)(x)| = \left| \int_a^x k(x,t)\, (K^n f)(t)\, dt \right|$$

$$\le \frac{M^{n+1} \|f\|_\infty}{n!} \int_a^x (t-a)^n\, dt = M^{n+1} \|f\|_\infty \frac{(x-a)^{n+1}}{(n+1)!} .$$

which concludes the induction.
Using this fact we obtain for all $n \in \mathbb{N}$

$$\|K^n f\| = \max\{|(K^n f)(x)| : a \le x \le b\} \le M^n \|f\|_\infty \frac{(b-a)^n}{n!}$$

so

$$\|K^n\| = \sup\{\|K^n\| : \|f\| \le 1\} \le M^n \frac{(b-a)^n}{n!} .$$

Since $(\frac{1}{n!})^{1/n} \to 0$ as $n \to \infty$ we conclude that $\limsup_{n\to\infty} \| K^n \|^{1/n} = 0$.

It follows that, for every $g \in \mathcal{C}[a,b]$ and parameter λ there exists a unique solution of the Volterra integral equation given by the Neumann series

$$g + \sum_{n=1}^{\infty} \lambda^n K^n g. \qquad \square$$

4.11.15 **Remark**. In AMS §5 we established the existence of a unique solution using Banach's Fixed Point Theorem and that theorem provided a similar way of finding a solution by iteration. □

Corollary 4.11.11 tells us that in any unital Banach algebra, $B(e; 1)$ is an open ball of regular elements. As we might expect this implies a topological property for the set of regular elements.

4.11.16 **Theorem**. *In any unital Banach algebra* $(A, \|\cdot\|)$, *the set of regular elements is open.*

Proof. Given a regular element $x \in A$ and $y \in A$,

$$\| e - x^{-1}y \| = \| x^{-1}(x-y) \| \le \| x^{-1} \| \, \| x-y \|.$$

So from Corollary 4.11.11 we have that $x^{-1}y$ is regular when $\| x-y \| < 1/\| x^{-1} \|$.
But in this case $y = xx^{-1}y$ is itself regular being as it is, the product of regular elements x and $x^{-1}y$. Therefore,

$$\{ y \in A : \| x-y \| < 1/\| x^{-1} \| \}$$

is a set of regular elements and this implies that the set of regular elements is open. □

4.11.17 **Corollary**. *In any unital Banach algebra the set of singular elements is closed.*

Another topological property relating to the regular elements is as follows.

4.11.18 **Theorem**. *On a unital normed algebra* $(A, \|\cdot\|)$, *the mapping* $x \mapsto x^{-1}$ *is a homeomorphism of the set of regular elements onto itself.*

Proof. Since the mapping $x \mapsto x^{-1}$ is its own inverse mapping it is sufficient to prove continuity. For regular elements $x, y \in A$,

$$\| x^{-1} - y^{-1} \| = \| x^{-1}(x-y) y^{-1} \| \le \| x^{-1} \| \, \| y^{-1} \| \, \| x-y \| .$$

But $\qquad \| y^{-1} \| \le \| x^{-1} - y^{-1} \| + \| x^{-1} \| \le \| x^{-1} \| \, \| y^{-1} \| \, \| x-y \| + \| x^{-1} \|.$

So for all regular $y \in A$ where $\| x-y \| < 1/2\| x^{-1} \|$ we have $\| y^{-1} \| < 2 \| x^{-1} \|$ and so

$$\| x^{-1} - y^{-1} \| < 2 \| x^{-1} \|^2 \| x-y \|,$$

which implies that the mapping $x \mapsto x^{-1}$ is continuous at x. □

4.12 EXERCISES

1. For the following continuous linear mappings T determine $\| T \|$.
   (i) Given $k \in \{1,2, \dots, n\}$, the projection $p_k: (\mathbb{R}^n, \|\cdot\|_2) \to \mathbb{R}$ where $p_k(\lambda_1, \lambda_2, \dots, \lambda_n) = \lambda_k$.
   (ii) $T: (\mathbb{R}^3, \|\cdot\|_2) \to (\mathbb{R}^3, \|\cdot\|_2)$ where $T(\lambda_1, \lambda_2, \lambda_3) = (\lambda_3, \lambda_1, \lambda_2)$.
   (iii) $D_0: (\mathcal{C}^1[0,1], \|\cdot\|') \to \mathbb{R}$ where $\| f \|' = \| f \|_\infty + \| f' \|_\infty$ and $D_0(f) = f'(0)$.
   (iv) the shift $S: (\ell_1, \|\cdot\|_1) \to (\ell_1, \|\cdot\|_2)$ where $S(\{\lambda_1, \lambda_2, \dots, \lambda_n, \dots\}) = \{0, (\lambda_1, \lambda_2, \dots, \lambda_n \dots\}$.

2. (i) Given a sequence $\{\alpha_1, \alpha_2, \dots, \alpha_n, \dots\}$ of scalars prove that $\{\alpha_1\lambda_1, \alpha_2\lambda_2, \dots, \alpha_n \lambda_n, \dots\} \in \ell_2$ for all $\{\lambda_1, \lambda_2, \dots, \lambda_n, \dots\} \in \ell_2$ if and only if $\{\alpha_1, \alpha_2, \dots, \alpha_n, \dots\}$ is bounded.
   (ii) A *diagonal* operator T on $(\ell_2, \|\cdot\|_2)$ is defined by a bounded sequence of scalars $\{\alpha_1, \alpha_2, \dots, \alpha_n, \dots\}$ where for $x \equiv \{\lambda_1, \lambda_2, \dots, \lambda_n, \dots\} \in \ell_2$, $Tx = \{\alpha_1\lambda_1, \alpha_2\lambda_2, \dots, \alpha_n\lambda_n, \dots\}$.
   (a) Prove that T is continuous and $\| T \| = \sup \{ | \alpha_n | : n \in \mathbb{N} \}$.
   (b) Prove that T is a topological isomorphism onto if and only if $\inf \{ | \alpha_n | : n \in \mathbb{N} \} > 0$.
   (iii) Deduce that $(m, \|\cdot\|_\infty)$ is isometrically isomorphic to a linear subspace of $(\mathcal{B}(\ell_2), \|\cdot\|)$ and that $(\mathcal{B}(\ell_2), \|\cdot\|)$ is not separable.

3. Given $y \equiv \{\mu_1, \mu_2, \dots, \mu_n, \dots\} \in \ell_1$ consider the functional f_y defined on ℓ_1 by $f_y(x) = \sum \lambda_n \overline{\mu}_n$ for all $x \equiv \{\lambda_1, \lambda_2, \dots, \lambda_n, \dots\} \in \ell_1$.
   (i) Prove that f_y is linear and continuous on ℓ_1 with respect to both the $\|\cdot\|_2$ and $\|\cdot\|_1$ norms.
   (ii) Determine $\| f_y \|$ in both $(\ell_1, \|\cdot\|_2)$ and $(\ell_1, \|\cdot\|_1)$ and find whether f_y attains its norm on the unit sphere in each case.

4. Consider the functional F defined on the real linear space $\mathcal{C}[-\pi,\pi]$ by

$$F(f) = \int_{-\pi}^{\pi} f(t) \sin t \, dt.$$

   (i) Prove that F is linear and determine whether F is continuous on $\mathcal{C}[-\pi,\pi]$ with respect to norms $\|\cdot\|_2$, $\|\cdot\|_\infty$ and $\|\cdot\|_1$.
   (ii) Wherever possible for $\mathcal{C}[-\pi,\pi]$ with norms $\|\cdot\|_2$, $\|\cdot\|_\infty$ and $\|\cdot\|_1$ determine $\| F \|$ in the corresponding dual space, and find whether F attains its norm on the unit sphere of the space.

5. (i) Consider a continuous linear mapping T of a Banach space $(X, \|\cdot\|)$ into a normed linear space $(Y, \|\cdot\|')$ where $T(X)$ is dense in $(Y, \|\cdot\|')$ and T^{-1} exists and is continuous on $(T(X), \|\cdot\|')$. Prove that T maps X onto Y and $(Y, \|\cdot\|')$ is complete.

(ii) Consider Banach spaces $(X, \|\cdot\|)$ and $(Y, \|\cdot\|')$ and a continuous linear mapping T from a dense linear subspace of $(X, \|\cdot\|)$ onto a dense linear subspace of $(Y, \|\cdot\|')$. Given $\tilde{T}$, the unique continuous linear extension of T mapping $(X, \|\cdot\|)$ into $(Y, \|\cdot\|')$, prove that

(a) if T is an isometric isomorphism then $\tilde{T}$ is an isometric isomorphism onto $(Y, \|\cdot\|')$,

(b) if T is a topological isomorphism then $\tilde{T}$ is a topological isomorphism onto $(Y, \|\cdot\|')$.

(iii) Deduce that

(a) isometrically isomorphic normed linear spaces have isometrically isomorphic completions,

(b) topologically isomorphic normed linear spaces have topologically isomorphic completions.

6. A real trigonometric polynomial t on $\mathbb{R}$ of degree n is of the form

$$t(\theta) = \sum_{k=0}^{n} (a_k \sin k\theta + b_k \cos k\theta) \text{ where } a_k, b_k \in \mathbb{R} \text{ for } k \in \{0, 1, \ldots, n\}.$$

Consider the real linear space $\mathcal{T}_n$ of trigonometric polynomials of degree less than or equal to n and the linear operator D on $\mathcal{T}_n$ defined by

$$D(t) = t'(\theta).$$

(i) Prove that D is continuous on $(\mathcal{T}_n, \|\cdot\|)$ for any norm $\|\cdot\|$ on $\mathcal{T}_n$.

(ii) For norm $\|\cdot\|_2$ on $\mathcal{T}_n$ where $\|\cdot\|_2$ is generated by the inner product

$$(t, s) = \int_{-\pi}^{\pi} ts(\theta)\, d\theta$$

prove that $\| D \| = n$.

(iii) For norm $\|\cdot\|_\infty$ on $\mathcal{T}_n$ where $\| t \|_\infty = \max \{ | t(\theta) | : \theta \in [-\pi,\pi]\}$ prove that $\| D \| = n$.

(This result was first proved by S. Bernstein about 1900.
For an elementary proof see R.P. Boas, *Math.Mag.* **42** (1969), 165–174.)

7. Consider $(\mathcal{C}[0,1], \|\cdot\|_\infty)$ and the set of continuous linear functionals $\{p_x : x \in [0,1] \text{ where } p_x(f) = f(x) \text{ for all } f \in \mathcal{C}[0,1]\}$.

(i) Prove that this set is contained in the unit sphere of $(\mathcal{C}[0,1], \|\cdot\|_\infty)^*$.

(ii) Show that $(\mathcal{C}[0,1], \|\cdot\|_\infty)^*$ is not rotund.

8. (i) Consider a normed linear space $(X, \|\cdot\|)$ with a Schauder basis $\{e_n\}$. Prove that $\{e_n\}$ is a monotone basis if and only if $\| T_n \| \leq 1$ for all $n \in \mathbb{N}$ where T_n is the finite rank operator defined for each

$$x \equiv \sum_{n=1}^{\infty} \lambda_n e_n \text{ by } T_n(x) = \sum_{k=1}^{n} \lambda_k e_k .$$

(ii) Prove that if a Hilbert space has a countable orthonormal basis then it has a monotone Schauder basis.

9. (i) A linear space X has norms $\|\cdot\|$ and $\|\cdot\|'$ where for some $M > 0$, $\| x \| \leq M \| x \|'$ for all $x \in X$. Prove that $(X, \|\cdot\|)^* \subseteq (X, \|\cdot\|')^*$.

(ii) Deduce that if $\|\cdot\|$ and $\|\cdot\|'$ are equivalent norms for X then

$$(X, \|\cdot\|)^* = (X, \|\cdot\|')^*$$

and $\|\cdot\|^*$ and $\|\cdot\|'^*$, norms on X^* induced by $\|\cdot\|$ and $\|\cdot\|'$, are equivalent norms on X^*.

(iii) For $\mathcal{C}[-\pi,\pi]$, prove that

$$(\mathcal{C}[-\pi,\pi], \|\cdot\|_1)^* \underset{\neq}{\subset} (\mathcal{C}[-\pi,\pi], \|\cdot\|_2)^* \underset{\neq}{\subset} (\mathcal{C}[-\pi,\pi], \|\cdot\|_\infty)^*.$$

10. (i) Given a linear space X with algebraic dual $X^\#$, prove that each $x \in X$ generates a linear functional $\hat{x}$ on $X^\#$ defined by

$$\hat{x}(f) = f(x) \quad \text{for all } f \in X^\#.$$

(ii) Hence, or otherwise prove that a linear space X is finite dimensional if and only if its algebraic dual $X^\#$ is finite dimensional.

(iii) A normed linear space $(X, \|\cdot\|)$ has a finite dimensional dual space X^*. Prove that X is finite dimensional.

11. Prove that $(\ell_1, \|\cdot\|_\infty)^*$ is isometrically isomorphic $(c_0, \|\cdot\|_\infty)^*$ and that $\big(\mathcal{B}((\ell_1, \|\cdot\|_\infty), (c_0, \|\cdot\|_\infty)), \|\cdot\|\big)$ is isometrically isomorphic to $(\mathcal{B}(c_0, \|\cdot\|_\infty), \|\cdot\|)$.

12. (i) Given a closed hyperplane $M \equiv \{x \in X: f(x) = \lambda\}$ in a real normed linear space $(X, \|\cdot\|)$, and a point $x_0 \in X$, prove that

(a) $d(0, M) = \dfrac{|\lambda|}{\| f \|}$,

(b) $d(x_0, M) = \dfrac{|\lambda - f(x_0)|}{\| f \|}$ and

(c) there exists a $y_0 \in M$ such that $\| x_0 - y_0 \| = d(x_0, M)$ if and only if there exists an $x \in X$ such that $f(x) = \| f \| \| x \|$.

(ii) Given closed hyperplanes $M_1 \equiv \{x \in X : f(x) = \lambda_1\}$ and $M_2 \equiv \{x \in X : f(x) = \lambda_2\}$ in $(X, \|\cdot\|)$, prove that

$$d(M_1, M_2) = \frac{|\lambda_1 - \lambda_2|}{\| f \|}.$$

13. (i) Given a real normed linear space $(X, \|\cdot\|)$, prove that a linear functional f is continuous if and only if for any scalar λ, the half-space $\{x \in X : f(x) \geq \lambda\}$ is closed.

(ii) Given an open ball $B(x_0; r)$ contained in a closed half-space $H \equiv \{x \in X : f(x) \geq \lambda\}$ prove that

$$d(B(x_0;\ r), H) = \frac{\inf f(B(x_0; r))-\lambda}{\| f \|} = \frac{f(x_0)-\lambda}{\| f \|} - r.$$

14. Given a nonempty subset A of a real normed linear space $(X, \|\cdot\|)$, a hyperplane $M \equiv \{x \in X : f(x) = \lambda\}$ is called a *hyperplane of support* for A if A is contained in one or other of the half-spaces $\{x \in X : f(x) \geq \lambda\}$ or $\{x \in X : f(x) \leq \lambda\}$ and there exists at least one $x_0 \in A$ such that $f(x_0) = \lambda$.

(i) Given a compact set K in a normed linear space $(X, \|\cdot\|)$ and a closed hyperplane $M \equiv \{x \in X : f(x) = 0\}$, prove that there exists an $x_0 \in K$ such that $d(x_0, M) = d(K, M)$ and that $M_0 \equiv \{x \in X : f(x) = f(x_0)\}$ is a hyperplane of support for K at x_0.

(ii) Consider the closed unit ball $B[0; 1]$ in $(X, \|\cdot\|)$. Prove that $M_1 \equiv \{x \in X : f(x) = f(x_1)\}$ is a hyperplane of support for $B[0; 1]$ at $x_1 \in B[0; 1]$, if and only if $| f(x_1) | = \| f \| \| x_1 \|$.

15. Consider the functional f defined on $(c_0, \|\cdot\|_\infty)$ for $x \equiv \{\lambda_1,\lambda_2, \dots, \lambda_n, \dots\}$ by

$$f(x) = \sum_{k=1}^{\infty} \frac{\lambda_k}{2^k}.$$

(i) Prove that f is linear and continuous and $\| f \| = 1$, but show that f does not attain its norm on the unit sphere.

(ii) Show that for any $x_0 \notin M \equiv \{x \in X : f(x) = 0\}$ there is no closest point to x_0 in M.

(iii) Show that the closed unit ball has no hyperplane of support of the form $\{x \in X : f(x) = \lambda\}$.

16. Given a normed linear space $(X, \|\cdot\|)$ and elements $x \in X$ and $f \in X^*$, an operator $x \otimes f$ is defined on X by

$$x \otimes f(y) = f(y)\, x \quad \text{for all } y \in X.$$

(i) Prove that $x \otimes f$ is a continuous linear operator on X and

$$\| x \otimes f \| = \| x \| \| f \|.$$

(ii) Prove that every continuous linear operator T on X with n-dimensional range T(X) has the form

$$T = x_1 \otimes f_1 + x_2 \otimes f_2 + \dots + x_n \otimes f_n$$

where $x_1,x_2, \dots, x_n \in X$ and $f_1,f_2, \dots, f_n \in X^*$.

(iii) Prove that X is finite dimensional if and only if every linear operator on X is continuous.

17. Given a normed linear space $(X, \|\cdot\|)$ with proper closed linear subspace M, consider the quotient space X/M with norm $\|\cdot\|': X/M \to \mathbb{R}$ defined by

$$\| [x] \|' = \inf \{\| x+m \| : m \in M\}.$$

(i) (a) Prove that the quotient mapping $\pi: X \to X/M$ defined by

$$\pi(x) = x + M$$

is a continuous linear mapping and $\| \pi \| = 1$.

(b) Prove that π attains its norm on the closed unit ball of $(X, \|\cdot\|)$ if and only if for each $x \in X$ there exists an $m \in M$ such that

$$\| x-m \| = d(x, M) .$$

(ii) The linear subspace M is said to have *codimension* n in the linear space X if there exists a linearly independent set $\{e_1, e_2, \ldots, e_n\}$ in $X \setminus M$ such that each $x \in X$ can be represented in the form

$$x = \lambda_1 e_1 + \lambda_2 e_2 + \ldots + \lambda_n e_n + z$$

for scalars $\{\lambda_1, \lambda_2, \ldots, \lambda_n\}$ and $z \in M$.

Prove that if M has codimension n in X then X/M has dimension n.

(iii) A linear functional f on X has $M \subseteq \ker f$.

(a) Define a linear functional F on X/M by

$$F(x+M) = f(x).$$

Prove that F is continuous on $(X/M, \|\cdot\|)$ if and only if f is continuous on $(X, \|\cdot\|)$ and $\| F \| = \| f \|$.

(b) Prove that if M has finite codimension in X then f is continuous on $(X, \|\cdot\|)$.

18. Given a normed linear space $(X, \|\cdot\|)$, prove that $\mathfrak{B}(X)$ is a commutative algebra if and only if X is one dimensional.

(Hint: If X has more than one dimension use Exercise16 to construct non–commutative operators with one dimensional range.)

19. Given a unital normed algebra $(A, \|\cdot\|)$ consider the *left regular representation* map $a \mapsto T_a$ of A into $\mathfrak{B}(A)$ defined by

$$T_a(x) = ax \qquad \text{for all } x \in A.$$

Prove that $(A, \|\cdot\|)$ is isometrically algebra isomorphic to a subalgebra of $(\mathfrak{B}(A), \|\cdot\|)$ under left regular representation.

20. Consider a unital Banach algebra $(A, \|\cdot\|)$.

(i) Prove that if $\| x \| < 1$ then

$$\| (e-x)^{-1} - (e-x) \| \leq \frac{\| x \|^2}{1-\| x \|}.$$

(ii) Prove that if $x \in A$ is regular and $y \in A$ is such that $\| yx^{-1} \| < 1$ then $x-y$ is regular and

$$(x-y)^{-1} = x^{-1} + \sum_{k=1}^{\infty} x^{-1}(yx^{-1})^k.$$

21. Given a unital Banach algebra A, prove that

(i) if x and xy are regular then y is also regular,

(ii) if xy and yx are regular then both x and y are regular,

(iii) if $xy = e$ and $yx = z \neq e$ then $z^2 = z$ where $z \neq 0, e$.

22. Solve the Volterra integral equations

(i) $f(x) = 1 + \lambda \int_0^x e^{x-t} f(t)\, dt$ for $0 \leq x \leq 1$,

(ii) $f(x) = 1 + \lambda \int_0^x (x-t) f(t)\, dt$ for $0 \leq x \leq 1$.

§5. THE SHAPE OF THE DUAL

To gain a firmer grasp of the concept of the dual of a normed linear space we need to examine the particular shape of the dual space for several example normed linear spaces.

5.1 Finite dimensional normed linear spaces

Given a n-dimensional linear space X_n over $\mathbb{C}$ (or $\mathbb{R}$) with basis $\{e_1, e_2, \dots, e_n\}$ we recall from Theorem 4.10.5 that the algebraic dual $X_n^{\#}$ is also a linear space with basis $\{f_1, f_2, \dots, f_n\}$ dual to $\{e_1, e_2, \dots, e_n\}$ where

$$f_k(e_j) = 1 \quad \text{when } j = k, \qquad = 0 \quad j \neq k.$$

Furthermore, since $f = \sum_{k=1}^{n} f(e_k)\, f_k$, $X_n^{\#}$ is isomorphic to $\mathbb{C}^n$ (or $\mathbb{R}^n$) under the mapping

$$f \mapsto (f(e_1), f(e_2), \dots, f(e_n)).$$

and every linear functional f on X_n is of the form

$$f(x) = \sum_{k=1}^{n} \lambda_k f(e_k) \quad \text{where } x \equiv \sum_{k=1}^{n} \lambda_k e_k.$$

Being an isomorphism onto $\mathbb{C}^n$ implies that every linear functional f on X_n has the form

$$f(x) = \sum_{k=1}^{n} \lambda_k \bar{\alpha}_k \qquad \text{for some } (\alpha_1, \alpha_2, \dots, \alpha_n) \in \mathbb{C}^n$$

$$\text{where } x \equiv \sum_{k=1}^{n} \lambda_k e_k.$$

Now given any norm $\|\cdot\|$ on X_n, we have from Theorem 2.1.12 that $X_n^* = X_n^{\#}$. However, the actual norm on X_n^* does depend on the norm given on X_n. To obtain some idea of the form of the actual dual $(X_n, \|\cdot\|)^*$ we determine the corresponding norm $\|\cdot\|$ on $\mathbb{C}^n$ (or $\mathbb{R}^n$) so that $(\mathbb{C}^n, \|\cdot\|)$ is isometrically isomorphic to $(X_n, \|\cdot\|)^*$ under the mapping

$$f \mapsto (f(e_1), f(e_2), \dots, f(e_n)).$$

5.1.1 **Example.** Consider the normed linear space $(\mathbb{C}^n, \|\cdot\|_\infty)$ with standard basis $\{e_1, e_2, \dots, e_n\}$ for $\mathbb{C}^n$ where

$$e_k \equiv (0, \dots, 0, 1, 0, \dots, 0).$$
kth place

Now every continuous linear functional f on $\mathbb{C}^n$ is of the form

$$f(x) = \sum_{k=1}^{n} \lambda_k f(e_k) \qquad \text{for } x \equiv (\lambda_1, \lambda_2, \dots, \lambda_n) = \lambda_1 e_1 + \lambda_2 e_2 + \dots + \lambda_n e_n.$$

So $$| f(x) | \leq \sum_{k=1}^{n} | \lambda_k | | f(e_k) |$$
$$\leq (\sum_{k=1}^{n} | f(e_k) |) \max \{| \lambda_k | : k \in \{1, 2, \ldots, n\}\}$$
$$= (\sum_{k=1}^{n} | f(e_k) |) \| x \|_\infty$$
so that $\| f \| \leq (\sum_{k=1}^{n} | f(e_k) |)$.

Consider the element $x_0 \equiv (\lambda_1^0, \lambda_2^0, \ldots, \lambda_n^0)$ where

$$\left.\begin{array}{ll} \lambda_k^0 = \dfrac{\overline{f(e_k)}}{| f(e_k) |} & \text{if } f(e_k) \neq 0 \\ = 0 & \text{otherwise} \end{array}\right\}$$

Then $\| x_0 \|_\infty \leq 1$ and
$$f(x_0) = \sum_{k=1}^{n} | f(e_k) |$$
so
$$\| f \| = \sum_{k=1}^{n} | f(e_k) |.$$
This tells us that the isomorphism
$$f \mapsto (f(e_1), f(e_2), \ldots, f(e_n))$$
of $\mathbb{C}^{n*}$ onto $\mathbb{C}^n$ is an isometric isomorphism of $(\mathbb{C}^n, \|\cdot\|_\infty)^*$ onto $(\mathbb{C}^n, \|\cdot\|_1)$. □

5.1.2 **Example**. Consider the normed linear space $(\mathbb{C}^n, \|\cdot\|_p)$ where $1 < p < \infty$ with standard basis as in Example 5.1.1. Again we have that every continuous linear functional f on $\mathbb{C}^n$ is of the form
$$f(x) = \sum_{k=1}^{n} \lambda_k f(e_k) \qquad \text{for } x \equiv (\lambda_1, \lambda_2, \ldots, \lambda_n) \equiv \lambda_1 e_1 + \lambda_2 e_2 + \ldots + \lambda_n e_n.$$
So $$| f(x) | \leq \sum_{k=1}^{n} | \lambda_k | | f(e_k) |$$
$$\leq (\sum_{k=1}^{n} | \lambda_k |^p)^{1/p} (\sum_{k=1}^{n} | f(e_k) |^q)^{1/q} \text{ by Hölder's inequality where } \frac{1}{p} + \frac{1}{q} = 1$$
$$= (\sum_{k=1}^{n} | f(e_k) |^q)^{1/q} \| x \|_p$$
so that $\| f \| \leq (\sum_{k=1}^{n} | f(e_k) |^q)^{1/q}$.

Consider the element $x_0 \equiv (\lambda_1^0, \lambda_2^0, \dots, \lambda_n^0)$ where

$$\left.\begin{aligned}\lambda_k^0 &= \overline{f(e_k)}\,|f(e_k)|^{q-2} \quad \text{if } f(e_k) \neq 0\\ &= 0 \quad \text{otherwise}\end{aligned}\right\}.$$

Then $\quad \|x_0\| = \left(\sum_{k=1}^{n} |f(e_k)|^q\right)^{1/p}$ since $p(q-1) = q$,

and $\quad f(x_0) = \sum_{k=1}^{n} |f(e_k)|^q \leq \|f\|\,\|x_0\|_p$

so $\quad \|f\| \geq \left(\sum_{k=1}^{n} |f(e_k)|^q\right)^{1/q}$ since $1 - \frac{1}{p} = \frac{1}{q}$.

Therefore, $\quad \|f\| = \left(\sum_{k=1}^{n} |f(e_k)|^q\right)^{1/q}$.

This tells us that the isomorphism

$$f \mapsto (f(e_1), f(e_2), \dots, f(e_n))$$

of $\mathbb{C}^{n*}$ onto $\mathbb{C}^n$ is an isometric isomorphism of $(\mathbb{C}^n, \|\cdot\|_p)^*$ onto $(\mathbb{C}^n, \|\cdot\|_q)$. □

5.2 Hilbert spaces

In a Hilbert space, the significant characterisation of orthogonality given in Theorem 2.2.15 and the existence of best approximation points to closed linear subspaces given in Theorem 2.2.19 enables us to establish a satisfactory representation of continuous linear functionals on such spaces.

In Example 4.10.6 we noted that the inner product generates continuous linear functionals in a natural way. The following theorem tells us that every continuous linear functional is generated in this manner by the inner product.

5.2.1 The Riesz Representation Theorem for Hilbert Space.

For any given continuous linear functional f *on a Hilbert space* H, *there exists a unique* $z \in H$ *such that*

$$f(x) = (x, z) \quad \textit{for all } x \in H.$$

Proof. Since f is a continuous linear functional, ker f is a closed linear subspace of H.

When ker f = H then $\quad f(x) = 0 \quad$ for all $x \in H$

so $\quad f(x) = (x, 0) \quad$ for all $x \in H$.

When ker f $\neq$ H then from Corollary 2.2.21 we have that there exists a nonzero element $z_0 \in H$ such that z_0 is orthogonal to ker f. From Lemma 1.24.14 we have that ker f is a linear subspace of codimension one; that is, every $x \in H$ can be represented uniquely in the form

$$x = \lambda z_0 + y \quad \text{where } \lambda \text{ is a scalar and } y \in \ker f.$$

Now $f(x) = \lambda f(z_0)$.

Scaling z_0 we put $z \equiv \dfrac{\overline{f(z_0)}}{\|z_0\|^2}\, z_0$.

Then $(z_0,z) = f(z_0)$ and

$$f(x) = \lambda f(z_0) = (\lambda z_0+y, z) = (x, z) \quad \text{for all } x \in H.$$

To show the uniqueness of z, consider any $z' \in H$ such that

$$f(x) = (x, z) = (x, z') \quad \text{for all } x \in H.$$

Then $(x, z-z') = 0$ for all $x \in H$ and, as noted in Remarks 2.2.14, this implies that $z' = z$. □

5.2.2 **Remark.** Now every continuous linear functional f on a Hilbert space H is of the form f_z where

$$f_z(x) = (x, z) \quad \text{for all } x \in H.$$

We notice from Example 4.10.6 that

$$f_z\left(\frac{z}{\| z \|}\right) = \| z \| = \| f_z \|$$

so every continuous linear functional attains its norm on the unit sphere at the unit vector from which it is generated by the inner product. □

It is of interest to follow the implications of this theorem for finite dimensional Hilbert spaces, because it helps reveal the rich linear space structure of such spaces which is so useful in developing tensor calculus.

5.2.3 **Application**. *Contravariant and covariant vectors.*

Given an n-dimensional linear space X_n with basis $\{e_1, e_2, \ldots, e_n\}$, we have seen that the dual space X_n^* is also an n-dimensional linear space with basis $\{f_1,f_2, \ldots, f_n\}$ where

$$f_i(e_j) = \delta^i_j \qquad \text{for all } i, j \in \{1, 2, \ldots, n\};$$

$$(\delta^i_j \text{ is the } \textit{Kronecker delta} \text{ where } \delta^i_j = 1 \quad i = j$$
$$= 0 \quad i \neq j).$$

When X_n is also an inner product space, it follows from the Riesz Representation Theorem 5.2.1 that for each $i \in \{1, 2, \ldots, n\}$, given $f_i \in X^{\#}_n$ there exists an element $e^i \in X_n$ such that

$$f_i(x) = (x, e^i) \quad \text{for all } x \in X_n .$$

Since $f_i(e_j) = \delta^i_j$, it follows that the set $\{e^1, e^2, \ldots, e^n\}$ is also a basis for X_n. Given $\{e_1, e_2, \ldots, e_n\}$ a basis for X_n, the set $\{e^1, e^2, \ldots, e^n\}$ is called the *dual basis* for X_n. From the definition of this set it is clear that, for each $i,j \in \{1, 2, \ldots, n\}$, e^i is orthogonal to e_j for all $i \neq j$ and f_i attains its norm at $\frac{e^i}{\| e^i \|}$.

Suppose that the inner product on X_n is defined by

$$(x, y) = \sum_{i,j=1}^{n} a_{ij} \lambda^i \overline{\mu^j}$$

where $x \equiv \lambda^1 e_1 + \lambda^2 e_2 + \ldots + \lambda^n e_n$ and $y \equiv \mu^1 e_1 + \mu^2 e_2 + \ldots + \mu^n e_n$, and (a_{ij}) is a positive definite hermitian matrix. Then we can determine the components of each element of the dual basis $\{e^1, e^2, \ldots, e^n\}$ in terms of the original basis $\{e_1, e_2, \ldots, e_n\}$.

Now $(e_i, e_j) = a_{ij}$.

Suppose that $e^i = a^{i1}e_1 + a^{i2}e_2 + \ldots + a^{in}e_n$.

Then
$$\begin{aligned} f_i(e_j) &= \delta^i_j = (e_j, e^i) \\ &= a^{i1}(e_j, e_1) + a^{i2}(e_j, e_2) + \ldots + a^{in}(e_j, e_n) \\ &= a^{i1}\overline{a}_{1j} + a^{i2}\overline{a}_{2j} + \ldots a^{in}\overline{a}_{nj} = \sum_{k=1}^{n} a^{ik}\,\overline{a}_{kj}. \end{aligned}$$

So $a^{ik} = \dfrac{\text{cofactor of } a_{ik}}{\det(a_{ij})}$ and (a^{ij}) is the inverse matrix to (a_{ij}).

Further, since $(e_j, e^i) = \delta^i_j$ we have that $(e^i, e^j) = a^{ij}$.

Given any vector $x \in X_n$ where
$$x \equiv \lambda^1 e_1 + \lambda^2 e_2 + \ldots + \lambda^n e_n$$
we call $(\lambda^1, \lambda^2, \ldots, \lambda^n)$ the *contravariant components* of x. But also
$$x \equiv \lambda_1 e^1 + \lambda_2 e^2 + \ldots + \lambda_n e^n$$
and we call $(\lambda_1, \lambda_2, \ldots, \lambda_n)$ the *covariant components* of x.

Since we can express the dual basis in terms of the original basis through the matrix (a_{ij}), we can relate the contravariant and covariant components of x in a similar way:
$$x = \sum_{i=1}^{n} \lambda^i e_i = \sum_{i=1}^{n} \lambda_i e^i = \sum_{i,j=1}^{n} \lambda_i a^{ik} e_k$$
so for each $k \in \{1, 2, \ldots, n\}$, $\lambda^k = (x, e^k) = \sum_{i=1}^{n} \lambda_i a^{ik}$

and of course
$$\lambda_k = (x, e_k) = \sum_{i=1}^{n} \lambda^i a_{ik}.$$
□

5.2.4 **Remark**. Given a Hilbert space H, the mapping $z \mapsto f_z$ of H into H* where
$$f_z(x) = (x, z) \qquad \text{for all } x \in H$$
is, by the Riesz Representation Theorem 5.2.1, one-to-one and onto. But further, this mapping is additive since
$$\begin{aligned} f_{z_1+z_2}(x) &= (x, z_1+z_2) = (x, z_1) + (x, z_2) \\ &= (f_{z_1} + f_{z_2})(x) \qquad \text{for all } x \in H. \end{aligned}$$

The mapping is also conjugate homogeneous because
$$\begin{aligned} f_{\alpha z}(x) &= (x, \alpha z) = \overline{\alpha}(x, z) \\ &= \overline{\alpha}\, f_z(x) \qquad \text{for all } x \in H. \end{aligned}$$
These properties, together with the norm preserving property $\| f_z \| = \| z \|$, imply that the mapping is an isometry of H onto H* since
$$\| f_{z_1} - f_{z_2} \| = \| f_{z_1 - z_2} \| = \| z_1 - z_2 \|.$$

For a complex Hilbert space this mapping is not linear. But for a real Hilbert space the mapping is linear and is an isometric isomorphism of H onto H*. □

Given a Hilbert space H, the mapping $z \mapsto f_z$ where

$$f_z(x) = (x, z) \qquad \text{for all } x \in X$$

is a conjugate linear isometry of H onto H*. However, a Hilbert space is self-dual by which we mean that a Hilbert space H is actually isometrically isomorphic to its dual H*. We establish this indirectly, using the mapping from the Riesz Representation Theorem 5.2.1 to show that H and H* have the same orthogonal dimension and so by Theorem 3.31 are isometrically isomorphic.

5.2.5. **Theorem**. *A Hilbert space* H *is isometrically isomorphic to its dual* H*.

Proof. Consider a maximal orthonormal set $\{e_\alpha\}$ in H and the set $\{f_\alpha\}$ in H* where for each α

$$f_\alpha(x) = (x, e_\alpha) \quad \text{for all } x \in H.$$

Now the mapping $e_\alpha \mapsto f_\alpha$ is one-to-one and onto and $\| f_\alpha \| = \| e_\alpha \| = 1$ for all α. But H* is an inner product space with inner product

$$(f_x, f_y) = (y, x) \quad \text{for all } x,y \in H,$$

so $$(f_\alpha, f_\beta) = (e_\beta, e_\alpha) = 0 \quad \text{for } \alpha \neq \beta$$

and $\{f_\alpha\}$ is an orthonormal set in H*.

If for $f \in H^*$, $(f, f_\alpha) = 0$ for all α then by the Riesz Representation Theorem 5.2.1 there exists an $e \in H$, $\| e \| = \| f \|$ such that

$$f(x) = (x, e) \qquad \text{for all } x \in H$$

so $$(e, e_\alpha) = 0 \qquad \text{for all } \alpha.$$

But $\{e_\alpha\}$ is a maximal orthonormal set so $e = 0$. Therefore, $f = 0$ and $\{f_\alpha\}$ is a maximal orthonormal set for H*.

We conclude that H and H* have the same orthogonal dimension and so are isometrically isomorphic. □

5.3 Infinite dimensional sequence spaces

For an infinite dimensional normed linear space $(X, \|\cdot\|)$ we saw in Remark 4.10.2 that the dual space X* is a proper linear subspace of $X^\#$. Its size as well as its norm is determined by the norm $\|\cdot\|$ on X. We will see how we have to take this into account as we modify the procedure followed with the finite dimensional examples in Section 5.1.

5.3.1 **Example**. Consider the normed linear space $(c_0, \|\cdot\|_\infty)$ with Schauder basis $\{e_1, e_2, \ldots, e_k, \ldots\}$ where

$$e_k \equiv \{0, \ldots, 0, \underset{\text{kth place}}{1}, 0, \ldots\} \qquad \text{for each } k \in \mathbb{N}.$$

Any $x \equiv \{\lambda_1, \lambda_2, \ldots, \lambda_k, \ldots\} \in c_0$ can be represented as an infinite series in $(c_0, \|\cdot\|_\infty)$,

$$x = \sum_{k=1}^{\infty} \lambda_k e_k, \text{ (see Example 1.25.10(iii)).}$$

Consider any continuous linear functional f on $(c_0, \|\cdot\|_\infty)$.

Writing $s_n \equiv \sum_{k=1}^{n} \lambda_k e_k$ we have, since f is linear

$$f(s_n) = \sum_{k-1}^{n} \lambda_k f(e_k),$$

and since f is continuous and $\| x - s_n \|_\infty \to 0$ as $n \to \infty$, then $f(s_n) \to f(x)$ as $n \to \infty$; that is, every continuous linear functional f on $(c_0, \|\cdot\|_\infty)$ can be represented as an infinite series of the form

$$f(x) = \sum_{k=1}^{\infty} \lambda_k f(e_k) \quad \text{for all } x \equiv \{\lambda_1, \lambda_2, \ldots, \lambda_k, \ldots\} \in c_0 .$$

We show that the mapping

$$f \mapsto \{f(e_1), f(e_2), \ldots, f(e_k), \ldots\}$$

is an isometric isomorphism of $(c_0, \|\cdot\|_\infty)^*$ onto $(\ell_1, \|\cdot\|_1)$.

We begin by showing that $\{f(e_1), f(e_2), \ldots, f(e_k), \ldots\} \in \ell_1$.

Given $n \in \mathbb{N}$, consider the element $x_0 \equiv \{\lambda_1^0, \lambda_2^0, \ldots, \lambda_k^0, \ldots\} \in c_0$ where

$$\left.\begin{aligned} \lambda_k^0 &= \frac{\overline{f(e_k)}}{|f(e_k)|} && \text{if } f(e_k) \neq 0 \text{ and } 1 \le k \le n \\ &= 0 && \text{otherwise} \end{aligned}\right\}.$$

Now $\| x_0 \|_\infty \le 1$ and

$$f(x_0) = \sum_{k=1}^{n} |f(e_k)| \le \| f \| \, \| x_0 \|_\infty \le \| f \|.$$

But this is true for all $n \in \mathbb{N}$, so $\sum |f(e_k)| < \infty$; that is,

$$\{f(e_1), f(e_2), \ldots, f(e_k), \ldots\} \in \ell_1.$$

But this also implies that

$$\sum_{k=1}^{\infty} |f(e_k)| \le \| f \|.$$

Now
$$\begin{aligned} |f(x)| &\le \sum_{k=1}^{\infty} |\lambda_k| \, |f(e_k)| \\ &\le \left(\sum_{k=1}^{\infty} |f(e_k)| \right) \sup \{ |\lambda_k| : k \in \mathbb{N} \} = \left(\sum_{k=1}^{\infty} |f(e_k)| \right) \| x \|_\infty \end{aligned}$$

so that
$$\| f \| \le \sum_{k=1}^{\infty} |f(e_k)|$$

and therefore,
$$\| f \| = \sum_{k=1}^{\infty} |f(e_k)|.$$

This tells us that the mapping

$$f \mapsto \{f(e_1), f(e_2), \ldots, f(e_k), \ldots\}$$

from $(c_0, \|\cdot\|_\infty)^*$ into $(\ell_1, \|\cdot\|_1)$ is norm preserving.

We now show that the mapping is onto.

For any $\{\alpha_1, \alpha_2, \ldots, \alpha_k, \ldots\} \in \ell_1$ and $x \equiv \{\lambda_1, \lambda_2, \ldots, \lambda_k, \ldots\} \in c_0$ we have that the series $\sum \lambda_k \alpha_k$ is absolutely convergent. So the linear functional f defined on c_0 by

$$f(x) = \sum_{k=1}^{\infty} \lambda_k \overline{\alpha}$$

satisfies $| f(x) | \le \sum_{k=1}^{\infty} | \alpha_k |$ when $\| x \|_\infty \le 1$, and so is continuous.

It is clear that the mapping

$$f \mapsto \{f(e_1), f(e_2), \ldots, f(e_k), \ldots\}$$

is linear so we conclude that it is an isometric isomorphism of $(c_0, \|\cdot\|_\infty)^*$ onto $(\ell_1, \|\cdot\|_1)$. Being an isomorphism onto ℓ_1 implies that every continuous linear functional f on $(c_0, \|\cdot\|_\infty)$ has the form

$$f(x) = \sum_{k=1}^{\infty} \lambda_k \overline{\alpha} \qquad \text{for some } \{\alpha_1, \alpha_2, \ldots, \alpha_k, \ldots\} \in \ell_1$$
$$\text{where } x \equiv \{\lambda_1, \lambda_2, \ldots, \lambda_k, \ldots\} \in c_0.$$ □

5.3.2 **Remark.** We note that since $(\ell_1, \|\cdot\|_1)$ is isometrically isomorphic to a dual space then it follows from Corollary 4.10.3 that $(\ell_1, \|\cdot\|_1)$ is complete. □

5.3.3 **Example.** Given $1 < p < \infty$ consider the normed linear space $(\ell_p, \|\cdot\|_p)$ with the same sequence $\{e_1, e_2, \ldots, e_k, \ldots\}$ in ℓ_p as in Example 5.3.1.
For any $x \equiv \{\lambda_1, \lambda_2, \ldots, \lambda_k, \ldots\} \in \ell_p$ we have that

$$\| x - \sum_{k=1}^{n} \lambda_k e_k \|_p = \left(\sum_{k=n+1}^{\infty} | \lambda_k |^p \right)^{1/p} \to 0 \qquad \text{as } n \to \infty \text{ since } x \in \ell_p\,;$$

that is, $$x = \sum_{k=1}^{\infty} \lambda_k e_k\,.$$

So the sequence $\{e_1, e_2, \ldots, e_k, \ldots\}$ is also a Schauder basis for $(\ell_p, \|\cdot\|_p)$.
Consider any continuous linear functional f on $(\ell_p, \|\cdot\|_p)$.

Writing $s_n \equiv \sum_{k=1}^{n} \lambda_k e_k$ we have, since f is linear

$$f(s_n) = \sum_{k=1}^{n} \lambda_k f(e_k)$$

and since f is continuous and $\| x - s_n \|_p \to 0$ as $n \to \infty$, then $f(s_n) \to f(x)$ as $n \to \infty$; that is, every continuous linear functional f on $(\ell_p, \|\cdot\|_p)$ can be represented as an infinite series of the form

$$f(x) = \sum_{k=1}^{\infty} \lambda_k f(e_k) \qquad \text{for all } x \equiv \{\lambda_1, \lambda_2, \ldots, \lambda_k, \ldots\} \in \ell_p.$$

We show that the mapping

$$f \mapsto \{f(e_1), f(e_2), \ldots, f(e_k), \ldots\}$$

is an isometric isomorphism of $(\ell_p, \|\cdot\|_p)^*$ onto $(\ell_q, \|\cdot\|_q)$ where $\frac{1}{p} + \frac{1}{q} = 1$.

We begin by showing that $\{f(e_1), f(e_2), \ldots, f(e_k), \ldots\} \in \ell_q$.

Given $n \in \mathbb{N}$, consider the element $x_0 \equiv \{\lambda_1^0, \lambda_2^0, \dots, \lambda_k^0, \dots\} \in \ell_q$ where

$$\left.\begin{aligned} \lambda_k^0 &= \overline{f(e_k)}\, |f(e_k)|^{q-2} \quad \text{if } f(e_k) \neq 0 \text{ and } 1 \le k \le n \\ &= 0 \quad \text{otherwise} \end{aligned}\right\}$$

Now $\| x_0 \|_p = \left(\sum_{k=1}^{n} |f(e_k)|^q \right)^{1/p}$ and $f(x_0) = \sum_{k=1}^{n} |f(e_k)|^q \le \| f \| \, \| x_0 \|_p$

so that $\left(\sum_{k=1}^{n} |f(e_k)|^q \right)^{1/q} \le \| f \|$ since $1 - \frac{1}{p} = \frac{1}{q}$.

But this is true for all $n \in \mathbb{N}$, so $\sum |f(e_k)|^q < \infty$; that is, $\{f(e_1), f(e_2), \dots, f(e_k), \dots\} \in \ell_q$.

But this also implies that

$$\left(\sum_{k=1}^{\infty} |f(e_k)|^q \right)^{1/q} \le \| f \|.$$

Now

$$\begin{aligned} |f(x)| &\le \sum_{k=1}^{\infty} |\lambda_k| \, |f(e_k)| \\ &\le \left(\sum_{k=1}^{\infty} |\lambda_k|^p \right)^{1/p} \left(\sum_{k=1}^{\infty} |f(e_k)|^q \right)^{1/q} \quad \text{by Hölder's inequality} \\ &= \left(\sum_{k=1}^{\infty} |f(e_k)|^q \right)^{1/q} \| x \|_p \end{aligned}$$

so that

$$\| f \| \le \left(\sum_{k=1}^{\infty} |f(e_k)|^q \right)^{1/q}$$

and therefore

$$\| f \| = \left(\sum_{k=1}^{\infty} |f(e_k)|^q \right)^{1/q}.$$

This tells us that the mapping

$$f \mapsto \{f(e_1), f(e_2), \dots, f(e_k), \dots\}$$

from $(\ell_p, \|\cdot\|_p)^*$ into $(\ell_q, \|\cdot\|)$ is norm preserving.

We now show that the mapping is onto.

For any $\{\alpha_1, \alpha_2, \dots, \alpha_k, \dots\} \in \ell_p$ we deduce from Hölder's inequality that the series $\sum \lambda_k \bar{\alpha}_k$ is absolutely convergent. So the linear function f defined on ℓ_p by

$$f(x) = \sum_{k=1}^{\infty} \lambda_k \bar{\alpha}_k$$

satisfies $|f(x)| \le \left(\sum_{k=1}^{\infty} |\alpha_k|^q \right)^{1/q}$ when $\| x \|_p \le 1$, and so is continuous.

It is clear that the mapping

$$f \mapsto \{f(e_1), f(e_2), \dots, f(e_k), \dots\}$$

is a linear mapping so we conclude that it is an isometric isomorphism of $(\ell_p, \|\cdot\|_p)^*$ onto $(\ell_q, \|\cdot\|_q)$.

Being an isomorphism onto ℓ_q implies that every continuous linear functional f on $(\ell_p, \|\cdot\|_p)$ has the form

$$f(x) = \sum_{k=1}^{\infty} \lambda_k \bar{\alpha}_k \qquad \text{for some } \{\alpha_1, \alpha_2, \ldots, \alpha_k, \ldots\} \in \ell_q$$
$$\text{where } x \equiv \{\lambda_1, \lambda_2, \ldots, \lambda_k, \ldots\} \in \ell_p. \qquad \square$$

5.3.4 **Remarks.** Since every $(\ell_p, \|\cdot\|_p)$ space where $1 < p < \infty$ is isometrically isomorphic to a dual space $(\ell_q, \|\cdot\|_q)^*$ where $\frac{1}{p} + \frac{1}{q} = 1$, we deduce from Corollary 4.10.3 that all the $(\ell_p, \|\cdot\|_p)$ spaces where $1 < p < \infty$ are complete.
When $p = 2$ we have classical Hilbert sequence space $(\ell_2, \|\cdot\|_2)$ and the mapping

$$f \mapsto \{f(e_1), f(e_2), \ldots, f(e_k), \ldots\}$$

is an isometric isomorphism of $(\ell_2, \|\cdot\|_2)$ onto its own dual $(\ell_2, \|\cdot\|_2)^*$; (see Remark 5.2.4). $\square$

5.4 The Banach Space $(\mathcal{C}[a,b], \|\cdot\|_\infty)$

The form of the continuous linear functionals on the function space $(\mathcal{C}[a.b], \|\cdot\|_\infty)$ is given in another important representation theorem due to F. Riesz. This result is significant in the development of functional analysis because of its generalisations and applications.

We will confine ourself to the real Banach space $(\mathcal{C}[a,b], \|\cdot\|_\infty)$ of continuous real functions [a,b]. The representation theorem shows that a continuous linear functional on $(\mathcal{C}[a,b], \|\cdot\|_\infty)$ can be represented as a Riemann–Stieltjes integral. We develop background theory sufficient to explain the theorem and refer the reader to the text T.M. Apostol, *Mathematical Analysis*, Addison Wesley, 1957, for a standard treatment of the theory of this integral.

5.4.1 **Definitions.** Given bounded real functions f and α on [a,b] and a partition P of [a,b] where

$$a \equiv t_0 < t_1 < \ldots < t_n \equiv b$$

a *Riemann–Stieltjes sum* is the real number

$$S(P, f, \alpha) = \sum_{k=1}^{n} f(\xi_k)\,(\alpha(t_k) - \alpha(t_{k-1}))$$

where $t_{k-1} \le \xi_k \le t_k$ for each $k \in \{1, 2, \ldots, n\}$.
We say that f is *Riemann–Stieltjes integrable* with respect to α if there exists a real number I and given $\varepsilon > 0$ there exists a partition P_ε of [a,b] such that

$$| S(P, f, \alpha) - I | < \varepsilon \qquad \text{for all partitions P, refinements of } P_\varepsilon.$$

In this case I is denoted by

$$I = \int_a^b f(t)\, d\alpha(t)$$

and is called the *Riemann–Stieltjes integral of* f *with respect to* α.

5.4.2 **Remarks.**

(i) We note that if $\alpha(t) = t$ for all $t \in [a,b]$ then $S(P,f,\alpha) = S(P,f)$ the usual Riemann sum for f with respect to P. Further, f is Riemann integrable on [a,b] if and only if f is Riemann–Stieltjes integrable with respect to such an α on [a,b].

(ii) If f is Riemann–Stieltjes integrable with respect to an α which has a continuous derivative α' on [a,b], then $f\alpha'$ is Riemann integrable and

$$\int_a^b f(t)\, d\alpha(t) = \int_a^b f(t)\, \alpha'(t)\, dt \,;$$

(see T.M. Apostol, *Mathematical Analysis,* p. 197).

(iii) Clearly, the following elementary linearity properties hold.

(a) If f and g are Riemann–Stieltjes integrable with respect to α on [a,b] then $\lambda f + \mu g$ is also Riemann–Stieltjes integrable with respect to α on [a,b] for any $\lambda,\mu \in \mathbb{R}$ and

$$\int_a^b (\lambda f+\mu g)(t)\, d\alpha(t) = \lambda \int_a^b f(t)\, d\alpha(t) + \mu \int_a^b g(t)\, d\alpha(t)$$

(b) If f is Riemann–Stieltjes integrable with respect to both α and β on [a,b] then f is Riemann–Stieltjes integrable with respect to $\lambda\alpha+\mu\beta$ on [a,b] for any $\lambda,\mu \in \mathbb{R}$ and

$$\int_a^b f(t)\, d(\lambda\alpha+\mu\beta)(t) = \lambda \int_a^b f(d)\, d\alpha(t) + \mu \int_a^b f(t)\, d\beta(t)$$

(See T.M. Apostol, *Mathematical Analysis,* p. 193).

(iv) For our purposes the following existence theorem is important. For any continuous function f and bounded monotone increasing function α on [a,b], f is Riemann–Stieltjes integrable on [a,b]; (see T.M. Apostol, *Mathematical Analysis,* p. 211). □

Among the continuous linear functionals on $(\mathfrak{C}[a,b], \|\cdot\|_\infty)$ we draw attention to the positive linear functionals.

5.4.3 **Definitions.** We define the set

$$\mathfrak{C}^+[a,b] \equiv \{f \in \mathfrak{C}[a,b] : f(t) \geq 0 \text{ for all } t \in [a,b]\}$$

and the partial order relation on $\mathfrak{C}[a,b]$

$$g \leq f \qquad \text{if } f - g \in \mathfrak{C}^+[a,b].$$

A linear functional F on $\mathfrak{C}$ [a,b] is said to be a *positive linear functional* if

$$F(f) \geq 0 \qquad \text{for all } f \in \mathfrak{C}^+[a,b].$$

It is remarkable that although such positive linear functionals are defined by purely algebraic properties, they are always continuous.

5.4.4 **Lemma.** *Every positive linear functional* F *on* $(\mathfrak{C}[a,b], \|\cdot\|_\infty)$ *is continuous and* $F(1) = \|F\|$.

Proof. Clearly for every $f \in \mathcal{C}[a,b]$, $\|f\| 1 \pm f \geq 0$ and so

$$\|f\| F(1) \pm F(f) = F(\|f\| 1 \pm f) \geq 0.$$

Since $F(1) \geq 0$ we have

$$|F(f)| \leq \|f\| F(1)$$

and so F is continuous and $\|F\| \leq F(1)$.
On the other hand since $\|1\| = 1$ we have $F(1) \leq \|F\|$. □

The continuous linear functionals on $(\mathcal{C}[a,b], \|\cdot\|_\infty)$ can be expressed as the difference of two positive linear functionals.

5.4.5 **Lemma**. *Given a continuous linear functional* F *on* $(\mathcal{C}[a,b], \|\cdot\|_\infty)$, *there exist positive linear functionals* F^+ *and* F^- *on* $\mathcal{C}[a,b]$ *such that* $F = F^+ - F^-$.

Proof. We define the real functional F^+ on $\mathcal{C}^+[a,b]$ as follows. For $f \in \mathcal{C}^+[a,b]$ put

$$F^+(f) = \sup\{F(g) : 0 \leq g \leq f\}.$$

Since $F(0) = 0$ we have $F^+(f) \geq 0$.
But also since F is a continuous linear functional

$$F^+(f) \leq \|F\| \|f\|_\infty.$$

Clearly $F^+(\lambda f) = \lambda F^+(f)$ for all $\lambda \geq 0$.
We show that F^+ is additive on $\mathcal{C}^+[a,b]$.
For $f_1,f_2 \in \mathcal{C}^+[a,b]$, and $0 \leq g_1 \leq f_1$ and $0 \leq g_2 \leq f_2$ we have $0 \leq g_1+g_2 \leq f_1+f_2$ and so

$$F^+(f_1+f_2) \geq F(g_1+g_2) = F(g_1) + F(g_2).$$

Therefore, $$F^+(f_1+f_2) \geq F^+(f_1) + F^+(f_2).$$

Conversely, for $0 \leq g \leq f_1+f_2$ we have
$0 \leq \min\{f_1, g\} \leq f_1$ and $0 \leq g - \min\{f_1, g\} \leq f_2$ so that

$$\begin{aligned} F^+(f_1+f_2) &= \sup\{F(g) : 0 \leq g \leq f_1+f_2\} \\ &\leq \sup\{F(\min\{f_1,g\})\} + \sup\{F(g-\min\{f_1,g\})\} \leq F^+(f_1) + F^+(f_2). \end{aligned}$$

Now F^+ can be extended to a linear functional on $\mathcal{C}[a,b]$ as follows.
For $f \in \mathcal{C}[a,b]$ we consider

$$f^+ = \frac{|f| + f}{2} \text{ and } f^- = \frac{|f| - f}{2}$$

and note that f^+ and f^- are in $\mathcal{C}^+[a,b]$ and $f = f^+ - f^-$.
So we define F^+ on $\mathcal{C}[a,b]$ by

$$F^+(f) = F^+(f^+) - F^+(f^-).$$

We note that $-f = f^- - f^+$ so $F^+(-f) = F^+(f^-) - F^+(f^+) = -F^+(f)$ and we conclude that F^+ is linear.
We define the linear function F^- on $\mathcal{C}[a,b]$ by

$$F^-(f) = F^+(f) - F(f).$$

Then F^- is also continuous. But also since F^+ is a positive linear functional

$$F^+(f) \geq F(f) \quad \text{for all } f \in \mathcal{C}^+[a,b],$$

so $$F^-(f) \geq 0 \quad \text{for all } f \in \mathcal{C}^+[a,b];$$

that is, F^- is also a positive linear functional. □

We establish the representation of the continuous linear functionals on $(\mathcal{C}[a,b], \|\cdot\|_\infty)$ by first finding a representation of the positive linear functionals on $\mathcal{C}[a,b]$.

5.4.6 **Theorem**. *For a positive linear functional* F *on the real Banach space* $(\mathcal{C}[a,b],\|\cdot\|_\infty)$, *there exists a bounded monotone increasing function* α *on* $[a,b]$ *such that*

$$F(f) = \int_a^b f(t)\, d\alpha(t) \quad \textit{for all } f \in \mathcal{C}[a,b].$$

Proof. We proceed to define the function α on $[a,b]$.
Given $t \in (a,b)$ and $n \in \mathbb{N}$ sufficiently large we define the function $\phi_{t,n}$ on $[a,b]$ as

$$\left.\begin{aligned} \phi_{t,n}(x) &= 1 && a \le x \le t \\ &= 1 - n(x-t) && t \le x \le t + \tfrac{1}{n} \\ &= 0 && t + \tfrac{1}{n} < x \le b \end{aligned}\right\}$$

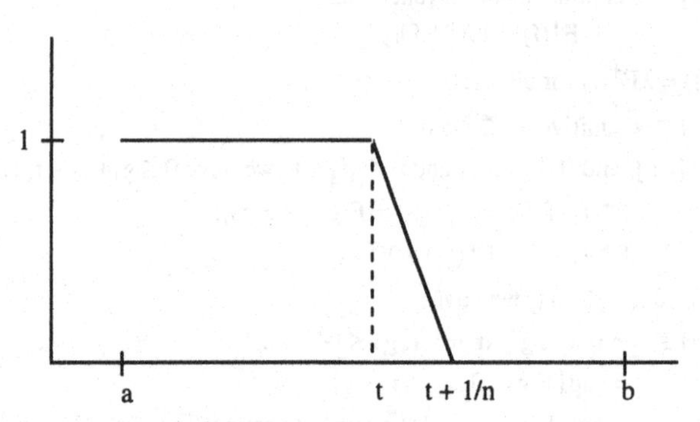

Figure 8. The graph of $\phi_{t,n}$.

Clearly for $m > n$

$$0 \le \phi_{t,m}(x) \le \phi_{t,n}(x) \le 1 \quad \text{for all } x \in [a,b].$$

Since F is a positive linear functional, the sequence $\{F(\phi_{t,n})\}$ is a decreasing sequence of real numbers bounded below by 0 and so is convergent to a real number which we denote by $\alpha(t)$.
Now for $a < t_1 < t_2 < b$ and $n \in \mathbb{N}$ we have $0 \le \phi_{t_1,n}(x) \le \phi_{t_2,n}(x) \le 1$ for all $x \in [a,b]$ and therefore

$$0 \le F(\phi_{t_1,n}) \le F(\phi_{t_2,n}) \le F(1)$$

and so $\alpha(t_1) \le \alpha(t_2)$.
We define $\alpha(a) = 0$ and $\alpha(b) = F(1)$ and then α is a monotone increasing function on $[a,b]$.
Given a continuous function f on $[a,b]$, then f is uniformly continuous on $[a,b]$, (see AMS §8). So given $\varepsilon > 0$ there exists a $\delta > 0$ such that

$$| f(t_1) - f(t_2) | < \varepsilon \quad \text{for } t_1,t_2 \in [a,b] \text{ when } | t_1-t_2 | < \delta .$$

Now f is Riemann–Stieltjes integrable with respect to α so there exists a partition P_ε of [a,b] such that

$$| S(P, f, \alpha) - \int_a^b f(t)\, d\alpha(t) | < \varepsilon$$

for all partitions P, refinements of P_ε. Consider a partition P of [a,b],

$$a \equiv t_0 < t_1 < \ldots < t_m \equiv b$$

which is a refinement of P_ε such that

$$\max\left\{ | t_k-t_{k-1} | : k \in \{1, 2, \ldots , m\} \right\} < \frac{\delta}{2}.$$

Choose $n \in \mathbf{N}$ sufficiently large that

$$\frac{2}{n} < \min\{ | t_k-t_{k-1} | : k \in \{1, 2, \ldots , m\} \}.$$

Then only consecutive intervals of

$$[t_0, t_1 + \frac{1}{n}], \ldots , [t_{k-1}, t_k + \frac{1}{n}], \ldots , [t_{m-1}, t_m] \tag{$*$}$$

may have points in common.

For each $k \in \{1, 2, \ldots , m\}$, the decreasing sequence of real numbers $\{F(\phi_{t_k,n})\}$ converges to $\alpha(t_k)$ and so we may choose $n \in \mathbf{N}$ sufficiently large that

$$\alpha(t_k) \le F(\phi_{t_k,n}) \le \alpha(t_k) + \frac{\varepsilon}{m} \quad \text{for all } k \in \{1, 2, \ldots , m\}.$$

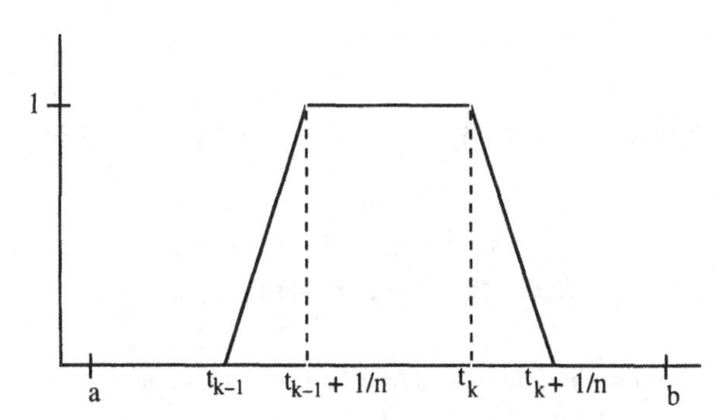

Figure 9. The graph of $\phi_{t_k,n} - \phi_{t_{k-1},n}$.

We now define the function f_0 on [a,b] by

$$f_0(x) = f(t_1)\, \phi_{t_1,n}(x) + \sum_{k=2}^{m} f(t_k)\, (\phi_{t_k,n} - \phi_{t_{k-1},n})(x) .$$

A point $x \in [a,b]$ belongs to one or two of the intervals from ($*$).

If x belongs to only one interval then either $t_0 \le x < t_1$ and $f_0(x) = f(t_1)$ or for some $k \in \{1, 2, \ldots , m\}$, $t_{k-1} + \frac{1}{n} < x \le t_k$ and $f_0(x) = f(t_k)$.

Then $$| f(x) - f_0(x) | < \varepsilon.$$

If x belongs to two intervals then for some $k \in \{1, 2, \ldots, m-1\}$
$t_k \le x < t_k + \frac{1}{n}$ and so $f_0(x) = f(t_k)\,(\phi_{t_k,n} - \phi_{t_{k-1},n})(x) + f(t_{k+1})(\phi_{t_{k+1},n} - \phi_{t_k,n})(x)$.
From the definition of the functions $\phi_{t,n}$ we have

$$f_0(x) = f(t_k)(1-n(x-t_k)) + f(t_{k+1})\, n(x-t_k).$$

Since $|\, x-t_k \,| < \delta$ and $|\, x-t_{k+1} \,| < \delta$ we have $|\, f(x)-f(t_k) \,| < \varepsilon$ and $|\, f(x)-f(t_{k+1}) \,| < \varepsilon$ and

$$|\, f(x)-f_0(x) \,| \le |\, f(x)-f(t_k) \,|\,(1-n(x-t_k)) + |\, f(x)-f(t_{k+1}) \,|\, n(x-t_k) < \varepsilon.$$

We conclude that for all $x \in [a,b]$

$$|\, f(x)-f_0(x) \,| < \varepsilon \quad \text{and so} \quad \|\, f-f_0 \,\|_\infty < \varepsilon.$$

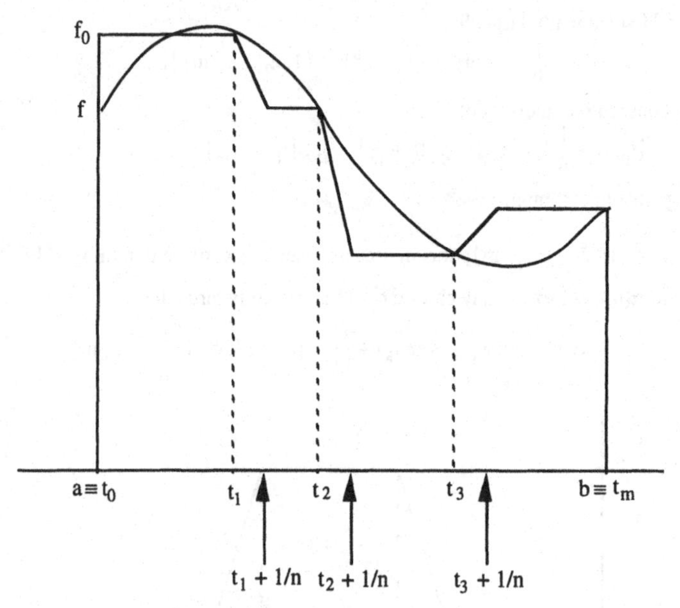

Figure 10. The graphs of f and f_0.

Now since F is a positive linear functional from Lemma 5.4.4 we have

$$|\, F(f) - F(f_0) \,| \le \|\, F \,\|\, \varepsilon.$$

From the definition of f_0 we have

$$F(f_0) = f(t_1)\, F(\phi_{t_1,n}) + \sum_{k=2}^{m} f(t_k)\,(F(\phi_{t_k,n}) - F(\phi_{t_{k-1},n}))\,.$$

But since for each $k \in \{1, 2, \ldots, m\}$, $0 \le F(\phi_{t_k,n}) - \alpha(t_k) < \varepsilon/m$, we have

$$\left|\, F(f_0) - \sum_{k=1}^{m} f(t_k)\,(\alpha(t_k) - \alpha(t_{k-1}))\,\right| < 2\,\varepsilon\,\|\, f \,\|$$

using the fact that $\alpha(t_0) = 0$.

But $\sum_{k=1}^{m} f(t_k)\,(\alpha(t_k) - \alpha(t_{k-1}))$ is a Riemann–Stieltjes sum $S(P, f, \alpha)$ for f with respect to α where P is a partition which is a refinement of P_ε. Therefore,

$$| F(f) - S(P, f, \alpha) | \le | F(f) - F(f_0) | + | F(f_0) - S(P, f, \alpha) | < (\| F \| + 2 \| f \|) \varepsilon$$

and we conclude that

$$F(f) = \int_a^b f(t)\, d\alpha(t) . \qquad \square$$

We now extend this result to the representation of continuous linear functionals on the space. To do so we introduce the following ideas.

5.4.7 **Definitions.** Given a real function α on $[a,b]$ and a partition P of $[a,b]$,

$$a \equiv t_0 < t_1 < \ldots < t_n \equiv b$$

we write

$$V(P, \alpha) = \sum_{k=1}^{n} | \alpha(t_k) - \alpha(t_{k-1}) |$$

The function α is said to be of *bounded variation* if the set of real numbers

$$\{ V(P, \alpha) : P \text{ any partition of } [a,b] \}$$

is bounded.

If α is of bounded variation

$$V(\alpha) \equiv \sup\{ V(P, \alpha) : P \text{ any partition of } [a,b] \}$$

is called the *total variation* of α on $[a,b]$.

5.4.8 **Remark.** Clearly any bounded monotone function on $[a,b]$ is of bounded variation. But further, functions of bounded variation have the following characterisation by monotone functions.

A function α on $[a,b]$ is of bounded variation if and only if α can be expressed as the difference of two monotone increasing functions; (see T.M. Apostol, *Mathematical Analysis*, p. 168). $\square$

We express our general representation theorem in terms of functions of bounded variation.

5.4.9. **The Riesz Representation Theorem for** $(\mathfrak{C}[a,b], \|\cdot\|_\infty)$.

For any continuous linear functional F *on a real Banach space* $(\mathfrak{C}[a.b], \|\cdot\|_\infty)$, *there exists a function* α *of bounded variation on* $[a,b]$ *such that*

$$F(f) = \int_a^b f(t)\, d\alpha(t) \qquad \textit{for all } f \in \mathfrak{C}[a,b],$$

and

$$V(\alpha) = \| F \|.$$

Proof. From Lemma 5.4.5 there exist positive linear functionals F^+ and F^- on $\mathcal{C}[a,b]$ such that $F = F^+ - F^-$.
By Theorem 5.4.6 there exist bounded monotone increasing functions α^+ and α^- on $[a,b]$ such that

$$F^+(f) = \int_a^b f(t)\, d\alpha^+(t) \quad \text{and} \quad F^-(f) = \int_a^b f(t)\, d\alpha^-(t) \quad \text{for all } f \in \mathcal{C}[a,b].$$

Then $$F(f) = F^+(f) - F^-(f) = \int_a^b f(t)\, d(\alpha^+ - \alpha^-)$$

by Remark 5.4.2(iii)(b).
But from Remark 5.4.8, $\alpha \equiv \alpha^+ - \alpha^-$ is of bounded variation and we have

$$F(f) = \int_a^b f(t)\, d\alpha(t) \quad \text{for all } f \in \mathcal{C}[a,b].$$

Since F has this representation, given $f \in \mathcal{C}[a,b]$ and $\varepsilon > 0$ there exists a partition P of $[a,b]$, $a \equiv t_0 < t_1 < \ldots < t_n \equiv b$ such that

$$| F(f) - \sum_{k=1}^{n} f(t_k)\,(\alpha(t_k) - \alpha(t_{k-1})) | < \varepsilon.$$

So $$| F(f) | < \varepsilon + \sum_{k=1}^{n} | f(t_k) |\, | \alpha(t_k) - \alpha(t_{k-1}) | \le \varepsilon + \| f \|_\infty \sum_{k=1}^{n} | \alpha(t_k) - \alpha(t_{k-1}) |$$
$$= \varepsilon + \| f \|_\infty V(\alpha).$$

It follows that

$$| F(f) | \le V(\alpha) \| f \|_\infty \quad \text{for all } f \in \mathcal{C}[a,b]$$

so $$\| F \| \le V(\alpha).$$

Following the pattern of the proof of Theorem 5.4.6, given $\varepsilon > 0$ consider a partition P of $[a,b]$, $a \equiv t_0 < t_1 < \ldots < t_m \equiv b$ such that

$$| V(\alpha) - \sum_{k=1}^{m} | \alpha(t_k) - \alpha(t_{k-1}) | | < \varepsilon .$$

Since the sequence $\{F^+(\phi_{t_k,n})\}$ converges to $\alpha^+(t_k)$ and the sequence $\{F^-(\phi_{t_k,n})\}$ converges to $\alpha^-(t_k)$, the sequence $\{F(\phi_{t_k,n})\}$ converges to $\alpha(t_k)$. So we choose $n \in \mathbb{N}$ sufficiently large that

$$| F(\phi_{t_k,n}) - \alpha(t_k) | < \varepsilon/m \quad \text{for all } k \in \{1, 2, \ldots, m\}.$$

Consider the continuous function f_0^* on [a, b] defined by

$$f_0^*(x) = \varepsilon\phi_{t_1,n}(x) + \sum_{k=2}^{m} \varepsilon_k(\phi_{t_k,n} - \phi_{t_{k-1},n})(x)$$

where

$$\left.\begin{aligned} \varepsilon_k &= 1 \quad \text{if} \quad \alpha(t_k) - \alpha(t_{k-1}) \ge 0 \\ &= -1 \quad \text{if} \quad \alpha(t_k) - \alpha(t_{k-1}) < 0 \end{aligned}\right\} \text{for } k \in \{1, 2, \ldots, m\}.$$

Now $$F(f_0^*) = \varepsilon_1 F(\phi_{t_1,n}) + \sum_{k=2}^{m} \varepsilon_k\big(F(\phi_{t_k,n}) - F(\phi_{t_{k-1},n})\big).$$

So we have $\quad |F(f_0^*) - \sum_{k=1}^{m} \varepsilon_k (\alpha(t_k) - \alpha(t_{k-1}))| < 2\varepsilon$ using the fact that $\alpha(t_0) = 0$,

then $\quad |F(f_0^*) - \sum_{k=1}^{m} |(\alpha(t_k) - \alpha(t_{k-1}))| < 2\varepsilon$

and we deduce that

$$|F(f_0^*) - V(\alpha)| < 3\varepsilon.$$

Therefore $\quad V(\alpha) - 3\varepsilon < |F(f_0^*)| \le \|F\| \le V(\alpha)$

and we conclude that $V(\alpha) = \|F\|$. □

5.4.10 **Remarks.**

(i) There is an alternative method of proving Theorem 5.4.9 using the Hahn–Banach Theorem of the next chapter. However, the Hahn–Banach Theorem establishes properties for normed linear spaces in general and its proof depends on the Axiom of Choice. Properties of particular spaces such as that given in Theorem 5.4.9 can be developed directly. We have chosen to present the direct proof even at the cost of a slight increase in technicality. For the alternative method the reader should consult A.L. Brown and A. Page, *Elements of Functional Analysis*, Van Nostrand Reinhold, 1970, p. 202.

(ii) In earlier examples where we determine the shape of the dual we find an isometric isomorph of the dual. Theorem 5.4.9 does not go so far as this. Given a continuous linear function F on $(\mathcal{C}[a,b], \|\cdot\|_\infty)$ there is no unique function of bounded variation α on [a,b] such that

$$F(f) = \int_a^b f(t)\, d\alpha(t)\,.$$

We can add an arbitrary constant to α and we can alter α at its points of discontinuity in (a,b) without altering the value of the integral.

A function α of bounded variation on [a,b] is said to be ***normalised*** if $\alpha(a) = 0$ and α is continuous on the right at all points of (a,b). The set $\mathcal{NBV}[a,b]$ of such normalised functions of bounded variation is a Banach space with norm $\|\alpha\| = V(\alpha)$. It can be shown that $(\mathcal{C}[a,b], \|\cdot\|_\infty)^*$ is isometrically isomorphic to $(\mathcal{NBV}[a,b], \|\cdot\|)$. (See A.E. Taylor and D.C. Lay, *Introduction to Functional Analysis*, Kreiger, 1980, p. 150.)

(iii) It is not difficult to extend Theorem 5.4.9 to complex Banach spaces $(\mathcal{C}[a,b], \|\cdot\|_\infty)$. □

5.5 EXERCISES

1. Prove that the dual space of $(\mathbb{C}^n, \|\cdot\|_1)$ is isometrically isomorphic to $(\mathbb{C}^n, \|\cdot\|_\infty)$ by showing that for every linear functional f on $\mathbb{C}^n$ and $x_0 \equiv (\lambda_1^0, \lambda_2^0, \dots, \lambda_n^0)$ where

$$\left.\begin{aligned} \lambda_k^0 &= \frac{\overline{f(e_k)}}{|f(e_k)|} \quad \text{if } f(e_k) \neq 0 \text{ and k is the first member of } \{1, 2, \dots, n\} \\ &\qquad\qquad \text{where } |f(e_k)| = \max\{|f(e_k)| : k \in \{1, 2, \dots, n\}\} \\ &= 0 \qquad \text{otherwise} \end{aligned}\right\}$$

we have $f(x_0) = \|f\| \|x_0\|_1$.

2. Consider a Hilbert space with a maximal orthonormal sequence $\{e_n\}$. Prove that for any $f \in H^*$, $\|f\| = 1$, the series $\sum \overline{f(e_n)}\, e_n$ is convergent to some $x \in H$ and $\|x\| = 1$ and $f(x) = 1$.

3. Prove that the dual space of $(\ell_1, \|\cdot\|_1)$ is isometrically isomorphic to $(m, \|\cdot\|_\infty)$ by considering the sequence $\{e_1, e_2, \dots, e_k, \dots\}$ in ℓ_1 where

$$e_k \equiv \{0, \dots, 0, \underset{\text{kth place}}{1}, 0, \dots\} \quad \text{for each } k \in \mathbf{N},$$

and for every continuous linear functional f on ℓ_1, and any $n \in \mathbf{N}$, an element $x_0 \equiv \{\lambda_1^0, \lambda_2^0, \dots, \lambda_k^0\} \in \ell_1$ where

$$\left.\begin{aligned} \lambda_k^0 &= \frac{\overline{f(e_k)}}{|f(e_k)|} \quad \text{if } f(e_k) \neq 0 \text{ and k is the first member of } \mathbf{N} \text{ where} \\ &\qquad\qquad |f(e_k)| = \max\{|f(e_j)| : j \in \{1, 2, \dots, n\}\} \\ &= 0 \qquad \text{otherwise} \end{aligned}\right\}.$$

Deduce that every continuous linear functional f on $(\ell_1, \|\cdot\|_1)$ has the form

$$f(x) = \sum_{k=1}^{\infty} \lambda_k \bar{\alpha}_k \quad \text{for some } \{\alpha_1, \alpha_2, \dots, \alpha_k, \dots\} \in m$$

$$\text{where } x \equiv \{\lambda_1, \lambda_2, \dots, \lambda_k, \dots\} \in \ell_1.$$

4. (i) Using the representation of continuous linear functionals on $(c_0, \|\cdot\|_\infty)$ and $(\ell_1, \|\cdot\|_1)$, exhibit in each case a continuous linear functional which does not attain its norm on the closed unit ball.

   (ii) Using the representation of continuous linear functionals on $(\ell_p, \|\cdot\|_p)$ where $1 < p < \infty$, prove that every continuous linear functional on such a space attains its norm on the closed unit ball.

5. Determine the form of the dual of the normed linear spaces $(E_0, \|\cdot\|_\infty)$, $(E_0, \|\cdot\|_1)$ and $(E_0, \|\cdot\|_2)$. Exhibit in each case a continuous linear functional on the space which does not attain its norm on the closed unit ball.

6. The real Banach space $(c_0, \|\cdot\|_\infty)$ is a proper closed linear subspace of the Banach space $(c, \|\cdot\|_\infty)$. However, by considering the mapping $y \mapsto f$ of ℓ_1 into c^* defined by

$$f(x) = \mu_1 \lim_{n\to\infty} \lambda_n + \sum_{k=1}^{\infty} \mu_k \lambda_{k-1},$$

where $x \equiv \{\lambda_1, \lambda_2, \ldots, \lambda_k, \ldots\} \in c$ and $y \equiv \{\mu_1, \mu_2, \ldots, \mu_k, \ldots\} \in \ell_1$, or otherwise, prove that the dual spaces $(c_0, \|\cdot\|_\infty)^*$ and $(c, \|\cdot\|_\infty)^*$ are isometrically isomorphic.

7. In a Hilbert space H, given $z \in H$ consider the continuous linear functional f_z defined by

$$f_z(x) = (x, z) \qquad \text{for all } x \in H.$$

(i) Prove that the dual space H^* is a Hilbert space with inner product

$$(f_x, f_z) = (z, x).$$

(ii) Hence, or otherwise, prove that H and H^{**}, the dual of H^*, are isometrically isomorphic.

8. (i) Prove that $(E_0, \|\cdot\|_\infty)^*$ is isometrically isomorphic to $(c, \|\cdot\|_\infty)^*$.

(ii) Prove that $(\mathcal{P}[a,b], \|\cdot\|_\infty)^*$ is isometrically isomorphic to $(\mathcal{C}[a,b], \|\cdot\|_\infty)^*$ and deduce that $(\mathcal{P}[a,b], \|\cdot\|_\infty)^*$ is a Hilbert space.

(Hint: Use Weierstrass' Approximation Theorem, AMS §9.)

9. (i) Given a bounded monotone increasing function α on $[a,b]$ prove that the function F on $(\mathcal{C}[a,b], \|\cdot\|_\infty)$ defined by

$$F(f) = \int_a^b f(t)\, d\alpha(t)$$

is a positive linear functional and $\| F \| = V(\alpha)$.

(ii) Given a function α of bounded variation on $[a,b]$ prove that the function F on $(\mathcal{C}[a,b], \|\cdot\|_\infty)$ defined by

$$F(f) = \int_a^b f(t)\, d\alpha(t)$$

is a continuous linear functional and $\| F \| = V(\alpha)$.

(iii) Given a continuous function α on [a,b], prove that the function F on $(\mathcal{C}[a,b], \|\cdot\|_\infty)$ defined by

$$F(f) = \int_a^b f(t)\, \alpha(t)\, dt$$

is a continuous linear functional and $\| F \| = \int_a^b | \alpha(t) |\, dt$.

(iv) For the linear functionals F_1 and F_2 on $(\mathcal{C}[a,b], \| \|_\infty)$ where

$$F_1 = \int_0^1 t^2 f(t)\, dt \text{ and } F_2 = \int_0^1 (1-2t)\, f(t)\, dt$$

prove that both F_1 and F_2 are continuous and determine $\| F_1 \|$ and $\| F_2 \|$.

III. THE EXISTENCE OF CONTINUOUS LINEAR FUNCTIONALS

Given any linear space X, it follows from the existence of a Hamel basis for X and the fact that any linear functional is determined by its values on the Hamel basis, that the algebraic dual $X^{\#}$ is generally a "substantial" space. We know, from Remark 4.10.2, that for an infinite dimensional normed linear space $(X, \|\cdot\|)$, the dual X^* is a proper linear subspace of $X^{\#}$.

For the development of a theory of normed linear spaces in general, quite apart from particular examples or classes of examples, it is important to know that given any normed linear space $(X, \|\cdot\|)$, its dual X^* is also "substantial enough" and by this we mean that we have a dual which generalises sufficiently the properties we are accustomed to associate with the dual of a Euclidean space or indeed, with the duals of the familiar example spaces.

We now use the Axiom of Choice in the form of Zorn's Lemma, (see Appendix A.1), to prove the Hahn–Banach Theorem, an existence theorem which is crucial for the development of our general theory. The theorem assures us that for any nontrivial normed linear space there is always an adequate supply of continuous linear functionals.

The immediate application of this result is in the study of the structure of the second dual X^{**} of a normed linear space $(X, \|\cdot\|)$ and of the relation between the space X and its duals X^* and X^{**}.

§6. THE HAHN–BANACH THEOREM

There are several forms of the Hahn–Banach Theorem, some more general than others. The form sufficient for our purpose asserts the existence of norm preserving extensions of continuous linear functionals from a linear subspace of a normed linear space to the whole space. It is an interesting exercise to generalise this form, (see Exercise 6.13.5), but we will make no use of the generalisation. In fact we mainly use our restricted form of the theorem through its corollaries. The proof of the Hahn–Banach Theorem is an application of Zorn's Lemma but we isolate the computational first step of that proof in the following lemma.

6.1 **Lemma.** *Given a normed linear space* $(X, \|\cdot\|)$, *consider a continuous linear functional* f *defined on a proper linear subspace* M *of* X. *Given* $x_0 \in X \setminus M$, f *can be extended to a continuous linear functional* f *on* $M_0 \equiv sp\{x_0, M\}$ *such that* $\| f_0 \|$ *on* M_0 *is equal to* $\| f \|$ *on* M.

Proof. We may suppose that $\| f \| = 1$.

Case 1:* X *a real linear space

Since M is of codimension 1 in M_0, any $x \in M_0$ can be represented uniquely in the form

$$x = \lambda x_0 + y \qquad \text{where } \lambda \in \mathbb{R} \text{ and } y \in M.$$

For f_0 to be a linear extension of f on M we must have

$$f_0(y) = f(y) \qquad \text{for all } y \in M$$

and

$$\begin{aligned} f_0(x) &= \lambda f_0(x_0) + f_0(y) \\ &= \lambda f_0(x_0) + f(y) \end{aligned}$$

and we are free to choose a value for $f_0(x_0)$.

We show that a value for $f_0(x_0)$ can be chosen such that f_0 is continuous on M_0 and $\| f_0 \| = 1$.

Now for f_0 continuous on M_0,

$$\begin{aligned} \| f_0 \| &= \sup\{ | f(x) | : \| x \| \le 1, x \in M_0 \} \\ &\ge \sup\{ | f(y) | : \| y \| \le 1, y \in M \} = \| f \| = 1. \end{aligned}$$

But f_0 is continuous and $\| f_0 \| = 1$ if

$$| f_0(x) | \le \| x \| \qquad \text{for all } x \in M_0.$$

Now this will happen if the value $f_0(x_0)$ satisfies

$$- \| \lambda x_0 + y \| \le \lambda f_0(x_0) + f(y) \le \| \lambda x_0 + y \|;$$

that is,

$$- f(y/\lambda) - \| x_0 + y/\lambda \| \le f_0(x_0) \le - f(y/\lambda) + \| x_0 + y/\lambda \| \quad \text{when } \lambda \ne 0. \qquad (*)$$

(Notice that when $\lambda < 0$, we have

$$- f(y/\lambda) + \frac{1}{\lambda} \| \lambda x_0 + y \| \le f_0(x_0) \le - f(y/\lambda) - \frac{1}{\lambda} \| \lambda x_0 + y \|$$

$$- f(y/\lambda) - \| - (x_0 + y/\lambda) \| \le f_0(x_0) \le - f(y/\lambda) + \| - (x_0 + y/\lambda) \| .)$$

Now for any two elements $y_1, y_2 \in M$, we have

$$f(y_2) - f(y_1) = f(y_2 - y_1) \le \| y_2 - y_1 \|.$$

But $\qquad \| y_2 - y_1 \| \le \| x_0 + y_2 \| + \| x_0 + y_1 \|.$

Therefore $\qquad -f(y_1) - \| x_0 + y_1 \| \le - f(y_2) + \| x_0 + y_2 \|.$

So $\qquad \sup\{ - f(y) - \| x_0 + y \| : y \in M \}$ and

$\qquad \inf\{ - f(y) + \| x_0 + y \| : y \in M \}$ both exist

and $\qquad \sup\{ - f(y) - \| x_0 + y \| : y \in M \} \le \inf\{ - f(y) + \| x_0 + y \| : y \in M \}$.

So it is possible to choose a value for $f_0(x_0)$ between these two and then that value will satisfy inequality (*) which gives $\| f_0 \| = 1$ as required.

Case 2:* X *a complex linear space

We have seen from Theorems 4.10.15 and 4.10.16 that a complex normed linear space $(X, \|\cdot\|)$ can be regarded as a real normed linear space $(X_{\mathbb{R}}, \|\cdot\|)$ and there is a one-to-one norm preserving correspondence $f \mapsto f_{\mathbb{R}}$ of X^* onto $(X_{\mathbb{R}})^*$ where f and $f_{\mathbb{R}}$ are related by

$$f(x) = f_{\mathbb{R}}(x) - i\, f_{\mathbb{R}}(ix) \qquad \text{for } x \in X.$$

Now $M_0 \equiv \text{sp}\{x_0, M\}$ as a real normed linear space is $M_{0_{\mathbb{R}}} = \text{sp}\{x_0, ix_0, M_{\mathbb{R}}\}$. From Case 1 applied twice $f_{\mathbb{R}}$ on $M_{\mathbb{R}}$ can be extended to a continuous linear functional $f_{0_{\mathbb{R}}}$ on $M_{0_{\mathbb{R}}}$ such that $\| f_{0_{\mathbb{R}}} \| = 1$.

But then f_0 given by

$$f_0(x) = f_{0_{\mathbb{R}}}(x) - i\, f_{0_{\mathbb{R}}}(ix) \qquad \text{for all } x \in M_0$$

extends f on M as a continuous linear functional f_0 on M_0 such that $\| f_0 \| = 1$. □

The Hahn–Banach Theorem uses Zorn's Lemma to carry extensions, like those achieved in Lemma 6.1 for one extra dimension, to the whole space.

6.2 The Hahn–Banach Theorem.

For a normed linear space $(X, \|\cdot\|)$, *consider a continuous linear functional* f *defined on a proper linear subspace* M *of* X. *Then* f *can be extended as a continuous linear functional* f_0 *on* X *such that* $\| f_0 \|$ *on* X *is equal to* $\| f \|$ *on* M.

Proof. Consider the set $\mathfrak{F}$ of all possible extensions of f as norm preserving continuous linear functionals on linear subspaces containing M, with partial order relation $\lesssim$ on $\mathfrak{F}$:

$$f_1 \lesssim f_2 \text{ if dom } f_1 \subseteq \text{dom } f_2 \text{ and } f_2 \text{ is an extension of } f_1.$$

From Lemma 6.1, we see that $\mathfrak{F}$ is nonempty.

Suppose that $\{f_\alpha\}$ is a totally ordered subset of $\mathfrak{F}$. We show that $\{f_\alpha\}$ has an upper bound in $\mathfrak{F}$. For $x \in \bigcup_\alpha \text{dom } f_\alpha$, there exists an α_0 such that $x \in \text{dom } f_{\alpha_0}$, so we define a functional f^1 on $\bigcup_\alpha \text{dom } f_\alpha$ by

$$f^1(x) = f_{\alpha_0}(x),$$

and since $\{f_\alpha\}$ is a totally ordered set, f^1 is well defined.

Now since $\{f_\alpha\}$ is a totally ordered subset, dom $f^1 = \bigcup_\alpha \text{dom } f_\alpha$ is a linear subspace of X containing M. But also from the definition we see that f^1 is a linear extension of f_α for all α. As such an extension,

$$\sup\{|f(x)| : \|x\| \le 1,\ x \in \text{dom } f^1\} \ge \|f\|.$$

If there exists an $x \in \text{dom } f^1$ where $\|x\| \le 1$ and $f^1(x) > \|f\|$ then, since there exists an α_0 such that $x \in \text{dom } f_{\alpha_0}$, we would have $f_{\alpha_0}(x) > \|f\|$ which contradicts the defining property for elements of $\mathfrak{F}$ that $\|f_{\alpha_0}\| = \|f\|$. So we conclude that $\|f^1\| = \|f\|$ and then $f^1 \in \mathfrak{F}$.

Since $f_\alpha \precsim f^1$ for all α, we have that f^1 is an upper bound for $\{f_\alpha\}$.

It now follows from Zorn's Lemma that $\mathfrak{F}$ has a maximal element f_0.
Now dom $f_0 = X$, for otherwise by Lemma 6.1, for $x_0 \in X \setminus \text{dom } f_0$ we could extend f_0 as a norm preserving continuous linear functional on sp $\{x_0, \text{dom } f_0\}$ but then f_0 would not be maximal in $\mathfrak{F}$. □

The Hahn–Banach Theorem has the following important corollary which guarantees the existence of certain continuous linear functionals which attain their norm on the closed unit ball, or considered geometrically guarantees the existence of tangent hyperplanes to the closed unit ball. It is this first corollary rather than the theorem itself which we will use in the subsequent development of our analysis.

Now the norm of a continuous linear functional f on a normed linear space $(X, \|\cdot\|)$ is given by
$$\|f\| = \sup\{|f(x)| : \|x\| \leq 1\}.$$
Given a continuous linear functional f there does not necessarily exist an $x \in X$, $\|x\| = 1$ such that $f(x) = \|f\|$. However, the corollary tells us that given an $x \in X$, $\|x\| = 1$ there does exist a continuous linear functional f such that $f(x) = \|f\|$.

6.3 **Corollary**. *Given a normed linear space* $(X, \|\cdot\|)$, *for each non–zero* $x_0 \in X$, *there exists a non–zero continuous linear function* f_0 *on* X *such that* $f_0(x_0) = \|f_0\| \|x_0\|$.

Proof. Consider the one dimensional linear subspace $M \equiv \text{sp}\{x_0\}$. Define the functional f on M by
$$f(\lambda x_0) = \lambda \|x_0\| \qquad \text{for scalar } \lambda.$$
Then f is a continuous linear functional on M such that $f(x_0) = \|x_0\|$ and $\|f\| = 1$. By the Hahn–Banach Theorem 6.2, f can be extended as a continuous linear functional f_0 on X such that $\|f_0\| = 1$. Then f_0 on X satisfies
$$f_0(x_0) = \|f_0\| \|x_0\|.$$
□

Let us explore the geometrical interpretation of this result.

6.4 **Remark**. Corollary 6.3 implies that given $x_0 \in X$, $\|x_0\| = 1$ there exists a continuous linear functional f_0 on X, $\|f_0\| = 1$ such that $f_0(x_0) = 1$. Geometrically this says that for each $x_0 \in X$ on the boundary of the closed unit ball B[0;1] there exists a closed tangent hyperplane $M_{f_0} \equiv \{x \in X : f_0(x) = 1\}$ to B[0; 1] at x_0 where $\|f_0\| = 1$.

Clearly,
$$d(0, M_{f_0}) = \frac{1}{\|f_0\|} = 1 = \|x_0\| \text{ and } x_0 \in M_{f_0}.$$
This is the sort of geometrical property we assume quite naturally in Euclidean space. □

We now show how Corollary 6.3 implies that X* is "substantial" in another sense.

6.5 **Definitions.**

(i) Given a linear space X, a linear subspace Y of $X^{\#}$ is said to be *total* on X if for for $x \in X$, $f(x) = 0$ for all $f \in Y$ implies that $x = 0$;
or contrapositively, for $x \neq 0$ there exists an $f \in Y$ such that $f(x) \neq 0$.

(ii) Given a linear space X, a linear subspace Y of $X^{\#}$ is said to *separate* the points of X if for $x, y \in X$, $x \neq y$ there exists an $f \in Y$ such that $f(x) \neq f(y)$.

Clearly, the fact that Y is a linear subspace of $X^{\#}$ implies that these properties are equivalent.

6.6 **Remark.** Given a normed linear space $(X, \|\cdot\|)$, Corollary 6.3 implies that the dual X^* is total on X or equivalently separates the points of X. □

The second corollary to the Hahn Banach Theorem implies that in general there are sufficient continuous linear functionals on a normed linear space, not only to separate points, but also to separate points from proper closed linear subspaces.

6.7 **Corollary.** *Given a normed linear space* $(X, \|\cdot\|)$, *for any proper closed linear subspace* M *and* $x_0 \in X \setminus M$, *there exists a continuous linear functional* f_0 *on* X *such that*

$$f_0(M) = 0, \quad f_0(x_0) = 1 \quad \textit{and} \quad \| f_0 \| = \frac{1}{d(x_0, M)}.$$

Proof. Consider $M_0 \equiv \text{sp}\{x_0, M\}$.
Now any $x \in M_0$ has the form

$$x = \lambda x_0 + y \text{ where } \lambda \text{ is a scalar and } y \in M.$$

Define a linear functional f on M_0 by $f(x) = \lambda$.
Then clearly $f(M) = 0$ and $f(x_0) = 1$.
Since $\ker f = M$ which is closed, then f is continuous on M_0. But also, $\| f \| = \dfrac{1}{d(0, M_f)}$.
However,

$$\begin{aligned} d(0, M_f) &\equiv \inf\{\| x \| : f(x) = 1 = f(x_0)\} = \inf\{\| x \| : f(x - x_0) = 0\} \\ &= \inf\{\| x_0 - y \| : f(y) = 0\} \equiv d(x_0, M). \end{aligned}$$

So $\| f \| = \dfrac{1}{d(x_0, M)}$.
By the Hahn–Banach Theorem there exists a continuous linear functional f_0 on X an extension of f on M_0 such that

$$\| f_0 \| = \| f \| = \frac{1}{d(x_0, M)}.$$ □

6.8 **Remark.** Corollary 6.7 implies that given a set A in a normed linear space $(X, \|\cdot\|)$, a point $x_0 \in \overline{\text{sp}A}$ if and only if for every continuous linear functional f on X where $f(A) = 0$ we have $f(x_0) = 0$. If for every continuous linear functional f on X where $f(A) = 0$ we have $f(x_0) = 0$ then $d(x_0, \overline{\text{sp}A}) = 0$, for otherwise by Corollary 6.7 there exists a continuous linear functional f_0 on X such that $f_0(\overline{\text{sp}A}) = 0$ and $f_0(x_0) = 1$.
The converse is immediate. □

The Hahn–Banach Theorem can be used to reveal a great deal about the relation between a normed linear space and its dual. To illustrate this we consider separability of the spaces.

The separability of a normed linear space does not necessarily imply the separability of its dual. For example the Banach space $(\ell_1, \|\cdot\|_1)$ is separable but its dual is isometrically isomorphic to $(m, \|\cdot\|_\infty)$ which is not separable (see Example 1.25.3 and Exercise 1.26.16). However, Corollary 6.7 enables us to establish the following relation.

6.9 **Theorem.** *A normed linear space* $(X, \|\cdot\|)$ *is separable if its dual* $(X^*, \|\cdot\|)$ *is separable.*

Proof. Consider a countable set $\{f_n \in X^* : n \in \mathbb{N}\}$ dense in $(X^*, \|\cdot\|)$.
For each $n \in \mathbb{N}$ there exists an $x_n \in X$, $\|x_n\| \leq 1$ such that $|f_n(x_n)| \geq \frac{1}{2}\|f_n\|$.
Consider $M \equiv \overline{sp}\,\{x_n : n \in \mathbb{N}\}$.
Suppose that $M \neq X$. Then for $x_0 \in X \setminus M$ we have from Corollary 6.7 that there exists an $f_0 \in X^*$ such that $f_0(M) = 0$ and $f_0(x_0) \neq 0$. But then

$$\tfrac{1}{2}\|f_n\| \leq |f_n(x_n)| = (f_n - f_0)(x_n)| \leq \|f_n - f_0\|$$

and $\qquad \|f_0\| \leq \|f_n - f_0\| + \|f_n\| \leq 3\|f_n - f_0\|$ for all $n \in \mathbb{N}$.
But this contradicts the density of $\{f_n \in X^* : n \in \mathbb{N}\}$ in $(X^*, \|\cdot\|)$.
So we conclude that $X = \overline{sp}\,\{x_n : n \in \mathbb{N}\}$; that is, $(X, \|\cdot\|)$ is separable. □

The Hahn–Banach Theorem is useful in determining the form of the dual for subspaces and quotient spaces of a normed linear space.

6.10 **Definition.** For a linear subspace M of a normed linear space $(X, \|\cdot\|)$ the *annihilator* of M is the subset

$$M^\perp \equiv \{f \in X^* : f(x) = 0 \quad \text{for all } x \in M\}.$$

It is evident that $M^\perp$ is always a closed linear subspace of $(X^*, \|\cdot\|)$; (see Exercise 6.13.8(i)).

6.11 **Theorem.** *Consider a linear subspace* M *of a normed linear space* $(X, \|\cdot\|)$.
(i) M^* *is isometrically isomorphic to* $X^*/M^\perp$.
(ii) *If* M *is a closed linear subspace, then* $(X/M)^*$ *is isometrically isomorphic to* $M^\perp$.

Proof.
(i) Given $f \in M^*$ we have from the Hahn–Banach Theorem 6.2 that there exists an extension $f_0 \in X^*$ and $\|f_0\| = \|f\|$.
Consider the mapping $T: M^* \to X^*/M^\perp$ defined by

$$T(f) = f_0 + M^\perp.$$

If $f_0^1 \in X^*$ is another extension of f then $f_0 - f_0^1 \in M^\perp$ and so this mapping is well defined. Clearly T is linear.

Since each element in $f_0 + M^\perp$ is an extension of f,

$$\| f \| \le \inf \{\| f_0+g \| : g \in M^\perp\} = \| f_0+M^\perp \| = \| T(f) \|.$$

But also since f_0 is a norm preserving extension of f,

$$\| f \| = \| f_0 \| \ge \inf \{\| f_0+g \| : g \in M^\perp\} = \| f_0+M^\perp \| = \| T(f) \|.$$

So T is an isometric isomorphism.

But since the restriction of any continuous linear functional on X is a continuous linear functional on M we conclude that T is onto.

(ii) Consider the quotient mapping $\pi: X \to X/M$ defined by $\pi(x) = x+M$ and the mapping $T: (X/M)^* \to X^*$ defined by

$$T(h) = h \circ \pi.$$

Clearly T is linear.

But also

$$| T(h)(x) | = | h(x+M) | \le \| h \| \| x+M \| \le \| h \| \| x \| \qquad \text{for all } x \in X$$

so $\qquad \| T(h) \| \le \| h \| \qquad$ for all $h \in (X/M)^*$.

If $x \in M$ then

$$T(h)(x) = h(x+M) = h(M) = 0$$

since M is the zero of X/M. Therefore T maps $(X/M)^*$ into $M^\perp$.

Further for $f \in M^\perp$ we can define unambiguously an $h \in (X/M)^*$ by

$$h(x+M) = f(x).$$

Clearly, such an h is a linear functional on X/M

and $\qquad | h(x+M) | = | f(x) | = | f(x+m) | \le \| f \| \| x+m \| \qquad$ for all $m \in M$

$$= \| f \| \| x+M \| \qquad \text{for all } x \in X.$$

So h is continuous on X/M and $\| h \| \le \| f \|$.

But then T maps $(X/M)^*$ onto $M^\perp$.

Moreover, T is one-to-one because if $T(h) = 0$ then $h(x+M) = 0$ for all $x \in M$ and so $h = 0$.

So the mapping T can be expressed as

$$T(h)(x) = h(x+M) = f(x) \qquad \text{for all } x \in X$$

and then $\qquad \| h \| \le \| f \| = \| T(h) \| \qquad$ for all $h \in (X/M)^*$.

We conclude that

$$\| T(h) \| = \| h \| \qquad \text{for all } h \in (X/M)^*$$

so T is an isometric isomorphism of $(X/M)^*$ onto $M^\perp$. □

6.12 **Remark**. The proofs of Theorem 6.11 can be given using the techniques of conjugate mappings presented in Chapter 5, Section 12. □

6.13 EXERCISES

1. A normed linear space $(X, \|\cdot\|)$ is said to be *smooth* if for each $x \in X$, $\|x\| = 1$, there exists only one continuous linear functional f on X where $\|f\| = 1$ and $f(x) = 1$.

   (i) Show that (a) $(\mathbb{R}^3, \|\cdot\|_1)$ is not smooth
   and (b) $(c_0, \|\cdot\|_\infty)$ is not smooth.

   (ii) Prove that Hilbert space is smooth.

2. (i) Prove that a normed linear space $(X, \|\cdot\|)$ is rotund if and only if for every $f \in X^*$ where $\|f\| = f(x) = f(y) = 1$ for $x,y \in X$ where $\|x\| = \|y\| = 1$ we have $x = y$.

   (ii) Prove that (a) if X^* is smooth then X is rotund
   and (b) if X^* is rotund then X is smooth.

3. Prove that a normed linear space $(X, \|\cdot\|)$ is smooth at $x \in X$, $\|x\| = 1$ if and only if $\displaystyle\lim_{\lambda \to 0} \frac{\|x+\lambda y\| - \|x\|}{\lambda}$ exists for all $y \in X$,
   and if it is smooth at x then this limit is $f(y)$ where $f \in X^*$, $\|f\| = 1$ and $f(x) = \|x\|$.

4. (i) Given a normed linear space $(X, \|\cdot\|)$, prove that for any $x \in X$,
   $$\|x\| = \sup\{|f(x)| : f \in X^*, \|f\| \le 1\}.$$

   (ii) A linear subspace Y of X^* is said to be *norming* if
   $$\|x\| = \sup\{|f(x)| : f \in Y, \|f\| \le 1\}.$$
   Prove that such a subspace Y is total on X.

   (iii) Prove that if a linear subspace Y of X^* is dense in X^* then Y is norming. But give an example of a linear subspace Y of X^* which is norming but not dense in X^*.

5. A *convex* functional ϕ on a linear space X is a real functional on X defined by
   $$\phi((1-\lambda)x+\lambda y) \le (1-\lambda)\,\phi(x) + \lambda\phi(y) \quad \text{for all } x,y \in X \text{ and } 0 \le \lambda \le 1.$$
   Prove the more general form of the Hahn–Banach Theorem.
   Consider a convex functional ϕ on a linear space X, a proper linear subspace M of X and a linear functional f on M such that
   $$\operatorname{Re} f(x) \le \phi|_M(x) \qquad \textit{for all } x \in M.$$
   Then there exists a linear functional f_0 on X an extension of f on M, such that
   $$\operatorname{Re} f_0(x) \le \phi(x) \qquad \textit{for all } x \in X.$$

6. Consider a normed linear space $(X, \|\cdot\|)$, a proper closed linear subspace M of X and $x_0 \in X \setminus M$.

(i) Prove that there exists a continuous linear functional f on X such that $\| f \| = 1$, $f(M) = 0$ and $f(x_0) = d(x_0, M)$.

(ii) Prove that there exists a closed hyperplane M_0 containing M such that
$$d(x_0, M_0) = d(x_0, M).$$

7. In the Banach space $(m, \|\cdot\|_\infty)$, consider m_0 the smallest closed linear subspace of $(m, \|\cdot\|_\infty)$ containing all sequences of the form
$$\{\lambda_1, \lambda_2-\lambda_1, \lambda_3-\lambda_2, \ldots, \lambda_n-\lambda_{n-1}, \ldots\}$$
where $x \equiv \{\lambda_1, \lambda_2, \lambda_3, \ldots, \lambda_n, \ldots\} \in m$.

(i) Show that $e \equiv \{1,1,1, \ldots\} \notin m_0$, and prove that there exists a continuous linear functional f^* on $(m, \|\cdot\|_\infty)$ such that $\| f^* \| = 1$, $f^*(e) = 1$ and $f^*(m_0) = 0$.

(ii) The *Banach limit* of a bounded sequence $x \equiv \{\lambda_1, \lambda_2, \ldots, \lambda_n, \ldots\}$ is defined by
$$\text{LIM}\ \lambda_n = f^*(x).$$
Prove that
(a) $\text{LIM}\ \lambda_n = \text{LIM}\ \lambda_{n+1}$.
(b) $\text{LIM}\ \lambda_n \geq 0$ if $\lambda_n \geq 0$ for all $n \in \mathbb{N}$.
(c) $\text{LIM}\ (\alpha\lambda_n+\beta\mu_n) = \alpha\ \text{LIM}\ \lambda_n + \beta\ \text{LIM}\ \mu_n$ for scalars α,β and $y \equiv \{\mu_1, \mu_2, \ldots, \mu_n, \ldots\} \in m$.
(d) $\liminf \lambda_n \leq \text{LIM}\ \lambda_n \leq \limsup \lambda_n$ if $\lambda_n \in \mathbb{R}$ for all $n \in \mathbb{N}$.
(e) If $x \in c$ then $\text{LIM}\ \lambda_n = \lim \lambda_n$.

8. (i) Given a nonempty subset M of a normed linear space $(X, \|\cdot\|)$, prove that $M^\perp$ is a closed linear subspace of $(X^*, \|\cdot\|)$.

(ii) Given a nonempty subset N of $(X^*, \|\cdot\|)$ we define the set
$$N_\perp \equiv \{x \in X : f(x) = 0 \text{ for all } f \in N\}.$$
Prove that $N_\perp$ is a closed linear subspace of $(X, \|\cdot\|)$.

(iii) Prove that $(M^\perp)_\perp = M$ if and only if M is a closed linear subspace of $(X, \|\cdot\|)$.

(iv) Prove that $N \subseteq (N_\perp)^\perp$ but by considering the linear space c_0 as a closed linear subspace of $(m, \|\cdot\|_\infty)$ show that
$$({c_0}_\perp)^\perp \neq c_0.$$

(v) Prove that if N is a finite dimensional linear subspace of $(X^*, \|\cdot\|)$ then
$$N = (N_\perp)^\perp.$$

9. Given a linear subspace M of a normed linear space $(X, \|\cdot\|)$, prove that for any $f \in X^*$,
$$d(f, M^\perp) = \| f|_M \|$$
and there exists an $f_0 \in M^\perp$ such that
$$\| f-f_0 \| = d(f, M^\perp).$$

§7. THE NATURAL EMBEDDING AND REFLEXIVITY

It is Corollary 6.3 to the Hahn–Banach Theorem which enables us to develop a significant theory of dual spaces and to define the important class of reflexive spaces.

7.1 **Definitions**. Given a normed linear space $(X, \|\cdot\|)$ and its dual X^* with norm $\|\cdot\|$ defined by

$$\| f \| = \sup \{ | f(x) | : \| x \| \leq 1 \},$$

the dual space $(X^*, \|\cdot\|)$ has its own dual $(X^*)^*$ usually written X^{**} and called the *second dual* space (or *second conjugate* space) of $(X, \|\cdot\|)$. We will usually denote elements of X by x,y,z, elements of X^* by f,g,h and elements of X^{**} by F,G,H.
The norm of X^{**} is of course defined by

$$\| F \| = \sup \{ | F(f) | : \| f \| \leq 1 \}.$$

The definition of the second dual $(X^{**}, \|\cdot\|)$ prompts us to enquire into its relation to the original space $(X, \|\cdot\|)$ from which it is generated.

For elements $x \in X$ and $f \in X^*$, we can consider f fixed and x varying over X as we do when we think of f as a functional on X. But alternatively, we can consider x fixed and f varying over X^*, and when we do this we have x acting as a functional on X^*.

7.2 **Theorem**. *Given a normed linear space* $(X, \|\cdot\|)$ *and* $x \in X$, *the functional* $\hat{x}$ *defined by*

$$\hat{x}(f) = f(x) \qquad \textit{for all } f \in X^*$$

is a continuous linear functional on X^* *and* $\hat{x}$ *as an element of* $(X^{**}, \|\cdot\|)$ *satisfies*

$$\| \hat{x} \| = \| x \|.$$

Proof. Clearly, $\hat{x}$ is linear:

$$\hat{x}(f+g) = (f+g)(x) = f(x) + g(x) = \hat{x}(f) + \hat{x}(g) \quad \text{for } f, g \in X^*$$

and

$$\hat{x}(\alpha f) = \alpha f(x) = \alpha \hat{x}(f) \quad \text{for scalar } \alpha \text{ and } f \in X^*,$$

using the fact that in X^* addition and multiplication by a scalar are defined pointwise.
But also $\hat{x}$ is continuous since

$$| \hat{x}(f) | = | f(x) | \leq \| x \| \| f \| \quad \text{for all } f \in X^*.$$

As a continuous linear functional on X^* we have that $\hat{x} \in X^{**}$.
But $\| \hat{x} \| = \sup \{ | f(x) | : \| f \| \leq 1 \} \leq \| x \|$.
However, by Corollary 6.3 applied to $(X, \|\cdot\|)$, for $x \in X$ there exists an $f \in X^*$ such that $\| f \| = 1$ and $f(x) = \| x \|$, so that

$$\| \hat{x} \| = \sup \{ | f(x) | : \| f \| \leq 1 \} \geq \| x \|.$$

Therefore, $\| \hat{x} \| = \| x \|$. □

Given a normed linear space $(X, \|\cdot\|)$ and $x \in X$ the identification given in Theorem 7.2 of $\hat{x}$ as an element of X^{**} suggests that we investigate the mapping $x \mapsto \hat{x}$ of X into X^{**}.

7.3 **Theorem**. *Given a normed linear space* $(X, \|\cdot\|)$, *the mapping* $x \mapsto \hat{x}$ *induced by the definition of* $\hat{x}$ *as*

$$\hat{x}(f) = f(x) \quad \textit{for all } f \in X^*$$

is an isometric isomorphism of X *into* X^{**}.

Proof. The mapping $x \mapsto \hat{x}$ is linear:

$$\widehat{x+y}(f) = f(x+y) = f(x) + f(y) = \hat{x}(f) + \hat{y}(f) = (\hat{x}+\hat{y})(f) \quad \text{for all } f \in X^*$$

so $\quad \widehat{x+y} = \hat{x} + \hat{y}$.

Also $\quad \widehat{\alpha x}(f) = f(\alpha x) = \alpha f(x) = \alpha\hat{x}(f) \quad$ for all scalar α and $f \in X^*$

so $\quad \widehat{\alpha x} = \alpha\hat{x}$

using the fact that f is linear.

From Theorem 7.2 we have that

$$\|\hat{x}\| = \|x\| \quad \text{for all } x \in X$$

so we conclude that the mapping $x \mapsto \hat{x}$ is an isometric isomorphism. □

This identification of the original space $(X, \|\cdot\|)$ as part of the second dual $(X^{**}, \|\cdot\|)$ suggests that we give a name to this mapping.

7.4 **Definition**. Given a normed linear space $(X, \|\cdot\|)$, the mapping $x \mapsto \hat{x}$ of X into X^{**} induced by the definition of $\hat{x}$ as $\hat{x}(f) = f(x)$ for all $f \in X^*$, is called the *natural embedding* of X into X^{**}. We denote by $\hat{X}$ the image of X in X^{**} under the natural embedding.

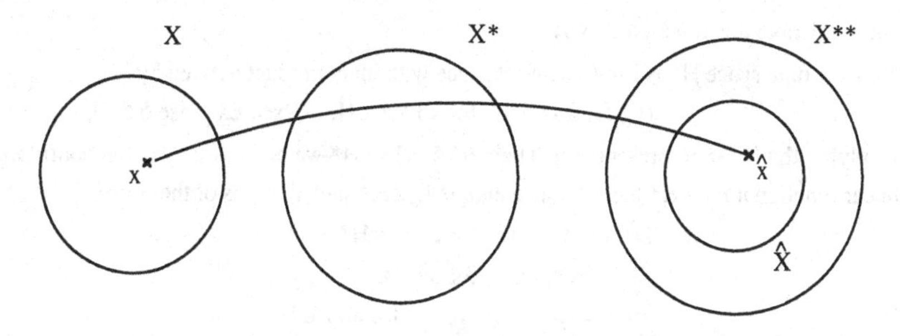

Figure 11. The natural embedding $x \mapsto \hat{x}$ of X into X^{**}.

Now it will be important to determine whether a normed linear space is isometrically isomorphic to its second dual under the natural embedding.

7.5 **Definition**. A normed linear space $(X, \|\cdot\|)$ where the natural embedding $x \mapsto \hat{x}$ maps X onto X^{**} is said to be *reflexive.*

7.6 **Remark**. Since a dual space is always complete we deduce that a normed linear space $(X, \|\cdot\|)$ is isometrically isomorphic to its second dual $(X^{**}, \|\cdot\|)$ only if $(X, \|\cdot\|)$ is a Banach space. So completeness is a necessary condition for a normed linear space to be reflexive. But it is not a sufficient condition as is shown in the examples below. □

7.7 Examples of reflexive spaces

We now give some examples of classes of normed linear spaces which are reflexive.

7.7.1 **Finite dimensional normed linear spaces**. A finite dimensional linear space is *algebraically reflexive.* Given an n-dimensional linear space X_n we have from Theorem 4.10.5 that the algebraic dual $X_n^{\#}$ is n-dimensional and so the second algebraic dual $X_n^{\#\#}$ is also n-dimensional. Now the mapping $x \mapsto \hat{x}$ of X_n into $X_n^{\#\#}$ induced by the definition of $\hat{x}$ as

$$\hat{x}(f) = f(x) \qquad \text{for all } f \in X_n^{\#}$$

is linear. But again from Theorem 4.10.5 we see that $X_n^{\#}$ is total on X_n and this implies that if $\hat{x}(f) = 0$ for all $f \in X_n^{\#}$ then $x = 0$, so the mapping $x \mapsto \hat{x}$ is one-to-one. But then $\hat{X}_n$, the image under this mapping, is n-dimensional so $\hat{X}_n = X_n^{\#\#}$. For an n-dimensional normed linear space $(X_n, \|\cdot\|)$ we have $X_n^{*} = X_n^{\#}$ so it is clear that $(X_n, \|\cdot\|)$ is reflexive. □

7.7.2 **Hilbert space**. From the Riesz Representation Theorem 5.2.1 we have that for any given continuous linear functional f on a Hilbert space H there exists a unique $z \in H$ such that f is of the form

$$f(x) = (x, z) \quad \text{for all } x \in H$$

and we denote this functional by f_z.

Now the dual space H^* is itself a Hilbert space with inner product defined by

$$(f_x, f_z) = (z, x) \quad \text{for all } x,z \in H; \qquad \text{(see Exercise 5.5.7).}$$

Applying the Riesz Representation Theorem 5.2.1 to H^* we have for any given continuous linear functional F on H^* there exists a unique $f_x \in H^*$ such that F is of the form

$$\begin{aligned} F(f_z) &= (f_z, f_x) \quad &&\text{for all } f_z \in H^* \\ &= (x, z) &&\text{for all } z \in H. \end{aligned}$$

But
$$\hat{x}(f_z) = f_z(x) = (x, z) \qquad \text{for all } z \in H$$

so
$$F(f) = \hat{x}(f) \qquad \text{for all } f \in H^* ;$$

that is,
$$F = \hat{x}$$

and we conclude that
$$\hat{H} = H^{**}.$$
□

7.7.3 **The $(\ell_p, \|\cdot\|_p)$ spaces where** $1 < p < \infty$. In Example 5.3.3. we showed that given $1 < p < \infty$ the dual space $(\ell_p, \|\cdot\|_p)^*$ is isometrically isomorphic to $(\ell_q, \|\cdot\|_q)$ where $\frac{1}{p} + \frac{1}{q} = 1$. We showed that every continuous linear functional f on $(\ell_p, \|\cdot\|_p)$ can be represented in the form

$$f(x) = \sum_{k=1}^{\infty} \lambda_k \overline{\mu}_k \quad \text{for} \quad x \equiv \{\lambda_1, \lambda_2, \dots, \lambda_k, \dots\} \in \ell_p$$
$$\text{and } \{\mu_1, \mu_2, \dots, \mu_k, \dots\} \in \ell_q.$$

Now defining for each $k \in \mathbb{N}$ the continuous linear functional f_k on ℓ_p by

$$f_k(x) = \lambda_k,$$

we can write

$$f(x) = \sum_{k=1}^{\infty} \overline{\mu}_k f_k(x)$$

so

$$f = \sum_{k=1}^{\infty} \overline{\mu}_k f_k .$$

For any given $F \in \ell_p^{**}$ we have

$$F(f) = \sum_{k=1}^{\infty} \overline{\mu}_k F(f_k) .$$

However, $(\ell_p, \|\cdot\|_p)^{**}$ is isometrically isomorphic to $(\ell_q, \|\cdot\|_q)^*$ and so every continuous linear functional F on $(\ell_q, \|\cdot\|_q)$ can be represented in the form

$$F(f) = \sum_{k=1}^{\infty} \overline{\mu}_k \lambda_k \qquad \text{for} \quad f \equiv \{\mu_1, \mu_2, \dots, \mu_k, \dots\} \in \ell_q$$
$$\text{and } \{\lambda_1, \lambda_2, \dots, \lambda_k, \dots\} \in \ell_p.$$

So $\{F(f_1), F(f_2), \dots, F(f_k), \dots\} \in \ell_p$.

But then

$$F(f) = \hat{x}(f) \text{ for all } f \in \ell_p^{**} \text{ where } x \equiv \{F(f_1), F(f_2), \dots F(f_k), \dots\} \in \ell_p;$$

that is, $\quad F = \hat{x}$

and we conclude that $\hat{\ell}_p = \ell_p^{**}$. □

7.8 **Remark**. It should be noted that it is not sufficient for reflexivity that a Banach space $(X, \|\cdot\|)$ be isometrically isomorphic to $(X^{**}, \|\cdot\|)$. R.C. James, *Proc. Nat. Acad. Sci. USA* **37** (1951), 174–177, has given an example of a nonreflexive Banach space with just this property. For reflexivity we must have $(X, \|\cdot\|)$ isometrically isomorphic to $(X^{**}, \|\cdot\|)$ under the natural embedding. That is why in Examples 7.7.2 and 7.7.3, although it is obvious from Example 5.3.3 that $(\ell_p, \|\cdot\|_p)$ is isometrically isomorphic to $(\ell_p, \|\cdot\|_p)^{**}$ and from Exercise 5.5.7 that a Hilbert space H is isometrically isomorphic to H^{**}, yet for reflexivity we need to establish the isometric isomorphism under the natural embedding and this requires a little more careful computation. □

7.9 Techniques to prove nonreflexivity

To prove nonreflexivity directly we have to show that the natural embedding is into but not onto. Such computation usually involves a knowledge of the form of the dual and the form of the second dual of the space and in other than a few cases this can be quite a complication. So the nonreflexivity of a normed linear space is usually established by indirect argument.

It is clear that if a normed linear space is not complete then it is not reflexive. But we can sometimes quickly determine non–reflexivity by an appeal to separability.

7.9.1 **Theorem.** *A separable Banach space* $(X, \|\cdot\|)$ *with a nonseparable dual* $(X^*, \|\cdot\|)$ *is not reflexive.*

Proof. It follows from Theorem 6.9 that the second dual $(X^{**}, \|\cdot\|)$ cannot be separable. But then $(X, \|\cdot\|)$ and $(X^{**}, \|\cdot\|)$ cannot be isometrically isomorphic, (see Exercise 1.26.15). □

7.9.2 **Example.** The Banach space $(c_0, \|\cdot\|_\infty)$ is separable, (see Example 1.25.2(ii)). But from Example 5.3.1, $(c_0, \|\cdot\|_\infty)^*$ is isometrically isomorphic to $(\ell_1, \|\cdot\|_1)$ and in Exercise 5.5.3 we prove that $(\ell_1, \|\cdot\|_1)^*$ is isometrically isomorphic to $(m, \|\cdot\|_\infty)$. So $(c_0, \|\cdot\|)^{**}$ is isometrically isomorphic to $(m, \|\cdot\|_\infty)$. But $(m, \|\cdot\|_\infty)$ is not separable, (see Example 1.25.3). Therefore we can conclude that $(c_0, \|\cdot\|_\infty)$ is not reflexive. □

A more generally applicable indirect method is a consequence of the following important property of reflexive spaces.

7.9.3 **Theorem.** *On a reflexive normed linear space* $(X, \|\cdot\|)$ *every continuous linear functional* f *attains its norm on the closed unit ball of* $(X, \|\cdot\|)$.

Proof. Applying Corollary 6.3 to $(X^*, \|\cdot\|)$ we have that for any given $f \in X^*$ there exists a continuous linear functional F on X^* such that $F(f) = \|f\|\,\|F\|$. But we are given that $\hat{X} = X^{**}$, so

$$F = \hat{x} \qquad \text{for some } x \in X.$$

Then

$$f(x) = \hat{x}(f) = \|f\|\,\|x\|.$$

□

7.9.4 **Remark.** It follows from Theorem 7.9.3 that if a Banach space has a continuous linear functional which does not attain its norm on the closed unit ball then the space is not reflexive. This provides a technique for proving nonreflexivity which does not even necessitate a knowledge of the form of the dual space. □

7.9.5 **Example**. Consider the Banach space $(\mathcal{C}[-\pi,\pi], \|\cdot\|_\infty)$. We exhibit a continuous linear functional on the space which does not attain its norm on the closed unit ball, (see Exercise 4.12.4).

Consider the linear functional F defined by

$$F(f) = \int_{-\pi}^{\pi} f(t) \sin t \, dt .$$

Now
$$|F(f)| \le \left(\int_{-\pi}^{\pi} |\sin t| \, dt \right) \| f \|_\infty \quad \text{for all } f \in \mathcal{C}[-\pi,\pi] .$$

So F is continuous and $\| F \| \le \int_{-\pi}^{\pi} |\sin t| \, dt = 4$. If we choose f_0 on $[-\pi,\pi]$ defined by

$$\left.\begin{aligned} f_0(t) &= -1 \quad -\pi \le t < 0 \\ &= 1 \quad\quad 0 \le t \le \pi \end{aligned}\right\}.$$

Then
$$F(f_0) = \int_{-\pi}^{\pi} |\sin t| \, dt = 4$$

and since $\| f_0 \|_\infty = 1$ we would have $\| F \| = 4$ and F attain its norm at f_0.

But $f_0 \notin \mathcal{C}[-\pi,\pi]$. However, we can modify f_0 to show that $\| F \| = 4$.

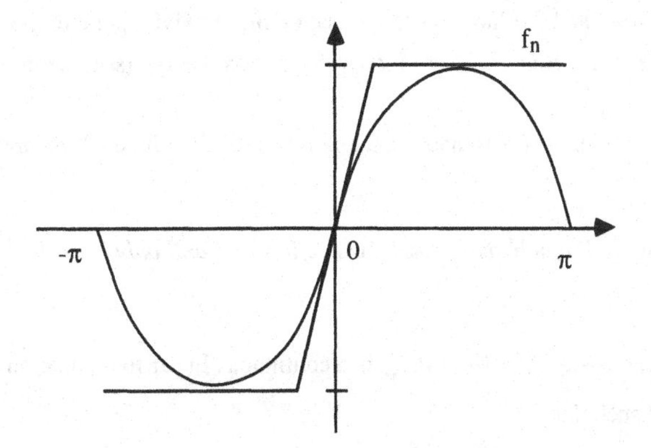

Figure 12. A modification of f_0 to give a sequence $\{f_n\}$ in $\mathcal{C}[-\pi,\pi]$ so that $F(f_n) \to 4$ as $n \to \infty$.

We define a sequence $\{f_n\}$ in $\mathcal{C}[-\pi,\pi]$ by

$$\left.\begin{aligned} f_n(t) &= -1 \quad\quad -\pi \le t < -\frac{1}{n} \\ &= nt \quad\quad -\frac{1}{n} \le t \le \frac{1}{n} \\ &= 1 \quad\quad\quad \frac{1}{n} < t \le \pi \end{aligned}\right\}.$$

Now $\| f_n \|_\infty = 1$ for all $n \in \mathbb{N}$.

But $F(f_n) = \int_{-\pi}^{\pi} f_n(t) \sin t \, dt > 4 - \frac{1}{n}$ for all $n \in \mathbb{N}$.

So $F(f_n) \to 4$ as $n \to \infty$ and therefore $\| F \| = 4$.

But it is clear that there is no $f \in \mathcal{C}[-\pi,\pi]$, $\| f \|_\infty = 1$ such that $F(f) = 4$. So we conclude that $(\mathcal{C}[-\pi,\pi], \|\cdot\|_\infty)$ is not reflexive. □

7.9.6 **Remark**. It is important to observe that one of the most significant theorems in this theory was proved by R.C. James, *Studia Math.* **23** (1964), 205–216. By proving the converse of Theorem 7.9.3 he established the following theorem.

The James characterisation of reflexivity.

A Banach space is reflexive if and only if every continuous linear functional attains its norm on the closed unit ball of the space.

Although this is an extremely useful tool in determining reflexivity, the proof is quite "deep". It is inappropriate to present it in this course and in fact we will not need to make use of it. □

Theorem 7.9.3 and the James characterisation of reflexivity point to the significance of the class of reflexive Banach spaces for approximation theory; (see Exercise 7.13.2).

We now show that if a Banach space is reflexive then its dual, or predual is also reflexive.

7.10 **Theorem**. *A Banach space* $(X, \|\cdot\|)$ *is reflexive if and only if its dual* $(X^*, \|\cdot\|)$ *is reflexive.*

Proof. Consider $\mathfrak{F} \in X^{***}$. Now $\mathfrak{F}|_{\hat{X}}$ is a continuous linear functional on $\hat{X}$, so there exists an $f \in X^*$ such that

$$\mathfrak{F}|_{\hat{X}}(\hat{x}) = \hat{f}(\hat{x}) \quad \text{for all } x \in X.$$

If X is reflexive then $\hat{X} = X^{**}$. So

$$\mathfrak{F}(F) = \hat{f}(F) \qquad \text{for all } F \in X^{**}.$$

Now $\mathfrak{F} = \hat{f}$ and we conclude that $\hat{X^*} = X^{***}$; that is, X^* is reflexive.

Conversely, suppose that $\hat{X}$ is a proper linear subspace of X^{**}. Since X is a Banach space, $\hat{X}$ is closed so from Corollary 6.7 there exists an $\mathfrak{F} \in X^{***}$, $\mathfrak{F} \neq 0$ such that $\mathfrak{F}(\hat{X}) = 0$.

If X^* is reflexive, $\mathfrak{F} = \hat{f}$ for some $f \in X^*$, so $\hat{f}(\hat{X}) = 0$ which implies that $f(X) = 0$; that is, $f = 0$. But this contradicts $\mathfrak{F} \neq 0$, so we conclude that $\hat{X} = X^{**}$; that is, X is reflexive. □

7.11 **Remark**. Theorem 7.10 implies that if a Banach space $(X, \|\cdot\|)$ is not reflexive then none of its duals or preduals is reflexive. □

7.12 The completion of a normed linear space

It is the theoretically powerful Hahn–Banach Theorem which enables us to establish in a painless fashion that with an incomplete normed linear space there is always a complete normed linear space which contains the original as a dense subspace.

7.12.1 **Definition**. The *completion* of an incomplete normed linear space $(X, \|\cdot\|)$ is a complete normed linear space $(\tilde{X}, \|\cdot\|)$ such that $(X, \|\cdot\|)$ is isometrically isomorphic to a dense subspace of $(\tilde{X}, \|\cdot\|)$.

The study of the second dual of a normed linear space enables us to define such a completion.

7.12.2 **Theorem.** *For any incomplete normed linear space* $(X, \|\cdot\|)$ *there exists a completion* $(\tilde{X}, \|\cdot\|)$ *which is unique up to isometric isomorphisms.*

Proof. Consider the second dual $(X^{**}, \|\cdot\|)$ of $(X, \|\cdot\|)$. By Corollary 4.10.3, $(X^{**}, \|\cdot\|)$ is complete so it contains $\overline{\hat{X}}$ the closure of the natural embedding of X. As a closed subspace of a complete space, $(\overline{\hat{X}}, \|\cdot\|)$ is complete by Proposition 1.13. Since X is isometrically isomorphic to $\hat{X}$ under the natural embedding, we have that $(\hat{X}, \|\cdot\|)$ is a completion of $(X, \|\cdot\|)$.
The uniqueness up to isometric isomorphisms follows from Theorem 4.8. □

7.12.3 **Corollary**. *Given an incomplete normed linear space* $(X, \|\cdot\|)$ *with completion* $(\tilde{X}, \|\cdot\|)$ *then* $(X, \|\cdot\|)^*$ *is isometrically isomorphic to* $(\tilde{X}, \|\cdot\|)^*$.

7.12.4 **Example**. E_0 and ℓ_1 are dense linear subspaces of $(c_0, \|\cdot\|_\infty)$. Then $(E_0, \|\cdot\|_\infty)$ and $(\ell_1, \|\cdot\|_\infty)$ have the same completion and $(E_0, \|\cdot\|_\infty)^*$ and $(\ell_1, \|\cdot\|_\infty)^*$ and $(c_0, \|\cdot\|_\infty)^*$ are isometrically isomorphic to $(\ell_1, \|\cdot\|_1)$. □

When the incomplete space is an inner product space then it is significant that there is a natural extension of the inner product to its completion.

7.12.5 **Theorem**. *For an incomplete inner product space* X *its completion* $\tilde{X}$ *is an inner product space with inner product* $(\,.\,,\,.\,)$ *defined by*

$$(\tilde{x}, \tilde{y}) = \lim_{n\to\infty} (x_n, y_n)$$

where $\{x_n\}$ and $\{y_n\}$ *are sequences in* X *convergent on* $\tilde{x}$ *and* $\tilde{y}$ *in* $\tilde{X}$.

Proof. Now $\{(x_n, y_n)\}$ is a Cauchy sequence of scalars so $\lim_{n\to\infty} (x_n, y_n)$ exists. It is not difficult to show that this limit is independent of the choice of sequences $\{x_n\}$ and $\{y_n\}$ in X converging to $\tilde{x}$ and $\tilde{y}$ in $\tilde{X}$. Properties (i)–(iv) of the inner product follow from those of the inner product in X and the limit properties. We check property (v).

If $(\tilde{x}, \tilde{x}) = 0$, then $\lim_{n\to\infty} \| x_n \|^2 = 0$ so (x_n) converges to 0 and then $\tilde{x} = 0$.

Clearly, the inner product on $\tilde{X}$ is an extension of that on X. □

7.12.6 **Example.** $(\mathcal{C}[a,b], \|\cdot\|_2)$ is an incomplete inner product space and its completion is a Hilbert space which contains $(\mathcal{C}[a,b], \|\cdot\|_2)$ as a dense linear subspace. But $(\mathcal{C}[a,b], \|\cdot\|_2)^*$ is isometrically isomorphic to $(\tilde{\mathcal{C}}[a,b], \|\cdot\|_2)^*$ and as it is a Hilbert space we have by Theorem 5.2.5 that $(\tilde{\mathcal{C}}[a,b], \|\cdot\|_2)$ is isometrically isomorphic to its dual. So the elements $\tilde{\mathcal{C}}[a,b]$ can be thought of as continuous linear functionals over $(\mathcal{C}[a,b], \|\cdot\|_2)$.

□

7.12.7 **Remark.** Some regard classical Hilbert space $(\mathcal{L}_2[a,b], \|\cdot\|_2)$ introduced in Example 2.2.12(iii) as $(\tilde{\mathcal{C}}_2[a,b], \|\cdot\|_2)$ the completion of $(\mathcal{C}_2[a,b], \|\cdot\|_2)$. □

7.13 **EXERCISES**

1. Determine whether the following normed linear spaces are reflexive.

   (i) $(E_0, \|\cdot\|_\infty)$ (ii) $(\ell_1, \|\cdot\|_1)$ (iii) $(m, \|\cdot\|_\infty)$

   (iv) $(c, \|\cdot\|_\infty)$ (v) $(\tilde{\mathcal{C}}[-\pi,\pi], \|\cdot\|_2)$ (vi) $(\mathcal{C}\mathcal{P}(2\pi), \|\cdot\|_\infty)$

2. Consider a reflexive Banach space $(X, \|\cdot\|)$.

   (i) Given any proper closed linear subspace M and $x_0 \in X \setminus M$, prove that there exists an element $y_0 \in M$ such that $\| x_0 - y_0 \| = d(x_0, M)$; (see Exercise 6.13.6).

   (ii) Given any proper closed convex set K and $x_0 \in X \setminus K$, is it true that there exists an element $y_0 \in K$ such that $\| x_0 - y_0 \| = d(x_0, K)$?

3. For a reflexive Banach space $(X, \|\cdot\|)$ prove that

   (i) X is smooth if and only if X* is rotund, and

   (ii) X is rotund if and only if X* is smooth.

4. Consider a linear space X with norms $\|\cdot\|$ and $\|\cdot\|'$.
   (i) Prove that $\|\cdot\|$ and $\|\cdot\|'$ are equivalent if and only if they generate equivalent norms $\|\cdot\|^*$ and $\|\cdot\|'^*$ on X^*; (see Exercise 4.12.9).
   (ii) Prove that a Banach space $(X, \|\cdot\|)$ is reflexive if and only if $(X, \|\cdot\|')$ is reflexive where $\|\cdot\|$ and $\|\cdot\|'$ are equivalent norms on X.

5. Using the information in Section 5 about the shape of the dual, exhibit on each space
   (i) $(c_0, \|\cdot\|_\infty)$, (ii) $(\ell_1, \|\cdot\|_1)$, (iii) $(\mathcal{C}[-1,1], \|\cdot\|_\infty)$.
   a continuous linear functional which does not attain its norm on the closed unit ball and deduce that none of these spaces is reflexive.

6. (i) Using the James characterisation of reflexivity, prove that a uniformly rotund Banach space is reflexive; (see Exercise 2.4.17).
   (ii) Hence, deduce that ℓ_p space $(1 < p < \infty)$ is reflexive.

7. Many of the properties of finite dimensional normed linear spaces can be derived from Corollary 6.3 to the Hahn–Banach Theorem without using compactness at all. Using such a method prove that
   (i) a finite dimensional normed linear space $(X_n, \|\cdot\|)$ is reflexive,
   (ii) every linear functional f on X_n is continuous,
   (iii) convergence in norm is equivalent to coordinatewise convergence, (that is, Corollary 2.1.11 holds),
   (iv) the closed unit ball is compact.

8. Given a linear space X consider the *algebraic embedding* $x \mapsto \hat{x}$ of X into $X^{\#\#}$ induced by the definition of $\hat{x}$ as

$$\hat{x}(f) = f(x) \qquad \text{for all } f \in X^{\#}.$$

   (i) Prove that $\hat{X}$ the image of X under the algebraic embedding $x \mapsto \hat{x}$ is an isomorphism of X into $X^{\#\#}$.
   (ii) Prove that $\hat{X} = X^{\#\#}$, (that is, X is algebraically reflexive), if and only if X is finite dimensional. (See Theorem 2.1.12).

§8. SUBREFLEXIVITY

In any discussion of approximation theory and reflexivity we notice the significance of continuous linear functionals which attain their norm on the closed unit ball of the space. A key result which gives useful information about the distribution of norm attaining continuous linear functionals in the dual was given by E. Bishop and R.R. Phelps, *Bull. Amer. Math. Soc.* **67** (1961), 97–98. They showed that completeness for a normed linear space implies that the set of norm attaining continuous linear functionals is dense in the dual. This result has been generalised out of all recognition and in its most general form is called the Ekeland Variational Principle. It has wide application in a remarkable number of areas as the survey article by I. Ekeland, *Bull. Amer. Math. Soc. N.S.* **1** (1979), 443–474 demonstrates.

We confine our attention to the result originally given by Bishop and Phelps and show its usefulness for some geometrical considerations. We begin by developing a geometrical lemma.

8.1 **Definition**. Given convex sets A and B in a linear space X the *convex hull* of A and B, denoted by co$\{A,B\}$ is the set $\{\lambda a + (1-\lambda)b : 0 \le \lambda \le 1 \text{ and } a \in A, b \in B\}$. Clearly co$\{A,B\}$ is a convex set in X.

8.2 **Notation**. It is convenient to denote by $\overline{B}$ the closed unit ball of $(X, \|\cdot\|)$.

8.3 **Lemma**. *Given a normed linear space* $(X, \|\cdot\|)$, *a nonzero continuous linear functional* f *on* X *and a constant* $k > 1$, *consider the sets*

$$T = \{x \in X : x \in \ker f \text{ and } 1 < \|x\| \le k\} \quad \textit{and}\; K \equiv \text{co}\,\{\overline{B}, T\}.$$

Then the functional $\|\cdot\|'$ *on* X *where*

$$\|x\|' = \inf\{\lambda > 0 : x \in \lambda K\}$$

is a norm equivalent to the given norm $\|\cdot\|$ *on* X.

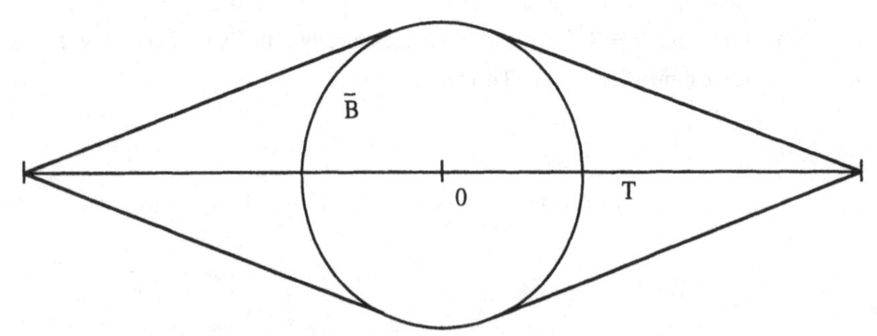

Figure 13. A pictorial representation of $K \equiv \text{co}\,\{\overline{B}, T\}$.

Proof. For $y \in \bar{B}$ and $x \in T$ we have $\alpha y \in \bar{B}$ and $\alpha x \in T$ for all $|\alpha| \leq 1$, so

$$\alpha(\lambda y + (1-\lambda)x) \in K \text{ for all } 0 \leq \lambda \leq 1 \text{ and } |\alpha| \leq 1.$$

Therefore, for $x \neq 0$, if $x \in \mu K$ for some $\mu > 0$ then $\lambda x \in |\lambda| \mu K$ for $\lambda \neq 0$ so

$$\| \lambda x \|' \leq |\lambda| \| x \|'.$$

For x substitute λx and for λ substitute $1/\lambda$ then we have

$$\| x \|' \leq \frac{1}{|\lambda|} \| \lambda x \|'$$

and we conclude that $\| \lambda x \|' = |\lambda| \| x \|'$ for $\lambda \neq 0$.

But also for $x,y \neq 0$, if $x \in \lambda K$ and $y \in \mu K$ for some $\lambda,\mu > 0$ then x/λ and $y/\mu \in K$.

Since K is convex,

$$\frac{\lambda}{\lambda+\mu}(x/\lambda) + \frac{\mu}{\lambda+\mu}(y/\mu) \in K;$$

that is, $\frac{x+y}{\lambda+\mu} \in K$ and so $x + y \in (\lambda+\mu)K$;

This implies that $\| x+y \|' \leq \| x \|' + \| y \|'$.

Since $\ker \|\cdot\|' = \{0\}$, $\|\cdot\|'$ is a norm for X. But also $\bar{B} \subseteq K \subseteq k\bar{B}$ which implies that

$$\| x \| \leq \| x \|' \leq \frac{1}{k} \| x \| \qquad \text{for all } x \in X$$

so $\|\cdot\|'$ and $\|\cdot\|$ are equivalent norms for X. □

We develop our theory for Banach spaces over the real numbers and we establish the Bishop–Phelps Theorem for Banach spaces over the real or complex numbers by appealing to Theorem 4.10.16.

The main argument of the Bishop–Phelps Theorem is based on the following lemma.

8.4 **Lemma.** *For a real Banach space* $(X, \|\cdot\|)$ *consider* $f \in X^*$, $\| f \| = 1$. *Given* $0 < k \leq 1$ *and* $u \in \bar{B}$ *where* $f(u) > 0$ *there exists an* $x_0 \in \bar{B}$ *where* $f(x_0) > 0$ *such that*

$$k \| x_0 - u \| \leq f(x_0) - f(u) \textit{ and } f(y) - f(x_0) < k \| y - x_0 \| \textit{ for all } y \in \bar{B},\ y \neq x_0.$$

Proof. We define a sequence $\{x_n\}$ in $\bar{B}$ inductively as follows.

Choose $x_1 = u$. For each $n \in \mathbb{N}$ consider the set

$$S_n \equiv \{y \in \bar{B} : f(y) - f(x_n) > K \| y - x_n \|\}.$$

If $S_n = \emptyset$, write $x_{n+1} = x_n$ and if $S_n \neq \emptyset$, choose $x_{n+1} \in S_n$ such that

$$f(x_{n+1}) \geq \frac{1}{2}(f(x_n) + \sup\{f(x) : x \in S_n\}). \tag{i}$$

For the sequence $\{x_n\}$ so defined

$$k \| x_n - x_{n-1} \| < f(x_n) - f(x_{n-1}) \qquad \text{for all } n \in \mathbb{N}$$

so

$$k \| x_m - x_n \| < f(x_m) - f(x_n) \qquad \text{for all } m > n. \tag{ii}$$

Now the sequence $\{f(x_n)\}$ is increasing and is bounded above, so it is convergent. But then this implies that the sequence $\{x_n\}$ is Cauchy. Since $(X, \|\cdot\|)$ is complete, the sequence $\{x_n\}$ is convergent to some $x_0 \in \overline{B}$.

Since f is continuous at x_0, inequality (ii) implies that

$$k \| x_0 - x_n \| \le f(x_0) - f(x_n) \quad \text{for all } n \in \mathbb{N}$$

and in particular $\quad k \| x_0 - u \| \le f(x_0) - f(u).$

Suppose there exists a $v \in \overline{B}$, $v \ne x_0$ such that

$$f(v) \ge f(x_0) + K \| v - x_0 \|.$$

Then $\quad f(v) > f(x_0) = \lim_{n\to\infty} f(x_n)$

since f is continuous at x_0. But then $v \in S_n$ for all $n \in \mathbb{N}$ which implies that

$$2\, f(x_{n+1}) - f(x_n) \ge \sup\{f(x) : x \in S_n\} \ge f(v)$$

and so $\quad f(x_0) \ge f(v).$

But this is a contradiction so

$$f(y) - f(x_0) < k \| y - x_0 \| \quad \text{for all } y \in \overline{B},\ y \ne x_0. \qquad \square$$

8.5 **Theorem**. *For a real Banach space* $(X, \|\cdot\|)$, *given* $f \in X^*$, $\| f \| = 1$ *and* $\varepsilon > 0$ *there exists a* $g \in X^*$, $\| g \| = 1$ *and* $x_0 \in X$, $\| x_0 \| = 1$ *such that*

$$g(x_0) = 1 \text{ and } | g(x) | \le \varepsilon \quad \text{for } x \in \ker f \text{ and } \| x \| \le 1.$$

Proof. We need to find a $g \in X^*$, $\| g \| = 1$ and $x_0 \in K$, $\| x_0 \| = 1$ such that $g(x_0) = 1$ and $| g(z) | \le 1$ for all $z \in T \equiv \{x \in X : x \in \ker f \text{ and } \| x \| \ge \frac{1}{\varepsilon}\}$. Consider $K \equiv \text{co}\ \{\overline{B}, T\}$.

From Lemma 8.3, K is an equivalent norm ball which contains $\overline{B}$.

If $x_0 \in \overline{B} \cap \text{bdy } K$, from Corollary 6.3 we see that there exists a $g \in X^*$, such that

$$g(x_0) = \sup\ \{| g(x) | : x \in K\} \ge \{| g(x) | : x \in \overline{B}\} \equiv \| g \| \ge g(x_0).$$

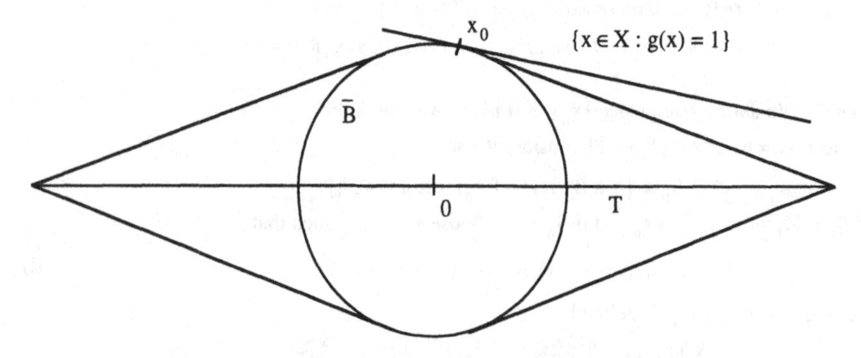

Figure 14. K is to one side of the hyperplane $\{x \in X : g(x) = 1\}$.

So to establish the theorem we need to show that $\bar{B} \cap \text{bdy } K \neq \emptyset$.

Consider $u \in \bar{B}$ but $u \notin \text{bdy } K$. We may consider $f(u) > 0$. There exists an $\alpha > 1$ such that $\alpha u \in K$. Then there exists $0 < \lambda < 1$ such that

$$\alpha u = \lambda x + (1-\lambda)z \qquad \text{for some } x \in \bar{B} \text{ and } z \in T.$$

So $\qquad x - u = (\alpha-1)u + (1-\lambda)(x-z)$

and $\qquad \| x-u \| \leq (\alpha-1) + (1-\lambda) \| x-z \| \leq \left(1 + \frac{1}{\varepsilon}\right)(\alpha-\lambda)$ since $\| x-z \| \leq 1 + \frac{1}{\varepsilon}$.

But also $f(u) < \alpha f(u) = \lambda f(x) < f(x)$.

So $\qquad f(x) - f(u) = (\alpha-1) f(u) + (1-\lambda) f(x) > (\alpha-\lambda) f(u) \geq \| x-u \| \dfrac{f(u)}{1+\frac{1}{\varepsilon}}$. $\qquad$ (i)

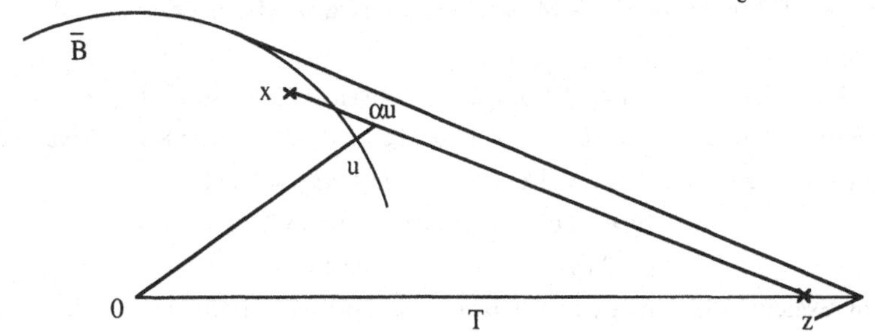

Figure 15. $u \in \bar{B}$ but $u \notin \text{bdy } K$ and $\alpha u \in K$ for $\alpha > 1$.

In Lemma 8.4, choosing $k = \dfrac{f(u)}{1+\frac{1}{\varepsilon}}$ we have that there exists an $x_0 \in \bar{B}$ where $f(x_0) > 0$,

such that $\qquad \dfrac{f(u)}{1+\frac{1}{\varepsilon}} \| x_0-u \| \leq f(x_0) - f(u)$ $\qquad$ (ii)

and $\qquad f(y) - f(x_0) < \dfrac{f(u)}{1+\frac{1}{\varepsilon}} \| y-x_0 \|$ for all $y \in \bar{B}$ and $y \neq x_0$. $\qquad$ (iii)

Comparing inequalities (i) and (iii) we see that $x_0 \in \text{bdy } K$. This completes the proof of the theorem. □

In order to show the full implications of this theorem we need the following lemma.

8.6 **The Parallel Hyperplane Lemma.**

For a real normed linear space $(X, \|\cdot\|)$, *given* $0 < \varepsilon \leq \frac{1}{2}$ *and* $f, g \in X^*$, $\| f \| = \| g \| = 1$ *and* $| g(x) | \leq \varepsilon$ *for* $x \in \ker f$ *and* $\| x \| \leq 1$ *then either* $\| f-g \| \leq 2\varepsilon$ *or* $\| f+g \| \leq 2\varepsilon$.

Proof. Consider $g|_{\ker f}$. Now $\| g|_{\ker f} \| \leq \varepsilon$ so by the Hahn–Banach Theorem 6.2 there exists an $h \in X^*$ an extension of $g|_{\ker f}$ to X such that $\| h \| \leq \varepsilon$.

Then $(g-h)(x) = 0$ for $x \in \ker f$, so $g-h = \alpha f$ for some real α.

Now $\quad |1-|\alpha|| = \|g\| - \|g-h\| \le \|h\| \le \varepsilon.$

So if $\alpha \ge 0$ then $\quad \|f-g\| = \|(1-\alpha)f-h\| \le |1-\alpha| + \|h\| \le 2\varepsilon.$

If $\alpha < 0$ then $\quad \|f+g\| = \|(1+\alpha)f+h\| \le |1+\alpha| + \|h\| \le 2\varepsilon.$ □

Theorem 8.5 and Lemma 8.6 have an immediate corollary, a gloss on Corollary 6.7 to the Hahn–Banach Theorem.

8.7 **Corollary**. *Consider a real Banach space* $(X, \|\cdot\|)$ *and any proper closed linear subspace* M. *Given* $\varepsilon > 0$ *there exists a* $g \in X^*$, $\|g\| = 1$ *and an* $x_0 \in X \setminus M$, $\|x_0\| = 1$ *such that* $g(x_0) = 1$ *and* $|g(x)| \le \varepsilon$ *for all* $x \in M$ *and* $\|x\| \le 1$ *and* $d(x_0, M) \ge 1-2\varepsilon$.

Proof. From the modification of Corollary 6.7 given in Exercise 6.13.6, there exists an $f \in X^*$, $\|f\| = 1$ such that $f(M) = 0$. Now consider ker f. From Theorem 8.5 we have that there exists a $g \in X^*$, $\|g\| = 1$ and $x_0 \in X$, $\|x_0\| = 1$ such that $g(x_0) = 1$ and

$$|g(x)| \le \varepsilon \qquad \text{for } x \in \ker f \text{ and } \|x\| \le 1$$

so

$$|g(x)| \le \varepsilon \qquad \text{for } x \in M \text{ and } \|x\| = 1.$$

Again by Exercise 6.13.6 there exists an α such that $|\alpha| = 1$ and $\alpha f(x_0) = d(x_0, \ker f)$. From Lemma 8.6 we may assume that $\|f-g\| \le 2\varepsilon$ so

$$1 = g(x_0) \le |g(x_0) - f(x_0)| + |f(x_0)| \le \|f-g\| + d(x_0, \ker f) \le 2\varepsilon + d(x_0, M).$$

Then $d(x_0, M) \ge 1 - 2\varepsilon$. □

8.8 **Definition**. A normed linear space $(X, \|\cdot\|)$ is said to be *subreflexive* if the set of continuous linear functionals which attain their norm on the closed unit ball is dense in the dual $(X^*, \|\cdot\|)$.

From Theorem 8.5 and Lemma 8.6 we deduce the key theorem of this section.

8.9 **The Bishop–Phelps Theorem**. *Every Banach space is subreflexive.*

Within a short time of publication of the original Bishop–Phelps Theorem 8.9 it was realised that the proof provided more information than had been claimed. So we present this form which has applications in its own right.

8.10 **The improved version of the Bishop–Phelps Theorem.**

Given a Banach space $(X, \|\cdot\|)$ *and an* $f \in X^*$, $\|f\| = 1$ *and* $0 < \varepsilon \le \frac{1}{2}$ *and* $u \in X$, $\|u\| = 1$ *such that* $|f(u) - 1| \le \varepsilon^2$ *then there exists a* $g \in X^*$, $\|g\| = 1$ *and an* $x_0 \in X$, $\|x_0\| = 1$ *such that*

$$g(x_0) = 1, \quad \|f-g\| \le 2\varepsilon \text{ and } \|x_0-u\| \le 2\varepsilon.$$

Proof. We firstly prove the theorem for a real Banach space $(X, \|\cdot\|)$. Within the proof of Theorem 8.5 inequality (ii) gives us that

$$\| x_0 - u \| \le \frac{(1+\frac{1}{\varepsilon})\, f(x_0-u)}{f(u)}$$

where $f(x_0) > 0$ and so $\| f+g \| > (f+g)(x_0) > 1$. Lemma 8.6 gives us that $\| f-g \| \le 2\varepsilon$.
But also we see that

$$0 \le f(x_0-u) \le 1 - f(u) \le \varepsilon^2 \qquad \text{so } \| x_0-u \| \le \frac{\varepsilon+\varepsilon^2}{1-\varepsilon^2}\ .$$

But since $0 < \varepsilon \le \frac{1}{2}$ we have $1 - \varepsilon^2 \le \frac{3}{4}$

so $$\| x_0-u \| \le \tfrac{4}{3}\,(\varepsilon+\varepsilon^2) \le \tfrac{4}{3}\,(\tfrac{3}{2}\,\varepsilon) = 2\varepsilon.$$

The theorem is extended to complex Banach spaces by Theorem 4.10.16 which establishes that there is a norm preserving one-to-one correspondence between the continuous complex linear functionals and the continuous real linear functionals on X. If $| f(u)-1 | < \varepsilon^2$ then $| f_{\mathbb{R}}(u)-1 | < \varepsilon^2$ so by Theorem 8.10 applied to $X_{\mathbb{R}}$ we have that there exists a $g_{\mathbb{R}} \in X_{\mathbb{R}}{}^*$, $\| g_{\mathbb{R}} \| = 1$ and an $x_0 \in X$, $\| x_0 \| = 1$ such that $g_{\mathbb{R}}(x_0) = 1$, $\| (f-g)_{\mathbb{R}} \| = \| f_{\mathbb{R}} - g_{\mathbb{R}} \| \le 2\varepsilon$ and $\| x_0-u \| \le 2\varepsilon$. Then for $g \in X^*$ defined as in Theorem 4.10.16 we have $\| g \| = 1$, $g(x_0) = 1$ and $\| f-g \| \le 2\varepsilon$ and $\| x_0-u \| \le 2\varepsilon$. □

We now show how the Bishop–Phelps Theorem can be applied. In Remark 7.9.6 we mentioned the James characterisation of reflexivity. The Bishop–Phelps Theorem 8.9 enables us to prove a special case of this result directly.

8.10 **Corollary**. *Consider a Banach space* $(X, \|\cdot\|)$ *where the dual space* X^* *is smooth. If every continuous linear functional attains its norm on the closed unit ball of the space then* X *is reflexive.*

Proof. By X^* being smooth we mean that to each $f \in X^*$, $\| f \| = 1$ there exists only one $F \in X^{**}$, $\| F \| = 1$ such that $F(f) = 1$.
However, for each $f \in X^*$, $\| f \| = 1$ there exists an $x \in X$, $\| x \| = 1$ such that $f(x) = 1$; that is, $\hat{x}(f) = 1$.
So the set of continuous linear functions on X^* which attain their norm on the closed unit ball of X^* is the set $\hat{X}$.
But by the subreflexivity of $(X^*, \|\cdot\|)$, the set $\hat{X}$ is dense in $(X^{**}, \|\cdot\|)$. But since $(X, \|\cdot\|)$ is complete, $\hat{X}$ is closed in $(X^{**}, \|\cdot\|)$. So $X^{**} = \hat{X}$; that is, X is reflexive. □

8.11 EXERCISES

1. Consider a real Banach space $(X, \|\cdot\|)$.

(i) Given $x \in X \setminus \{0\}$ we write

$D(x) \equiv \{f \in X^* : \| f \| = 1 \text{ and } f(x) = \| x \|\}$.

(a) Prove that given $x \in X \setminus \{0\}$ and any $y \in X$ and $\lambda > 0$

$$f_x(y) \leq \frac{\|x+y\| - \|x\|}{\lambda} \leq f_{x+\lambda y}(y)$$

for any $f_x \in D(x)$ and $f_{x+\lambda y} \in D(x+\lambda y)$ and that the inequality is reversed for $\lambda < 0$.

(b) Given $x \in X \setminus \{0\}$ deduce that if, for all $y \in X$ and $\lambda \neq 0$ and any selection $f_{x+\lambda y} \in D(x+\lambda y)$ we have

$f_{x+\lambda y}(y)$ converges as $\lambda \to 0$

then X is smooth at x.

(ii) Given $f \in X^*$, $\| f \| = 1$ a closed *slice* of the closed unit ball B(X) cut off by f is a subset

$\{x \in B(X) : f(x) \geq 1 - \delta\}$ for any $0 < \delta < 1$.

Given $x \in X$, $\| x \| = 1$, prove that for any $y \in X$, $\| y \| = 1$ and $0 < \lambda < \frac{\delta}{2}$,

$f_{x+\lambda y}(x) \geq 1 - \delta$ for all $f_{x+\lambda y} \in D(x+\lambda y)$.

(iii) Suppose that X has the property that every $f \in X^*$, $\| f \| = 1$ cuts off slices of B(X) of arbitrarily small diameter. Prove that

(a) X is rotund,

(b) every $f \in X^*$ attains its norm on B(X),

(Hint: Use Cantor's Intersection Theorem AMS §4.)

(c) X^* is smooth,

(d) X is reflexive.

(iv) Prove that if X is uniformly rotund then X is reflexive.

IV. THE FUNDAMENTAL MAPPING THEOREMS FOR BANACH SPACES

In addition to the Hahn–Banach Theorem the three mapping theorems, the Open Mapping Theorem, the Closed Graph Theorem and the Uniform Boundedness Theorem are vital for the development of any general theory of Banach spaces.

In these theorems we begin to appreciate the importance of the completeness condition. The proofs are based on Baire category arguments which reveal the implications of completeness for the metric topology. So we begin by developing this theory and demonstrate something of its force before applying it to establish the fundamental mapping theorems.

§9. BAIRE CATEGORY THEORY FOR METRIC SPACES

In complete metric spaces the metric topology has important characteristics and a knowledge of these is indispensable in establishing many significant results in the analysis of normed linear spaces and elsewhere.

We recall the following definition from the analysis of metric spaces.

9.1 **Definition**. Given a metric space (X, d), a subset A is said to be *dense* in (X, d) if its closure $\overline{A} = X$.
This means that A is dense in (X, d) if and only if every point of X is either a point of A or a cluster point of A. Equivalently, A is dense in (X, d) if and only if for every $x \in X$ and $\varepsilon > 0$, we have $B(x; \varepsilon) \cap A \neq \emptyset$.

The following concept related to density is used to partition metric spaces into disjoint classes.

9.2 **Definition**. Given a metric space (X, d), a subset A is said to be *nowhere dense* in X if int $\overline{A} = \emptyset$.

It is clear that A is nowhere dense in X if and only if $C(\overline{A})$ is dense in X; (see Exercise 9.19.1).

9.3 **Remark**. So a closed set A in (X, d), is nowhere dense if and only if C(A) is dense in X. A closed nowhere dense set A is its own boundary, bdy $A = \overline{C(A)} \cap A = A$. Further, the boundary of any open or closed set is nowhere dense since, for any open set G, bdy $G = \overline{G} \cap C(G)$ and C(bdy G) $= C(\overline{G}) \cup G$ from which it can be seen that C(bdy G) is dense in X. □

9.4 **Examples**.

(i) In any normed linear space $(X, \|\cdot\|)$,

(a) any finite subset is nowhere dense,

(b) any proper closed linear subspace is nowhere dense.

(ii) In $(\mathbb{R}, |\cdot|)$, the set of natural numbers $\mathbb{N}$ is nowhere dense.

(iii) In [0,1] with the usual metric, Cantor's Ternary Set K is closed and has empty interior so is nowhere dense.

(iv) In any discrete metric space (X, d), since every subset is both open and closed, every subset except the null set $\emptyset$ has nonempty interior so the only nowhere dense subset is $\emptyset$.□

It is clear that, whether or not a set is nowhere dense depends on the metric and the space.

9.5 **Examples**.

(i) A proper closed linear subspace Y of a normed linear space $(X, \|\cdot\|)$ is not nowhere dense in $(Y, \|\cdot\|_Y)$.

(ii) The set of natural numbers $\mathbb{N}$ as a subset of $(\mathbb{R}, |\cdot|)$ is a discrete metric space in its relative metric topology and so has no nonempty nowhere dense sets. □

9.6 **Remark**. Since the concept of nowhere denseness is defined in terms of interior and closure, concepts which are invariant under equivalent metrics, so this concept is also invariant under equivalent metrics. □

Now a finite union of nowhere dense sets is nowhere dense. This leads us to introduce the following definitions.

9.7 **Definitions.** Given a metric space (X, d), a subset A is said to be *first category* (or *meagre*) in X if A can be represented as the union of a countable number of nowhere dense sets. A subset A which is not of the first category in X is said to be of *second category* in X.

9.8 **Examples.**

(i) In any normed linear space $(X, \|\cdot\|)$,

(a) any countable set is first category,

(b) any countable union of first category sets is first category.

(ii) In $(\mathbb{R}, |\cdot|)$, the set of rational numbers $\mathbb{Q}$, being a countable set, is first category.

(iii) In [0,1] with the usual metric, Cantor's Ternary Set K is uncountable but is first category because it is nowhere dense.

(iv) In [0,1] with the usual metric, consider the set $\mathbb{K}$ constructed from Cantor's Ternary Set K in the following way:
In the interval $\left[\frac{1}{3}, \frac{2}{3}\right]$, consider the image of K under the mapping $f_1(t) = \frac{1}{3}(t+1)$. In the interval $\left[\frac{1}{9}, \frac{2}{9}\right]$, consider the image of K under mapping $f_{21}(t) = \frac{1}{9}(t+1)$, in the interval $\left[\frac{4}{9}, \frac{5}{9}\right]$, consider the image of K under the mapping $f_{22}(t) = \frac{1}{9}(t+4)$ and in the interval $\left[\frac{7}{9}, \frac{8}{9}\right]$, consider the image K under the mapping $f_{23}(t) = \frac{1}{9}(t+7)$. Continuing this process at the nth step we put a $\frac{1}{3^n}$ reduced copy of K into each of the intervals of length $\frac{1}{3^n}$ remaining in [0,1]. Now each copy of K is nowhere dense in [0,1] and there is a countable number of copies of K in the new set $\mathbb{K}$, so $\mathbb{K}$ is first category in [0,1] but it is uncountable and dense in [0,1]. □

9.9 **Definitions.** Given a metric space (X, d), a subset A is said to be *residual* in X if C(A) is first category. (X, d) is said to be a *Baire space* if every residual set in X is dense in X.

In a Baire space a first category subset has no interior so a residual subset is always dense and second category. A Baire space (X, d) is necessarily second category in itself, for if X were first category then the null set ∅ would be dense in X.

We develop the important characterisations for Baire spaces which we use in applications.

9.10 **Theorem**. *Given a metric space* (X, d), *the following are equivalent.*

(i) (X, d) *is a Baire space.*

(ii) *For every countable family of dense open sets* $\{G_n\}$ *we have* $\cap G_n$ *is dense in* X.

(iii) *For every countable family of closed sets* $\{F_n\}$ *such that* $X = \cup F_n$ *we have* $\cup \operatorname{int} F_n$ *is dense in* X.

Proof.

(i)⇒(iii) Since bdy F_n is nowhere dense, then $\cup$ bdy F_n is first category in X. But X is a Baire space so $C(\cup \text{ bdy } F_n)$ is dense in X. Now bdy $F_n = F_n \setminus \operatorname{int} F_n$ and since $X = \cup F_n$ we have that $C(\cup \text{ bdy } F_n) \subseteq \cup \operatorname{int} F_n$ and so $\cup \operatorname{int} F_n$ is dense in X.

(iii)⇒(ii) It is clear that $X = \overline{\cap G_n} \cup (\cup C(G_n))$, the union of a countable family of closed sets. So $\operatorname{int} \overline{\cap G_n} \cup (\cup \operatorname{int} C(G_n))$ is dense in X. But since G_n is dense in X, $\operatorname{int} C(G_n) = \emptyset$, and then $\operatorname{int} \overline{\cap G_n}$ is dense in X. But this implies that $\cap G_n$ is dense in X.

(ii)⇒(i) Consider a countable family $\{E_n\}$ of nowhere dense sets in X. Then $\cup E_n$ is first category in X. Since $C(\overline{E}_n)$ is open and dense in X so $\cap C(\overline{E}_n)$ is dense in X. But $\cap C(\overline{E}_n) \subseteq C(\cup E_n)$ so $C(\cup E_n)$ is dense in X. □

We are now ready to reveal the special properties of complete metric spaces.

9.11 **Baire's Theorem**. *A complete metric space* (X, d) *is a Baire space.*

Proof. Consider a countable family of dense open sets $\{G_n\}$ in X. We show that for any $x \in X$ and $r > 0$, we have $\cap G_n \cap B(x; r) \neq \emptyset$.

Since G_1 is dense in X there exists an $x_1 \in G_1 \cap B(x; r)$ such that

$$G_1 \cap B(x; r) \text{ contains } B[x_1; r_1] \text{ where } 0 < r_1 < \tfrac{r}{2}.$$

Since G_2 is dense in X there exists an $x_2 \in G_2 \cap B(x_1; r_1)$ such that

$$G_1 \cap G_2 \cap B(x; r) \cap B(x_1; r_1) \text{ contains } B[x_2; r_2] \text{ where } 0 < r_2 < \tfrac{r}{4}.$$

Continuing inductively we have that there exists an $x_n \in G_n \cap B(x_{n-1}; r_{n-1})$ such that

$$G_1 \cap G_2 \cap \ldots \cap G_n \cap B(x; r) \cap B(x_{n-1}; r_{n-1})$$

contains $B[x_n; r_n]$ where $0 < r_n < \frac{r}{2^n}$.

Now the sequence $\{B[x_n; r_n]\}$ is a nested sequence of closed sets whose diameters tend to zero. Since (X, d) is complete, we have by Cantor's Intersection Theorem, (see AMS §4), that there exists a $y \in \cap B[x_n; r_n]$ and so $y \in \cap G_n \cap B(x; r)$. □

Again whether or not a set is first or second category depends on the metric and the space.

9.12 **Example**. In any Banach space $(X, \|\cdot\|)$, a proper closed linear subspace Y is nowhere dense in X but by Theorem 9.11 is second category in $(Y, \|\cdot\|_Y)$. □

It is clear from Definitions 9.7 that a countable union of first category subsets is first category and we can use this fact to deduce the category of some subsets using Theorem 9.11.

9.13 **Examples**.
(i) As $(\mathbb{R}, |\cdot|)$ is complete it is a Baire space. The set of rationals $\mathbb{Q}$ is first category in $\mathbb{R}$ so we deduce that the set of irrationals $\mathbb{R} \setminus \mathbb{Q}$ is dense and second category in $\mathbb{R}$.
(ii) As [0,1] with the usual metric is complete it is a Baire space. Now the set $\mathbb{K}$ defined in Example 9.8(iv) is first category and so its complement is dense and second category in $\mathbb{R}$. □

The following theorem establishes a large class of sets as Baire spaces.

9.14 **Theorem**. *An open subset of a Baire space is a Baire space.*

Proof. Consider an open subset G of a metric space (X, d).
We show that if A is nowhere dense in G then A is nowhere dense in X. For any subset A of G, the closure of A in G is $\overline{A} \cap G$ and $\text{int}(\overline{A} \cap G) = \text{int}\,\overline{A} \cap G$. If $\text{int}(\overline{A} \cap G) = \emptyset$ then since A is a subset of G we have $\text{int}\,\overline{A} = \emptyset$.
Further, if a subset B is dense in X then $B \cap G$ is dense in G.
Therefore, for a Baire space (X, d), any first category subset A of G is first category in X and has C(A) dense in X which implies that $C(A) \cap G$ is dense in G and we conclude that $(G, d|_G)$ is a Baire space. □

9.15 **Remarks**. The particular properties from Baire Category Theory which are used widely in applications are as follows.
(i) A complete metric space (X, d) is second category and so for a countable family of closed sets $\{F_n\}$ such that $X = \bigcup F_n$, there exists at least one F_{n_0} which has nonempty interior. (Of course, since (X, d) is a Baire space we know that $\bigcup \text{int}\, F_n$ is dense in X but this property is not so often needed in applications.)
(ii) A complete metric space (X, d) is a Baire space and so for a countable family of dense open sets $\{G_n\}$ we have $\bigcap G_n$ is dense in X.
Baire Category Theory has a major use in providing nonconstructive existence proofs, showing that the desired examples occur as first or second category subsets in some complete metric space. □

We demonstrate the techniques of proof using Baire Category Theory and sample something of the results achieved by Baire category methods in the following three examples. These all concern properties of real functions on Baire spaces and their properties are involved in different ways. The first shows a limitation on the set of points of discontinuity of a real function continuous on a dense set, the second shows how a real lower semicontinuous function defined on an open set is continuous on a dense subset of its domain and the third shows how badly a continuous real function can behave from the point of view of its differentiability.

9.16 The set of points of discontinuity of a real function continuous on a dense set

We recall the following definitions from real analysis.

9.16.1 Definitions. Given a real function f on $\mathbb{R}$, for any bounded interval J we define $\omega(f, J)$, *the oscillation of* f *over* J by

$$\omega(f, J) \equiv \sup\{|f(x) - f(y)| : x,y \in J\},$$

and for $x_0 \in \mathbb{R}$ we define $\omega(f, x_0)$, the *oscillation of* f *at* x_0 by

$$\omega(f, x_0) \equiv \inf\{\omega(f, J) : \text{all such } J \text{ containing } x_0\}.$$

9.16.2 Theorem. *Given a real function f on* $\mathbb{R}$, *continuous at the points of a dense set, the set* D *of points of discontinuity of* f *is first category.*

Proof. It is clear that f is continuous at x_0 if and only if $\omega(f, x_0) = 0$.
So for each $n \in \mathbb{N}$, we define

$$E_n \equiv \{x \in \mathbb{R} : \omega(f, x) \geq \frac{1}{n}\}$$

and we show first that E_n is closed.
Consider a cluster point x_0 of E_n. Then for any bounded interval I containing x_0 there exists an $x \in E_n$ and so

$$\omega(f, I) \geq \omega(f, x) \geq \frac{1}{n}.$$

But then $\quad \omega(f, x_0) = \inf\{\omega(f, I) : \text{all such I containing } x_0\} \geq \frac{1}{n}.$
So $x_0 \in E_n$ and E_n is closed.
Now $D = \bigcup\{E_n : n \in \mathbb{N}\}$.
If for some E_n we have int $E_n \neq \emptyset$ then int $D \neq \emptyset$ which contradicts the fact that f is continuous on a dense set. Therefore, for each $n \in \mathbb{N}$, E_n is nowhere dense and we conclude that D is first category. □

9.16.3 **Remark**. The *ruler function* f defined on $\mathbb{R}$ by

$$\left.\begin{aligned} f(x) &= \frac{1}{q} \text{ for rational } x = \frac{p}{q} \neq 0, \quad (p,q \text{ mutually prime}) \\ &= 0 \text{ for irrational } x \\ f(0) &= 1 \end{aligned}\right\}$$

is continuous at irrational points and discontinuous at rational points. Now the set of rationals is first category in $\mathbb{R}$ so the ruler function is typical of Theorem 9.16.2. However, Theorem 9.16.2 tells us that it is not possible to have a real function on $\mathbb{R}$ continuous at rational points and discontinuous at irrational points, because the set of irrationals is second category in $\mathbb{R}$. □

9.17 The continuity of lower semicontinuous functions

9.17.1 **Definitions**. Given a metric space (X, d), a real function f on X is said to be *lower semicontinuous* at $x_0 \in X$ if given $\varepsilon > 0$ there exists a $\delta > 0$ such that

$$f(x) > f(x_0) - \varepsilon \quad \text{when } d(x, x_0) < \delta,$$

that is, if $$\liminf_{x \to x_0} f(x) \geq f(x_0).$$

A real function f on X is said to be *upper semicontinuous* at x_0 if $-f$ is lower semi–continuous at x_0; that is, f is upper semicontinuous at x_0 if given $\varepsilon > 0$ there exists a $\delta > 0$ such that

$$f(x) < f(x_0) + \varepsilon \quad \text{when } d(x, x_0) < \delta,$$

that is, if $$\limsup_{x \to x_0} f(x) \leq f(x_0).$$

Of course, f is continuous at x_0 if and only if f is both lower semicontinuous and upper semicontinuous at x_0.

We have the following global characterisation of lower semicontinuity.

9.17.2 **Proposition**. *Given a real function* f *on a metric space* (X, d), *the following are equivalent.*

(i) f *is lower semicontinuous on* X.

(ii) *For every* $\alpha \in \mathbb{R}$, *the set* $\{x \in X : f(x) \leq \alpha\}$ *is closed.*

(iii) *For every* $\alpha \in \mathbb{R}$, *the set* $\{x \in X : f(x) > \alpha\}$ *is open.*

Proof.

(i)⇒(ii) Given $\alpha \in \mathbb{R}$, consider x_0 a cluster point of the set $\{x \in X : f(x) \leq \alpha\}$. Since f is lower semi–continuous at x_0, given $\varepsilon > 0$ there exists a $\delta > 0$ such that

$$f(x) > f(x_0) - \varepsilon \quad \text{when } d(x, x_0) < \delta.$$

Since x_0 is a cluster point of $\{x \in X : f(x) \leq \alpha\}$ there exists an $x \in B(x_0; \delta)$ such that $\alpha \geq f(x) > f(x_0) - \varepsilon$ and so $f(x_0) \leq \alpha$ which implies that $\{x \in X : f(x) \leq \alpha\}$ is closed.

(ii)⇒(iii) This follows from the set property that for any subset A of $\mathbb{R}$,

$$X \setminus (f^{-1}(A)) = f^{-1}(\mathbb{R} \setminus A).$$

(iii)⇒(i) Given $x_0 \in X$ and $\varepsilon > 0$, (iii) implies that $\{x \in X : f(x) > f(x_0) - \varepsilon\}$ is open. But then there exists a $\delta > 0$ such that $B(x_0; \delta) \subseteq \{x \in X : f(x) > f(x_0) - \varepsilon\}$; that is,

$$f(x) > f(x_0) - \varepsilon \quad \text{when } d(x, x_0) < \delta$$

which implies that f is lower semicontinuous at x_0. □

How discontinuous can a lower semicontinuous function be? We use Baire Category Theory to show that if the lower semicontinuous function is defined on an open subset of a complete metric space then the function is actually continuous on a dense subset of its domain. This is an extremely useful result.

9.17.3 **Theorem.** *A real lower semicontinuous function* f *on a metric space* (X, d) *is continuous at the points of a residual subset of* X.

Proof. Consider $\{U_n\}$ a countable base for the usual topology on $\mathbb{R}$ and for each $n \in \mathbb{N}$ the set $D_n \equiv f^{-1}(U_n) \setminus \text{int}\, \overline{f^{-1}(U_n)}$ which is nowhere dense in X. Then $D \equiv \bigcup \{D_n : n \in \mathbb{N}\}$ is first category in X.

We show that f is continuous at the points of $X \setminus D$. Consider $x_0 \in X \setminus D$ and $k \in \mathbb{R}$ such that $f(x_0) < k$ and U_n such that $U_n \subseteq (-\infty, k)$ and $f(x_0) \in U_n$. Since $x_0 \notin D_n$ and $f(x_0) \in U_n$ we have $x_0 \in \text{int}\, \overline{f^{-1}(U_n)}$.

To show that f is continuous at x_0 it is sufficient to show that $f(x) \leq k$ for all $x \in \text{int}\, \overline{f^{-1}(U_n)}$.

Suppose there exists some $x_1 \in \text{int}\, \overline{f^{-1}(U_n)}$ with $f(x_1) > k$. By the lower semicontinuity of f at x_1 there exists an open neighbourhood V of x_1 such that $f(x) > k$ for all $x \in V$. Now $V \cap f^{-1}(U_n) \neq \emptyset$ and for $x_2 \in V \cap f^{-1}(U_n)$ we have $k < f(x_2) < k$ which is a contradiction.

So we conclude that f is continuous at the points of $X \setminus D$. □

9.17.4 **Remarks.**

(i) It is clear that a similar theorem holds for an upper semicontinuous function f by applying Theorem 9.17.3 to the function –f.

(ii) Such a result is of significance when (X, d) is a Baire space because then it tells us that the lower semicontinuous function is continuous at the points of a dense residual subset of X. □

9.18 The existence of continuous nowhere differentiable functions

Since most of the continuous real functions we manipulate in calculus have only a finite number of points in their domains where they are not differentiable, it is surprising to learn that there exist continuous functions which are nowhere differentiable. The first example of such a function was constructed by Karl Weierstrass in the nineteenth century. But Baire Category Theory shows that "most" continuous functions are of this type by showing that the class of functions with at least one derivative is "small". So Baire Category Theory reveals much more about the nature of continuous functions than Weierstrass' example and by a nonconstructive method.

9.18.1 **Theorem**. *In the Banach space* $(\mathcal{C}[0,1], \|\cdot\|_\infty)$ *the subset* $\mathcal{C}_1$ *of functions which have at least one derivative at a point of* $[0,1]$ *is first category.*

Proof. For each $n \in \mathbb{N}$ and $0 < h < \frac{1}{n}$ we define

$$E_{n,h} \equiv \left\{ f \in \mathcal{C}[0,1] : \text{for some } t \in [0,1-\tfrac{1}{n}],\ \left|\frac{f(t+h)-f(t)}{h}\right| \le n \right\}.$$

We show first that $E_{n,h}$ is closed.
Consider a cluster point f_0 of $E_{n,h}$ in $(\mathcal{C}[0,1], \|\cdot\|_\infty)$. Then given $\varepsilon > 0$ there exists an $f \in E_{n,h}$ such that $\| f_0 - f \|_\infty < \varepsilon$. But there exists a $t \in [0, 1-\frac{1}{n}]$ such that

$$\left|\frac{f_0(t+h)-f_0(t)}{h}\right| \le \frac{2\varepsilon}{h} + n$$

$$\le n \quad \text{since } \varepsilon > 0 \text{ is arbitrary.}$$

So $f_0 \in E_{n,h}$.
But then $E_n \equiv \bigcap \{E_{n,h} : \text{all } 0 < h < \frac{1}{n}\}$ is also closed.
We show that E_n is nowhere dense in $(\mathcal{C}[0,1], \|\cdot\|_\infty)$ by showing that that $C(E_n)$ is dense in $(\mathcal{C}[0,1], \|\cdot\|_\infty)$.
Given $f \in \mathcal{C}[0,1]$ and $\varepsilon > 0$ choose a polynomial p such that $\| f-p \|_\infty < \varepsilon$. This is possible because Weierstrass' Approximation Theorem, (AMS, §9), gives us that the polynomials are dense in the continuous functions on $[0,1]$. Then choose a continuous piecewise linear function q on $[0,1]$ of the following form.

$$\left.\begin{aligned} q(t) &= 2.5^m t && 0 \le t \le \frac{1}{2.10^m} \\ &= \frac{2}{2^m} - 2.5^m t && \frac{1}{2.10^m} < t \le \frac{1}{10^m} \\ q(t) &= q(t - \frac{1}{10^m}) && \frac{1}{10^m} < t \le 1 \end{aligned}\right\}$$

where we choose m sufficiently large that $\frac{1}{2^m} < \varepsilon$ and $2.5^m > n + 2\, \| p' \|_\infty$.

Then $\| q \|_\infty = \frac{1}{2^m} < \varepsilon$ and for all $t \in [0, 1]$ and h sufficiently small

$$\left|\frac{q(t+h)-q(t)}{h}\right| = 2.5^m > n + 2\, \| p' \|_\infty.$$

For the continuous function $g = p+q$ we have

$$\| g-q \|_\infty \le \| f-p \|_\infty + \| q \|_\infty < 2\varepsilon.$$

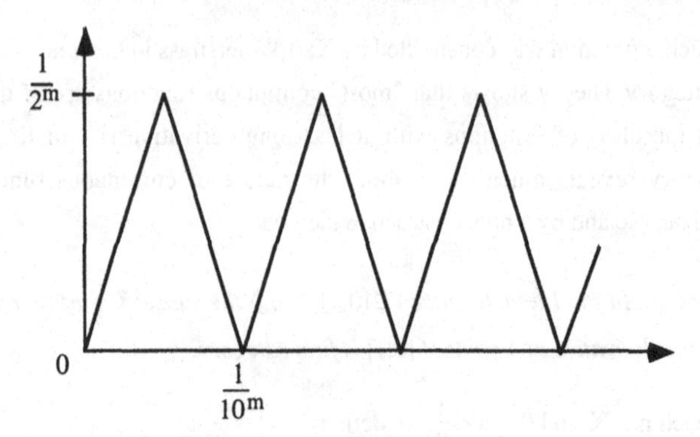

Figure 16. The "sawtooth" function q.

But also for all $t \in [0,1)$ and h sufficiently small

$$\left|\frac{g(t+h)-g(t)}{h}\right| \geq \left|\frac{q(t+h)-q(t)}{h}\right| - \left|\frac{p(t+h)-p(t)}{h}\right| > n$$

which implies that $g \in C(E_n)$.
We conclude that $C(E_n)$ is dense in $(\mathcal{C}[0,1], \|\cdot\|_\infty)$. We now have that $\cup\{E_n : n \in \mathbb{N}\}$ is first category in $(\mathcal{C}[0,1], \|\cdot\|_\infty)$, but $\mathcal{C}_1 \subseteq \cup\{E_n : n \in \mathbb{N}\}$. □

9.18.2 **Remark.** Since $(\mathcal{C}[0,1], \|\cdot\|_\infty)$ is a Baire space we conclude that the set of continuous nowhere differentiable functions on [0,1] is a residual set in $(\mathcal{C}[0,1], \|\cdot\|_\infty)$. Now although Weierstrass' Approximation Theorem implies that $\mathcal{C}^\infty[0,1]$ is also a dense subset of $(\mathcal{C}[0,1], \|\cdot\|_\infty)$, it follows from Theorem 9.18.1 that it is a first category subset. The set of nowhere differentiable functions is also dense but is a second category subset. □

9.19. Residual subsets of a Baire space.

9.19.1 **Definitions.** Given a metric space (X, d), a subset which can be represented as a countable intersection of open sets is called a G_δ *subset* and a subset which can be represented as a countable union of closed sets is called an F_σ *subset.*
The families of G_δ subsets and F_σ subsets are extensions of the families of open and closed sets.

9.19.2 **Remark.** It follows from de Morgan's Theorem that in any metric space (X, d), a subset A is a G_δ subset if and only if C(A) is an F_σ subset. So in $(\mathbb{R}, |\cdot|)$, the set of rationals $\mathbb{Q}$ as a countable set, is an F_σ subset and so the set of irrationals is a G_δ subset. □

In a metric space the open and closed sets have the following special properties.

9.19.3 **Proposition**. *In a metric space* (X, d),

(i) *every closed set* F *is a* G_δ *subset and*

(ii) *every open set* G *is an* F_σ *subset.*

Proof.

(i) For each $n \in \mathbb{N}$ consider $G_n \equiv \{B(x; \frac{1}{n}) : x \in F\}$.

For each $n \in \mathbb{N}$, G_n is open and $F \subseteq G_n$ so $F \subseteq \bigcap G_n$.

Consider $y \in \bigcap G_n$. For each $n \in \mathbb{N}$, there exists an $x \in F$ such that $y \in B(x; \frac{1}{n})$, so $y \in F$ or is a cluster point of F. But F is closed so $y \in F$ and we conclude that $\bigcap G_n \subseteq F$.

Therefore $F = \bigcap G_n$.

(ii) This follows by de Morgan's Theorem. □

The following theorem gives us further insight into the nature of residual subsets.

9.19.4 **Theorem.** *Consider a metric space* (X, d).

(i) *A residual subset* E *contains a* G_δ *subset of* X.

(ii) *A subset* E *which contains a dense* G_δ *subset of* X *is residual.*

(iii) *When* (X, d) *is a Baire space, a subset* E *is residual if and only if* E *contains a dense* G_δ *subset of* X.

Proof.

(i) Now C(E) is first category and so $C(E) = \bigcup_1^\infty E_n$ where E_n is nowhere dense for each $n \in \mathbb{N}$. So $E \supseteq \bigcap_1^\infty C(\overline{E}_n)$ a G_δ subset of X.

(ii) If E contains a dense G_δ subset of X then $E \supseteq \bigcap_1^\infty G_n$ where G_n is dense and open for each $n \in \mathbb{N}$. So $C(E) \subseteq \bigcup_1^\infty C(G_n)$ where int $C(G_n) = \emptyset$ for each $n \in \mathbb{N}$.

This implies that C(E) is first category and so E is residual.

(iii) From (i) we have that if E is residual then $E \supseteq \bigcap_1^\infty C(\overline{E}_n)$ where E_n is nowhere dense for each $n \in \mathbb{N}$. Then $C(\overline{E}_n)$ is dense for each $n \in \mathbb{N}$.

Since (X, d) is a Baire space, it follows from Theorem 9.10(ii) that $\bigcap_1^\infty C(\overline{E}_n)$ is also dense.

□

9.19.5 **Remark**. The information in Theorem 9.19.4 provides us with an important technique:

In any metric space it is obvious that the intersection of two dense subsets is not necessarily dense; in fact it could be empty. But in a complete metric space, if we have two dense subsets which are also residual then their intersection is also dense and residual. □

The following example illustrates how this technique works.

9.19.6 **Example**. Given a separable Banach space $(X, \|\cdot\|)$ and an open subset A of X, consider a real function f on $A \times X$ with the properties that for each $y \in X$, f(x,y) is lower semicontinuous in X and for each $y \in X$, f(x,y) is continuous in y. Then there exists a dense G_δ subset D of A such that f(x,y) is continuous at the points of D for every $y \in X$.

Proof. Now $(X, \|\cdot\|)$ is a Baire space and by Theorem 9.10, A is a Baire space. Consider $\{y_n\}$ a countable dense subset of X. Given $n \in \mathbb{N}$, there exists a dense G_δ subset D_n of A such that $f(x,y_n)$ is continuous at the points of D_n. By Theorem 9.19.4, $D = \bigcap_1^\infty D_n$ is a dense G_δ subset of A and $f(x,y_n)$ is continuous at the points of D for all $n \in \mathbb{N}$. But given $x \in X$, f(x,y) is continuous in y so we conclude that f(x,y) is continuous at the points of D for every $y \in X$. □

9.20. EXERCISES

1. Consider a metric space (X, d) and a subset A of X.
   (i) Prove that int $C(A) = C(\overline{A})$.
   (ii) Prove that $A \setminus \text{int}\,\overline{A}$ is nowhere dense in X.
   (iii) Prove that the following are equivalent
      (a) A is nowhere dense in X,
      (b) $C(\overline{A})$ is dense in X,
      (c) for any open subset G of X there exists an $x \in G$ and an $r > 0$ such that $B(x; r) \cap A = \varnothing$.

2. Consider a metric space (X, d) and a subset E of X.
   (i) Prove that if a subset A of E is nowhere dense in E then it is nowhere dense in X.
   (ii) Prove that if a subset A of E is first category in E then it is first category in X.
   (iii) Show that if a subset A of E is second category in E then it is not necessarily second category in X.
   (iv) Prove that if a subset A of E is second category in E which is second category in X then A is second category in X.

3. (i) For a metric space (X, d) with the property that every $x \in X$ is a cluster point of X, prove that singleton sets are nowhere dense in X.
   (ii) Hence, using Baire's Theorem prove that both $\mathbb{R}$ and Cantor's Ternary Set K are uncountable.

4. (i) (a) Show that the set of rationals $\mathbb{Q}$ is not a G_δ subset of $(\mathbb{R}, |\cdot|)$.
   (b) Show that in general for a metric space the families of G_δ subsets and F_σ subsets do not contain each other.
   (ii) (a) Prove that for a metric space (X, d) a real lower semicontinuous function f on X has the property that for every open set G in $\mathbb{R}$, $f^{-1}(G)$ is an F_σ subset of X.
   (b) Give an example to show that the converse is not true in general.

5. Consider a normed linear space $(X, \|\cdot\|)$. Prove that if X is of second category then the unit open ball is of second category and every nonempty open subset of X is second category.

6. (i) Using the fact that norms $\|\cdot\|_\infty$ and $\|\cdot\|_1$ are not equivalent on the linear space $\mathcal{C}[0,1]$, deduce that the normed linear space $(\mathcal{C}[0,1], \|\cdot\|_1)$ is not second category and so not complete.
   (ii) Given a Banach space $(X, \|\cdot\|)$, prove that every norm $\|\cdot\|'$ on X which is lower semicontinuous with respect to the $\|\cdot\|$-norm is continuous with respect to the $\|\cdot\|$-norm.

7. For the real function f on $\mathbb{R}$ defined by

$$f(x) = \left.\begin{array}{ll} 1 & x \text{ irrational} \\ 0 & x \text{ rational} \end{array}\right\}$$

prove that there is no sequence $\{f_n\}$ of continuous real functions pointwise convergent to f.
(Hint: Suppose that there exists some such sequence $\{f_n\}$ of continuous real functions pointwise convergent to f.
(i) For each $n \in \mathbb{N}$, define $E_n \equiv \{x \in \mathbb{R} : f_n(x) \geq \frac{1}{2}\}$ and prove that E_n is closed.
(ii) For each $n \in \mathbb{N}$, define $F_n \equiv \bigcap \{E_k : k \geq n\}$ and prove that F_n is closed.
(iii) Prove that x is irrational if and only if there exists a $\nu \in \mathbb{N}$ such that $x \in F_\nu$.
(iv) Prove that the set of irrationals is $\bigcup \{F_n : n \in \mathbb{N}\}$.
(v) Prove that (iv) implies that the set of irrationals is first category in $\mathbb{R}$ which it is not.)

8. Prove that on a complete metric space any real function which is the pointwise limit of a sequence of continuous real functions is itself continuous on a residual set. But give an example to show that the pointwise limit may still be discontinuous at the points of a dense set.

9. Prove that on a complete metric space any real function which is the pointwise limit of a monotone decreasing sequence of real lower semicontinuous functions is itself upper semicontinuous at the points of a residual subset.

10. Consider an infinite dimensional linear space X where X is the union of countably many finite dimensional linear subspaces.
    (i) Prove that with any norm $\|\cdot\|$, $(X, \|\cdot\|)$ is first category in itself.
    (ii) Prove that no infinite dimensional Banach space $(X, \|\cdot\|)$ has a countable Hamel basis.
    (iii) Show that the linear space E_0 cannot be given a norm $\|\cdot\|$ to make $(E_0, \|\cdot\|)$ complete.

11. Consider an infinite dimensional Banach space $(X, \|\cdot\|)$. Prove that X contains a proper linear subspace of codimension one which is of second category in X and which is not complete.

§10. THE OPEN MAPPING AND CLOSED GRAPH THEOREMS

When we study continuous mappings between metric spaces we are naturally led to examine the continuity of inverse mappings. It is useful to have criteria under which we can tell that a continuous one-to-one onto mapping is a homeomorphism without having to test the inverse mapping for continuity. Compactness provides a simple criterion in that a continuous one-to-one mapping from a compact metric space onto a metric space is a homeomorphism, (see AMS §9). However, it is reasonable to ask whether there is some simple criterion which applies to continuous linear mappings between normed linear spaces.

We should notice at the outset that not every continuous linear one-to-one onto mapping is a homeomorphism.

10.1 **Example**. Consider the identity mapping *id* of $(\mathcal{C}[0,1], \|\cdot\|_\infty)$ onto $(\mathcal{C}[0,1], \|\cdot\|_1)$. This mapping is continuous linear one-to-one onto but its inverse is not continuous, (see AMS §7). But we also note that $(\mathcal{C}[0,1], \|\cdot\|_\infty)$ is complete and $(\mathcal{C}[0,1], \|\cdot\|_1)$ is not complete so the normed linear spaces cannot be topologically isomorphic, (see Theorem 1.24.9). □

The Open Mapping Theorem provides as a corollary an important criterion for a continuous linear mapping between normed linear spaces to be a topological isomorphism.

10.2 **Definition**. A mapping T from a metric space (X, d) into a metric space (Y, d') is said to be an *open mapping* if for every open set G in (X,d), we have that T(G) is an open set in (Y, d').

10.3 **Remark**. Not all continuous mappings are open mappings and not all open mappings are continuous. The identity mapping between $\mathbb{R}$ with its usual metric and $\mathbb{R}$ with the discrete metric is a simple example illustrating both these facts. However, a continuous one-to-one onto mapping is a homeomorphism if and only if it is an open mapping, (see AMS §10). It is from this result that we develop our interest in open mappings. □

The Open Mapping Theorem gives us conditions under which a linear mapping between normed linear spaces is an open mapping.

For convenience we introduce the following notation.

10.4 **Notation**. Denoting by B the open unit ball in the normed linear space $(X, \|\cdot\|)$ and by B' the open unit ball in the normed linear space $(Y, \|\cdot\|')$, we use the following notation.

$rB \equiv B(0; r)$ and $x+rB \equiv B(x; r)$ in $(X, \|\cdot\|)$,

$rB' \equiv B(0; r)$ and $x+rB' \equiv B(0; r)$ in $(Y, \|\cdot\|')$.

We approach the Open Mapping Theorem through the following lemmas.

10.5 **Lemma**. *For a linear mapping* T *of a normed linear space* $(X, \|\cdot\|)$ *onto a normed linear space* $(Y, \|\cdot\|')$ *of second category,* $\overline{T(B)}$ *is a neighbourhood of* 0 *in* $(Y, \|\cdot\|')$.

Proof. Since T is onto, $Y = \bigcup\{nT(B) : n \in \mathbb{N}\}$. However, $(Y, \|\cdot\|')$ is second category so there exists some $n_0 \in \mathbb{N}$ such that $n_0\overline{T(B)}$ has nonempty interior. Any open set in $n_0\overline{T(B)}$ contains points of $n_0T(B)$ so there exists a $y_0 \in n_0T(B)$ such that

$$y_0 \in \text{int } n_0\overline{T(B)} .$$

But since translation is a homeomorphism,

$$0 \in \text{int } (n_0\overline{T(B)} - y_0).$$

Since $y_0 \in n_0T(B)$ there exists an $x_0 \in n_0 B$ such that $y_0 = Tx_0$ and since T is linear,

$$n_0T(B) - y_0 = T(n_0B - x_0).$$

Now for any $x \in n_0 B$, we have $\| x-x_0 \| < 2n_0$ so

$$n_0T(B) - y_0 \subseteq 2n_0T(B).$$

Again since translation is a homeomorphism

$$n_0\overline{T(B)} - y_0 = \overline{n_0T(B) - y_0}.$$

So $0 \in \text{int } (n_0\overline{T(B)} - y_0) \subseteq \text{int } 2n_0\overline{T(B)}$.

Since multiplication by a nonzero scalar is a homeomorphism, we conclude that

$$0 \in \text{int } \overline{T(B)} ;$$

that is, $\overline{T(B)}$ is a neighbourhood of 0 in $(Y, \|\cdot\|')$. □

Notice that in this lemma there are very few restrictions on the linear mapping T; it is onto but not necessarily continuous. Lemma 10.5 reveals the surprising consequences of the range space being second category. The next lemma shows that if we assume continuity of the linear mapping from a complete domain space onto a second category range space, we are able to refine the results of Lemma 10.5 further. We frame the statement of the lemma to exhibit the implications of completeness for continuous linear mappings.

10.6 **Lemma**. *For a continuous linear mapping* T *of a Banach space* $(X, \|\cdot\|)$ *into a normed linear space* $(Y, \|\cdot\|')$, if $\overline{T(B)}$ *is a neighbourhood of* 0 *in* $(Y, \|\cdot\|')$, *then* T(B) *is also a neighbourhood of* 0 *in* $(Y, \|\cdot\|')$.

Proof. If $0 \in \text{int}\, \overline{T(B)}$ then there exists a $\delta > 0$ such that $\delta B' \subseteq \overline{T(B)}$. For $y \in \delta B'$ we have $y \in \overline{T(B)}$ and so there exists an $x_1 \in B$ such that $y_1 = Tx_1$ and $\| y - y_1 \| < \frac{\delta}{2}$. Now $\frac{\delta}{2} B' \subseteq \overline{T(\frac{1}{2}B)}$ so there exists an $x_2 \in \frac{1}{2} B$ such that $y_2 = Tx_2$ and $\| (y-y_1) - y_2 \| < \frac{\delta}{2}$.
By induction we obtain a sequence $\{x_n\}$ in X such that $x_n \in \frac{1}{2^{n-1}} B$, $y_n = Tx_n$ and

$$\| y - (y_1 + y_2 + \ldots + y_n) \| < \frac{\delta}{2^n}.$$

We write $s_n \equiv \sum_{k=1}^{n} x_k$. Since $\| x_n \| < \frac{1}{2^{n-1}}$, we have

$$\| s_m - s_n \| \leq \sum_{k=n+1}^{m} \| x_k \| < \frac{1}{2^{n-1}} \qquad \text{for all } m > n.$$

So $\{s_n\}$ is a Cauchy sequence in $(X, \|\cdot\|)$. But also $\| s_n \| \leq \sum_{k=1}^{n} \| x_k \| < 2$.
Now $(X, \|\cdot\|)$ is complete so there exists an $x \in X$ such that $\{s_n\}$ is convergent to x and $\| x \| \leq 2$. But

$$\| Tx - y \|' \leq \| Tx - Ts_n \|' + \| Ts_n - \sum_{k=1}^{n} y_k \|' + \| \sum_{k=1}^{n} y_k - y \|'$$

and since T is continuous and linear we deduce that $y = Tx$, so $y \in 3T(B)$.
Then $\delta B' \subseteq 3T(B)$ and so $\frac{\delta}{3} B' \subseteq T(B)$; that is, $T(B)$ is a neighbourhood of 0 in $(Y, \|\cdot\|')$. □

We are now in a position to state and prove the Open Mapping Theorem.

10.7 **The Open Mapping Theorem.**
A continuous linear mapping T *of a Banach space* $(X, \|\cdot\|)$ *onto a Banach space* $(Y, \|\cdot\|')$ *is an open mapping.*

Proof. Consider an open set G in $(X, \|\cdot\|)$ and $y \in T(G)$. We show that $y \in \text{int}\, T(G)$.
Let $x \in G$ be such that $y = Tx$. Since G is open there exists an $r > 0$ such that $x + rB \subseteq G$.
Now Baire's Theorem 9.11 gives us that $(Y, \|\cdot\|')$ is second category so from Lemmas 10.5 and 10.6 we deduce that there exists a $\delta > 0$ such that $\delta B' \subseteq r\,T(B)$.
Now $y + \delta B' \subseteq y + rT(B) = T(x + rB) \subseteq T(G)$ so $y \in \text{int}\, T(G)$.
This implies that $T(G)$ is open in $(Y, \|\cdot\|')$. □

10.8 **Remark.** We note that Theorem 10.7 could be generalised to have range space $(Y, \|\cdot\|')$ a normed linear space of second category. However, we have given the most common form of the statement of the theorem and it is generally in this context where it has application. □

Most applications of the Open Mapping Theorem are derived from the following corollary which gives a simple criterion for a continuous linear mapping to be a homeomorphism.

10.9 **Corollary**. *A continuous linear one-to-one mapping* T *of a Banach space* $(X, \|\cdot\|)$ *onto a Banach space* $(Y, \|\cdot\|')$ *is a topological isomorphism.*

Proof. As T is one-to-one and onto its inverse T^{-1} exists and is a linear mapping from $(Y, \|\cdot\|')$ onto $(X, \|\cdot\|)$. From the Open Mapping Theorem 10.7, T is an open mapping so it is also a homeomorphism. □

Corollary 10.9 has a particularly useful application in determining whether norms are equivalent.

10.10 **Corollary**. *For a linear space* X, *if* $\|\cdot\|$ *and* $\|\cdot\|'$ *are norms such that both* $(X, \|\cdot\|)$ *and* $(X, \|\cdot\|')$ *are complete and there exists a* $K > 0$ *such that*

$$\|x\| \leq K\|x\|' \quad \textit{for all } x \in X$$

then $\|\cdot\|$ *and* $\|\cdot\|'$ *are equivalent norms for* X.

Proof. Consider the identity mapping *id*: $(X, \|\cdot\|') \to (X, \|\cdot\|)$. The inequality

$$\|x\| \leq K\|x\|' \quad \text{for all } x \in X$$

implies that *id* is continuous. But *id* is linear one-to-one and onto so by Corollary 10.9, it is a topological isomorphism which implies that $\|\cdot\|$ and $\|\cdot\|'$ are equivalent norms for X. □

The second mapping theorem which is in many cases easier to apply than the Open Mapping Theorem is the Closed Graph Theorem and its proof can be derived directly from the Open Mapping Theorem.

10.11 **Definitions**. Given metric spaces (X, d) and (Y, d') the *product metric* d_π for $X \times Y$ is defined by

$$d_\pi((x,y), (x',y')) = \max\{d(x, x'), d'(y, y')\}$$

and we call $(X \times Y, d_\pi)$ the *product space* of (X, d) and (Y, d'). It is easy to see that a sequence $\{(x_n,y_n)\}$ is convergent to (x,y) in $(X \times Y, d_\pi)$ if and only if $\{x_n\}$ is convergent to x in (X, d) and $\{y_n\}$ is convergent to y in (Y, d').
Given normed linear spaces $(X, \|\cdot\|)$ and $(Y, \|\cdot\|')$ over the same scalar field, $X \times Y$ is a linear space, the *product norm* $\|\cdot\|_\pi$ for $X \times Y$ is defined by

$$\|(x,y)\|_\pi = \max\{\|x\|, \|y\|'\}$$

and we call $(X \times Y, \|\cdot\|_\pi)$ the *product space* of $(X, \|\cdot\|)$ and $(Y, \|\cdot\|')$. The product norm generates the product metric.

Given a mapping T of a set X into a set Y the *graph* of T is the subset G_T of $X \times Y$ defined by

$$G_T \equiv \{(x,y) : y = Tx,\ x \in X\}.$$

When X and Y are linear spaces over the same scalar field then T is linear if and only if G_T is a linear subspace of $X \times Y$.

We say that a mapping T of a metric space (X, d) into a metric space (Y, d') has a *closed graph* if G_T is closed in $(X \times Y, d_\pi)$.

For mappings between metric spaces there is a close relation between continuity and having closed graph. We will explore this relation but first we give a more convenient way of expressing the fact that a mapping has a closed graph.

10.12 **Lemma**. *A mapping* T *of a metric space* (X, d) *into a metric space* (Y, d') *has a closed graph if and only if for every sequence* $\{x_n\}$ *in* X *where* $\{x_n\}$ *is convergent to* x *in* (X, d) *and* $\{Tx_n\}$ *is convergent to* y *in* (Y, d'), *we have* y = Tx.

Proof. Suppose that G_T is closed in $(X \times Y, d_\pi)$ and that $\{x_n\}$ is convergent to x in (X, d) and $\{Tx_n\}$ is convergent to y in (Y, d'). Then $\{(x_n, Tx_n)\}$ is convergent to (x,y) in $(X \times Y, d_\pi)$. But since G_T is closed, $(x,y) \in G_T$ and therefore y = Tx.

Conversely, if (x,y) is a cluster point of G_T in $(X \times Y, d_\pi)$ then there exists a sequence $\{(x_n, Tx_n)\}$ in G_T which is convergent to (x,y) in $(X \times Y, d_\pi)$. Then $\{x_n\}$ is convergent to x in (X, d) and $\{Tx_n\}$ is convergent to y in (Y, d'). □

From Lemma 10.12 it is clear that any continuous mapping between metric spaces always has a closed graph. But further, it is not difficult to prove that a mapping T from a metric space (X, d) into a compact metric space (Y, d') with closed graph is continuous, (see AMS, §8). It is reasonable to ask whether there is a simple criterion for a linear mapping between normed linear spaces with a closed graph, to be continuous.

We should note that not every linear mapping between normed linear spaces with closed graph is necessarily continuous.

10.13 **Example**. The differential operator D from $(\mathcal{C}^1[0,1], \|\cdot\|_\infty)$ into $(\mathcal{C}[0,1], \|\cdot\|_\infty)$ defined by

$$D(f) = f'$$

has a closed graph; (this is the classical uniform convergence theorem for differentiation, see AMS §8). However, D is not continuous, We note that $(\mathcal{C}[0,1], \|\cdot\|_\infty)$ is complete but $(\mathcal{C}^1[0,1], \|\cdot\|_\infty)$ is not complete, (see AMS, §4). □

The Closed Graph Theorem provides a criterion for the continuity of linear mappings between normed linear spaces. Its proof is an elegant application of the Open Mapping Theorem.

10.14 **The Closed Graph Theorem.**
A linear mapping T *of a Banach space* $(X, \|\cdot\|)$ *into a Banach space* $(Y, \|\cdot\|')$ *with a closed graph, is continuous.*

Proof. Consider X renormed with norm $\|\cdot\|_1$ defined by

$$\| x \|_1 = \| x \| + \| Tx \|'.$$

Then $$\| Tx \|' \leq \| x \| + \| Tx \|' = \| x \|_1 \quad \text{for all } x \in X$$
so T is a continuous linear mapping from $(X, \|\cdot\|_1)$ into $(Y, \|\cdot\|')$.
We show that $\|\cdot\|_1$ and $\|\cdot\|$ are equivalent norms for X.
Now $$\| x \| \leq \| x \| + \| Tx \|' = \| x \|_1 \quad \text{for all } x \in X$$
so if we show that $(X, \|\cdot\|_1)$ is complete then from Corollary 10.10 to the Open Mapping Theorem we will have that $\|\cdot\|$ and $\|\cdot\|'$ are equivalent.
We show that $(X, \|\cdot\|_1)$ is complete.
Consider a Cauchy sequence $\{x_n\}$ in $(X, \|\cdot\|_1)$. Then since

$$\| x_m - x_n \|_1 = \| x_m - x_n \| + \| Tx_m - Tx_n \|'$$

we have that $\{x_n\}$ is a Cauchy sequence in $(X, \|\cdot\|)$ and $\{Tx_n\}$ is a Cauchy sequence in $(Y, \|\cdot\|')$. As both $(X, \|\cdot\|)$ and $(Y, \|\cdot\|')$ are complete there exist an $x \in X$ such that $\{x_n\}$ is convergent x in $(X, \|\cdot\|)$ and a $y \in Y$ such that $\{Tx_n\}$ is convergent to y in $(Y, \|\cdot\|')$. But T has closed graph so $y = Tx$. Then

$$\| x_n - x \|_1 = \| x_n - x \| + \| Tx_n - Tx \|'$$

so $\{x_n\}$ is convergent to x in $(X, \|\cdot\|_1)$, and we conclude that $(X, \|\cdot\|_1)$ is complete. □

10.15 **EXERCISES**

1. Consider an open mapping T of a metric space (X, d) into a metric space (Y, d').
   (i) Prove that T maps compact sets in (X, d) to compact sets in (Y, d').
   (ii) Show that T does not necessarily map closed sets in (X, d) to closed sets in (Y, d').
2. (i) Consider the Banach space $(\mathcal{C}[0,1], \|\cdot\|_\infty)$ and $\mathcal{C}[0,1]$ with norm $\|\cdot\|_1$ and norm $\|\cdot\|_2$. Deduce from the Open Mapping Theorem that $(\mathcal{C}[0,1], \|\cdot\|_1)$ and $(\mathcal{C}[0,1], \|\cdot\|_2)$ are not complete.
   (ii) A mapping T on $\mathcal{C}[0,1]$ is defined by

$$T f(t) = \int_0^t f(s)\, ds.$$

(a) Prove that T is a continuous one-to-one mapping of $(\mathcal{C}[0,1], \|\cdot\|_\infty)$ onto the linear subspace $\mathcal{C}_0^1[0,1] \equiv \{f \in \mathcal{C}^1[0,1] : f(0) = 0\}$.

(b) Show that T^{-1} is not continuous and explain why this does not contradict the Open Mapping Theorem.

3. M and N are closed linear subspaces of a Banach space $(X, \|\cdot\|)$ such that $X = M \oplus N$. Prove that the norm $\|\cdot\|'$ on X defined for $z \in X$ where $z = x + y$, $x \in M$ and $y \in N$ by

$$\|z\|' = \|x\| + \|y\|$$

is an equivalent norm for X.

4. Consider the finite dimensional linear space X_m over $\mathbb{R}$ with basis $\{e_1, e_2, \ldots, e_m\}$ and any norm on X_m. Assuming that $(X_m, \|\cdot\|)$ is second category and using the Open Mapping Theorem with the mapping

$$x \mapsto (\lambda_1, \lambda_2, \ldots, \lambda_m)$$

of X_m into $\mathbb{R}^m$ where $x \equiv \lambda_1 e_1 + \lambda_2 e_2 + \ldots + \lambda_m e_m$, prove that $(X_m, \|\cdot\|)$ is topologically isomorphic to $(\mathbb{R}^m, \|\cdot\|_2)$.

5. (i) Consider a linear mapping T of a normed linear space $(X, \|\cdot\|)$ into a normed linear space $(Y, \|\cdot\|')$.

   (a) Prove that if T has closed graph then T has closed kernel.

   (b) But show that a linear mapping with a closed kernel does not necessarily have a closed graph; (see AMS §7).

   (ii) (a) Prove that a linear functional on a normed linear space is continuous if it has a closed graph.

   (b) Show that a linear mapping of a normed linear space into a finite dimensional normed linear space with closed graph is not necessarily continuous.

6. For the mapping T of $(\mathcal{C}^1[0,1], \|\cdot\|_\infty)$ into $(\mathcal{C}[0,1], \|\cdot\|_\infty)$ defined by

$$T(f) = f' + f$$

prove that

(i) T has closed graph,

(ii) T is not continuous,

(iii) $(\mathcal{C}^1[0,1], \|\cdot\|_\infty)$ is not complete.

7. A linear space X is a Banach space with respect to both norms $\|\cdot\|$ and $\|\cdot\|'$ and has the property that if a sequence $\{x_n\}$ in X is convergent with respect to both norms $\|\cdot\|$ and $\|\cdot\|'$ then the limit point is unique. Prove that norms $\|\cdot\|$ and $\|\cdot\|'$ are equivalent.

8. (i) For a linear mapping T of a Banach space $(X, \|\cdot\|)$ into a Banach space $(X, \|\cdot\|')$, prove that if $f \circ T \in X^*$ for all $f \in Y^*$ then T is continuous.

   (ii) For linear operators T and S on a Hilbert space H we have

   $$(Tx, y) = (x, Sy) \text{ for all } x,y \in H.$$

   Prove that both T and S are continuous.

   (iii) For a linear operator T on a Hilbert space H, prove that the functional ϕ_T on H defined by

   $$\phi_T(x) = (Tx, x)$$

   is continuous if and only if T is continuous.

9. (i) Show that Lemma 10.6 can be generalised to the following statement.
   For a linear mapping T *from a Banach space* $(X, \|\cdot\|)$ *into a normed linear space* $(Y, \|\cdot\|')$ *with closed graph, if* $\overline{T(B)}$ *is a neighbourhood of* 0 *in* $(Y, \|\cdot\|')$ *then* $T(B)$ *is also a neighbourhood of* 0 *in* $(Y, \|\cdot\|')$.

   (ii) Hence, show that the Open Mapping Theorem can be generalised to the following statement.
   A linear mapping T *from a Banach space* $(X, \|\cdot\|)$ *onto a Banach space* $(Y, \|\cdot\|')$ *with a closed graph is an open mapping.*

   (iii) Further, prove the following statement.
   A linear one-to-one mapping T *from a Banach space* $(X, \|\cdot\|)$ *onto a Banach space* $(Y, \|\cdot\|')$ *with a closed graph is a homeomorphism.*

10. (i) Consider a normed linear space $(X, \|\cdot\|)$, a proper closed linear subspace N and the quotient space $(X/N, \|\cdot\|)$. Prove that the quotient mapping $\pi : X \to X/N$ defined by $\pi(x) = x + N$.
   is a continuous, linear and open mapping.

   (ii) Given a continuous mapping T of a Banach space $(X, \|\cdot\|)$ onto a Banach space $(Y, \|\cdot\|')$, prove that the mapping $\tilde{T}$ of the quotient space $(X/\ker T, \|\cdot\|)$ onto $(Y, \|\cdot\|')$ defined by

   $$\tilde{T}(x+\ker T) = Tx$$

   is a topological isomorphism.

(iii) Show that the Open Mapping Theorem can be generalised to the following statement

A continuous linear mapping T *of a Banach space* (X, ||·||) *into a normed linear space* (Y, ||·||') *where* T(X) *is second category in* (Y, ||·||'), *is an open onto mapping and* (Y, ||·||') *is complete.*

11. We saw how the Closed Graph Theorem was deduced from the Open Mapping Theorem. However, show that the Open Mapping Theorem can be deduced from the Closed Graph Theorem.

(Hint: Consider a continuous linear mapping T of a Banach space (X, ||·||) onto a Banach space (Y, ||·||'). As in Exercise 10, consider the mapping

$$\tilde{T}: \frac{X}{\ker T} \to Y \text{ defined by } \tilde{T}(k+\ker T) = Tx.$$

Show that $\tilde{T}$ is one-to-one, onto and has a closed graph and therefore $\tilde{T}$ has a closed graph. Apply the Closed Graph Theorem to give that $\tilde{T}$ is continuous and deduce that T is an open mapping.)

12. (i) Consider a complex Banach algebra A. A linear functional ϕ on A is said to be *multiplicative* if also

$$\phi(xy) = \phi(x)\,\phi(y) \text{ for all } x,y \in A.$$

Prove that every multiplicative linear functional ϕ on A is continuous and $\|\phi\| \le 1$.

(Hint: Suppose on the contrary, that there exists an $x_0 \in A$, $\|x_0\| < 1$ and $\phi(x_0) = 1$. Consider $y_0 = \sum_{k=1}^{\infty} x_0^n$ and show that $x_0 + x_0y_0 = y_0$ a contradiction to the contrary hypothesis.)

(ii) A commutative Banach algebra is said to be *semisimple* if the set of multiplicative linear functions is total on A.

Prove that every algebra homomorphism from a complex Banach algebra A into a semisimple complex commutative Banach algebra B is continuous.

(iii) Consider a semisimple complex commutative algebra A.

Prove that all norms under which A is complete are equivalent.

13. Consider a separable Banach space $(X, \|\cdot\|)$ with a Schauder basis $\{e_n\}$. Consider X with norm $\|\cdot\|': X \to \mathbb{R}$ where for $x \equiv \sum_{k=1}^{\infty} \lambda_k e_k$,

$$\| x \|' = \sup\{\| \sum_{k=1}^{n} \lambda_k e_k \| : n \in \mathbb{N}\}$$

(i) Prove that $\|\cdot\|$ and $\|\cdot\|'$ are equivalent norms for X.

(ii) Hence, deduce that for any separable Banach space with a Schauder basis, the coordinate functionals are continuous.

(Hint: See Exercise 1.26.17 and Theorem 1.25.16)

§11. THE UNIFORM BOUNDEDNESS THEOREM

Our motivation for the study of uniform convergence is to determine the properties of the limit mapping of a pointwise convergent sequence of scalar mappings on a metric space. The three classical theorems on uniform convergence of sequences of real functions on an interval of the real line show that uniform convergence conditions are sufficient to guarantee the "good behaviour" of the limit function. In general, there is particular interest in the continuity of the limit mapping of a pointwise convergent sequence of continuous mappings.

In Dini's Theorem (AMS §9) the compactness of the domain space ensures that a pointwise convergent monotone sequence of continuous mappings with a continuous limit mapping is uniformly convergent.

As an extension of this type of investigation it is reasonable to ask, for a pointwise convergent sequence of continuous linear mappings between normed linear spaces, whether the limit mapping is continuous and whether convergence must necessarily be stronger than pointwise convergence. We will derive a useful result in this line as a consequence of a more general question concerning boundedness of a set of continuous linear mappings.

11.1 **Definitions**. Consider a set $\mathfrak{T}$ of continuous linear mappings from a normed linear space $(X, \|\cdot\|)$ into a normed linear space $(Y, \|\cdot\|')$.
We say that the set $\mathfrak{T}$ is *pointwise bounded* if for each $x \in X$ the set $\{Tx: T \in \mathfrak{T}\}$ is a bounded set in $(Y, \|\cdot\|')$.
We say that the set $\mathfrak{T}$ is *uniformly bounded* if $\mathfrak{T}$ is a bounded set in $(\mathfrak{B}(X,Y), \|\cdot\|)$ the normed linear space of continuous linear mappings from X into Y.

11.2 **Remark**. If the set $\mathfrak{T}$ is uniformly bounded then there exists an $M > 0$ such that

$$\|T\| \leq M \quad \text{for all } T \in \mathfrak{T}.$$

Therefore, for any $x \in X$,

$$\|Tx\|' \leq \|T\|\|x\| \leq M\|x\| \quad \text{for all } T \in \mathfrak{T},$$

so $\mathfrak{T}$ is pointwise bounded. □

However, the converse is not true in general.

11.3 **Example**. Consider the set of continuous linear functionals $\{f_n: n \in \mathbb{N}\}$ on $(E_0, \|\cdot\|_\infty)$ defined, for $x \equiv \{\lambda_1, \lambda_2, \ldots, \lambda_n, \ldots,\}$, by

$$f_n(x) = n\lambda_n.$$

Now for any given $x_0 \in X$ there exists an $n_0 \in \mathbb{N}$ such that $\lambda_n = 0$ for all $n > n_0$, so

$$|f_n(x_0)| \leq n_0 \|x_0\|_\infty \quad \text{for all } n \in \mathbb{N};$$

that is, $\{f_n : n \in \mathbb{N}\}$ is pointwise bounded. Nevertheless,

$$\| f_n \| = n \quad \text{for every } n \in \mathbb{N},$$

so $\{f_n : n \in \mathbb{N}\}$ is not uniformly bounded. But notice that the normed linear space $(E_0, \|\cdot\|_\infty)$ is not complete. □

On the other hand, for a set of continuous linear mappings on a Banach space, pointwise boundedness does imply uniform boundedness. This is the content of the Uniform Boundedness Theorem. This theorem, like the Open Mapping Theorem uses Baire category arguments and in particular Baire's Theorem 9.11 that a complete normed linear space is second category.

11.4 **The Uniform Boundedness Theorem.**
If a set $\mathfrak{T}$ *of continuous linear mappings from a Banach space* $(X, \|\cdot\|)$ *into a normed linear space* $(Y, \|\cdot\|')$ *is pointwise bounded then it is uniformly bounded.*

Proof. For each $n \in \mathbb{N}$, write

$$F_n \equiv \{x \in X : \| Tx \|' \leq n \text{ for all } T \in \mathfrak{T}\}.$$

Since each $T \in \mathfrak{T}$ is continuous, $T^{-1}(B'[0; n])$ is closed and so F_n is closed.
But also since $\mathfrak{T}$ is pointwise bounded, for each $x \in X$ there exists an $n \in \mathbb{N}$ such that $x \in F_n$; that is, $X = \cup\{F_n : n \in \mathbb{N}\}$. Now $(X, \|\cdot\|)$ is complete, so by Baire's Theorem 9.11 there exists some $n_0 \in \mathbb{N}$ such that F_{n_0} has nonempty interior. Therefore there exists some $x_0 \in X$ and $r > 0$ such that $x_0 + rB \subseteq F_{n_0}$, so

$$\| T(x_0+rB) \|' \leq n_0 \qquad \text{for all } T \in \mathfrak{T}.$$

Now since T is linear, $T(rB) = T(x_0+rB) - Tx_0$, so

$$\| Tx \|' \leq 2n_0 \qquad \text{for all } x \in rB$$

which implies that

$$\| Tx \|' \leq \frac{2n_0}{r} \qquad \text{for all } x \in B \text{ and all } T \in \mathfrak{T}.$$

Therefore, $\| T \| \leq \dfrac{2n_0}{r}$ for all $T \in \mathfrak{T}$; that is, $\mathfrak{T}$ is uniformly bounded. □

11.5 **Remark**. We note that, as with the Open Mapping Theorem, the Uniform Boundedness Theorem could be generalised to have domain space $(X, \|\cdot\|)$ a normed linear space of second category. However, the form given in Theorem 11.4 is generally sufficient for our purposes. □

One of the most important consequences of this theorem concerns the problem we mentioned in introducing the Uniform Boundedness Theorem, the continuity of the limit of a pointwise convergent sequence of continuous linear mappings.

11.6 **Definition**. We say that a sequence $\{T_n\}$ of continuous linear mappings of a normed linear space $(X, \|\cdot\|)$ into a normed linear space $(Y, \|\cdot\|')$ is *pointwise convergent* to a mapping T if for each $x \in X$, $\{T_n(x)\}$ is convergent to Tx in $(Y, \|\cdot\|')$.

Of course the pointwise limit of a sequence of continuous linear mappings is itself linear. But it need not be continuous.

11.7 **Example**. Consider the normed linear space $(\ell_1, \|\cdot\|_\infty)$ and the sequence $\{f_n\}$ of linear functionals on ℓ_1 defined, for $x \equiv \{\lambda_1, \lambda_2, \ldots, \lambda_n, \ldots\}$, by

$$f_n(x) = \sum_{k=1}^{n} \lambda_k .$$

Now $|f_n(x)| \leq n \|x\|_\infty$ for all $x \in X$, so for each $n \in \mathbb{N}$, f_n is continuous.
But $\{f_n\}$ is pointwise convergent to the linear functional f on ℓ_1 where

$$f(x) = \sum_{k=1}^{\infty} \lambda_k$$

since $x \in \ell_1$ and $|f_n(x) - f(x)| \leq \sum_{k=n+1}^{\infty} |\lambda_k|$ for each $n \in \mathbb{N}$.
However, for $x_n \equiv \{1, 1, \ldots, \underset{\text{nth place}}{1}, 0, \ldots\}$, $\|x_n\|_\infty = 1$ but $f(x_n) = n$ so f is not continuous on $(\ell_1, \|\cdot\|_\infty)$.
But notice again that the normed linear space $(\ell_1, \|\cdot\|_\infty)$ is not complete. □

11.8 **Remark**. We should notice that the pointwise convergence of a sequence of continuous linear mappings $\{T_n\}$ to a continuous linear mapping T does not necessarily imply that the set $\{T_n: n \in \mathbb{N}\}$ is uniformly bounded. In Example 11.3 we have a sequence $\{f_n\}$ which is not uniformly bounded. But for any given $x_0 \in E_0$,
$x_0 \equiv \{\lambda_1^0, \lambda_2^0, \ldots, \lambda_n^0 \ldots\}$ there exists an $n_0 \in \mathbb{N}$ such that $\lambda_n^0 = 0$ for all $n > n_0$.
Then $f_n(x_0) = 0$ for all $n > n_0$, so $\{f_n\}$ is pointwise convergent to the zero functional. □

However, for a sequence of continuous linear mappings on a Banach space we have the following significant corollary to the Uniform Boundedness Theorem.

11.9 **The Banach–Steinhaus Theorem**.
If $\{T_n\}$ *is a sequence of continuous linear mappings of a Banach space* $(X, \|\cdot\|)$ *into a normed linear space* $(Y, \|\cdot\|')$, *pointwise convergent to a mapping* T, *then* T *is a continuous linear mapping of* $(X, \|\cdot\|)$ *into* $(Y, \|\cdot\|')$.

Proof. Since $\{T_n\}$ is pointwise convergent then the set $\{T_n : n \in \mathbb{N}\}$ is pointwise bounded. From the Uniform Boundedness Theorem 11.4, $\{T_n : n \in \mathbb{N}\}$ is uniformly bounded; that is, there exists an $M > 0$ such that

$$\| T_n \| \leq M \qquad \text{for all } n \in \mathbb{N}.$$

Therefore, $\quad \| T_n(x) \|' \leq M \| x \| \quad$ for all $n \in \mathbb{N}$ and $x \in X$.

But for each $x \in X$, $\quad \| Tx \|' \leq \| Tx - T_n(x) \|' + \| T_n(x) \|'$,

so as $\{T_n\}$ is pointwise convergent to T we conclude that

$$\| Tx \|' \leq M \| x \| \qquad \text{for all } x \in X;$$

that is, T is continuous. □

However, the Banach–Steinhaus Theorem does not imply that pointwise convergence of a sequence of continuous linear mappings on a Banach space is any stronger than pointwise convergence; it does not provide a generalisation of Dini's Theorem.

11.10 **Example**. Consider the Banach space $(c_0, \|\cdot\|_\infty)$ and the sequence of continuous linear functionals $\{f_n\}$ on c_0 where for $x \equiv \{\lambda_1, \lambda_2, \ldots, \lambda_n, \ldots\}$,

$$f_n(x) = \lambda_n .$$

Now $\{f_n\}$ is pointwise convergent to the zero functional.

From the Uniform Boundedness Theorem 11.4, the set $\{f_n : n \in \mathbb{N}\}$ is uniformly bounded. But for $e_n \equiv \{0, \ldots, 0, 1, 0, \ldots\}$,

nth place

$$\| f_n \| \geq f_n(e_n) = 1$$

so $\{f_n\}$ is not convergent to the zero functional under the dual norm on $(c_0, \|\cdot\|_\infty)$. □

Another important application of the Uniform Boundedness Theorem is in characterising the boundedness of a set in a normed linear space.

11.11 **Definitions**. A nonempty set A in a normed linear space $(X, \|\cdot\|)$ is said to be *weakly bounded* if f(A) is a bounded set of scalars for each $f \in X^*$.

A nonempty set B in the dual $(X^*, \|\cdot\|)$ is said to be *weak * bounded* if $\hat{x}(B)$ is a bounded set of scalars for each $x \in X$.

There are occasions when we deduce the boundedness of a set from its weak boundedness or weak* boundedness, which is sometimes easier to test.

11.12. **Theorem**.

(i) *A nonempty set* A *in a normed linear space* $(X, \|\cdot\|)$ *is bounded if and only if it is weakly bounded.*

(ii) *If* $(X, \|\cdot\|)$ *is complete, a nonempty set* B *in the dual* $(X^*, \|\cdot\|)$ *is bounded if and only if it is weak * bounded.*

Proof. For any $x \in X$ and $f \in X^*$, $| f(x) | \leq \| f \| \| x \|$.
So if A is bounded in $(X, \|\cdot\|)$ then it is weakly bounded and if B is bounded in $(X^*, \|\cdot\|)$ then it is weak * bounded.

Conversely,
(i) $f(A) \equiv \{f(x) : x \in A\} = \{\hat{x}(f) : x \in A\}$ is bounded for each $f \in X^*$; that is, the set $\{\hat{x} : x \in A\}$ is pointwise bounded on $(X, \|\cdot\|)$.
But $(X^*, \|\cdot\|)$ is always complete so from the Uniform Boundedness Theorem, the set $\{\hat{x} : x \in A\}$ is uniformly bounded on $(X^*, \|\cdot\|)$, which implies that A is bounded in $(X, \|\cdot\|)$.
(ii) $\hat{x}(B) \equiv \{f(x) : x \in B\}$ is bounded for each $x \in X$; that is the set $\{f: f \in B\}$ is pointwise bounded on $(X, \|\cdot\|)$.
But here we assume that $(X, \|\cdot\|)$ is complete so from the Uniform Boundedness Theorem the set $\{f : f \in B\}$ is uniformly bounded on $(X, \|\cdot\|)$, which implies that B is bounded in $(X^*, \|\cdot\|)$. □

11.13 **Remark**. In Example 11.3 we have a set in the dual of an incomplete normed linear space which is weak * bounded but not bounded. So the completeness condition in Theorem 11.12(ii) is significant. □

Another application of the Uniform Boundedness Theorem concerns bilinear mappings.

11.14 **Definitions**. Given linear spaces X, Y and Z over the same scalar field, a mapping β of $X \times Y$ into Z is said to be a *bilinear mapping* if
for given $x \in X$, the associated mapping β_x: $Y \to Z$ defined by $\beta_x(y) = \beta(x,y)$ and
for given $y \in Y$, the associated mapping β_y: $Y \to Z$ defined by $\beta_y(y) = \beta(x,y)$ are
both linear.
If $(X, \|\cdot\|)$, $(Y, \|\cdot\|')$ and $(Z, \|\cdot\|'')$ are normed linear spaces and for given $x \in X$, β_x is continuous and for given $y \in Y$, β_y is continuous then β is said to be *separately continuous*.
If β is continuous on $X \times Y$ with the product norm then β is said to be *jointly continuous*.

If a bilinear mapping is jointly continuous then clearly it is separately continuous. The Uniform Boundedness Theorem enables us to establish a converse result.

11.15 **Theorem**. *Consider normed linear spaces* $(X, \|\cdot\|)$, $(Y, \|\cdot\|')$ *and* $(Z, \|\cdot\|'')$ *and a bilinear mapping* β *of* $X \times Y$ *into* Z *which is separately continuous. If* $(X, \|\cdot\|)$ *is complete then* β *is jointly continuous.*

Proof. For each $y \in Y$, $\| y \| = 1$ consider the associated continuous linear mapping β_y of $(X, \|\cdot\|)$ into $(Z, \|\cdot\|'')$. Then there exists a $K_y > 0$ such that

$$\| \beta_y(x) \|'' \leq K_y \| x \| \quad \text{for all } x \in X.$$

But then the set $\{\beta_y : y \in Y, \| y \| = 1\}$ is pointwise bounded on $(X, \|\cdot\|)$. Since $(X, \|\cdot\|)$ is complete, by the Uniform Boundedness Theorem the set is uniformly bounded on $(X, \|\cdot\|)$; that is, there exists a $K > 0$ such that $\| \beta_y(x) \|'' \leq K\| x \|$ for all $x \in X$ and $y \in Y$, $\| y \| = 1$. But since β is bilinear this implies that

$$\| \beta(x,y) \|'' \leq K\| x \| \| y \| \quad \text{for all } x \in X \text{ and } y \in Y$$

and this inequality gives us that β is jointly continuous. □

We now present an interesting application which exhibits the power of the Uniform Boundedness Theorem. Although we have seen, (from Example 3.16) that every $f \in \mathfrak{C}[-\pi,\pi]$ has mean square representation by its Fourier series, it was shown by du Bois Reymond in 1876 that the Fourier series for such a function may actually fail to converge pointwise to the function. He actually constructed a continuous function whose Fourier series is divergent at 0. The proof given here is an existence proof and it does also give us more information about the pointwise convergence of a Fourier series, (see Exercise 11.17.5).

11.16 **The du Bois Reymond Theorem.**
There exists an $f \in \mathfrak{C}[-\pi,\pi]$ *whose Fourier series is divergent at* 0.

Proof. Given $f \in \mathfrak{C}[-\pi,\pi]$ we write

$$S_n(t; f) \equiv \frac{\alpha_0}{2} + \sum_{k=1}^{n} (\alpha_k \cos kt + \beta_k \sin kt)$$

the nth partial sum of the Fourier series for f, where α_0, α_k and β_k are the Fourier coefficients for f defined with respect to the orthonormal set

$$\left\{\frac{1}{\sqrt{2\pi}}, \frac{1}{\sqrt{\pi}} \cos nt, \frac{1}{\sqrt{\pi}} \sin nt : n \in \mathbb{N}\right\}.$$

We show that $\{S_n(0; f)\}$ is divergent for some $f \in \mathfrak{C}[-\pi,\pi]$.
Suppose that $\{S_n(0; f)\}$ converges for every $f \in \mathfrak{C}[-\pi,\pi]$. Now for any $f \in \mathfrak{C}[-\pi,\pi]$,

$$S_n(0; f) = \frac{\alpha_0}{2} + \sum_{k=1}^{n} \alpha_k.$$

From the formula for the Fourier coefficients we have

$$S_n(0; f) = \frac{1}{\pi} \int_{-\pi}^{\pi} f(t)\, D_n(t)\, dt$$

where $D_n(t) = \frac{1}{2} + \sum_{k=1}^{n} \cos kt.$

But $$D_n(t) = \frac{\sin (n+\frac{1}{2})t}{2 \sin \frac{t}{2}} \quad \text{for } t \in (-\pi, \pi)\backslash\{0\}.$$

Notice that $\phi_n(f) \equiv S_n(0; f)$ is a linear functional on $\mathcal{C}[-\pi,\pi]$ and

$$|\phi_n(f)| \leq \frac{1}{\pi} \|f\|_\infty \int_{-\pi}^{\pi} |D_n(t)|\, dt.$$

So ϕ_n is a continuous linear functional on $(\mathcal{C}[-\pi,\pi], \|\cdot\|_\infty)$.
Now we supposed that $\{\phi_n(f)\}$ converges for every $f \in \mathcal{C}[-\pi,\pi]$, so by the Uniform Boundedness Theorem, the sequence $\{\|\phi_n\|\}$ is bounded. However, we have that

$$\|\phi_n\| = \frac{1}{\pi} \int_{-\pi}^{\pi} |D_n(t)|\, dt.$$

But $$\int_{-\pi}^{\pi} |D_n(t)|\, dt = 2\int_0^{\pi} |D_n(t)|\, dt \geq 2\int_0^{\pi} \frac{|\sin(n+\frac{1}{2})t|}{t} dt = 2\int_0^{\frac{\pi}{2}} \frac{|\sin(2n+1)t|}{t} dt$$

$$\geq 2\sum_{k=0}^{n-1} \int_{\frac{k\pi}{2n+1}}^{\frac{(k+1)\pi}{2n+1}} \frac{|\sin(2n+1)t|}{t} dt \geq 2\sum_{k=0}^{n-1} \frac{2n+1}{(k+1)\pi} \int_{\frac{k\pi}{2n+1}}^{\frac{(k+1)\pi}{2n+1}} |\sin(2n+1)t|\, dt$$

$$= \frac{4}{\pi}\sum_{k=0}^{n-1} \frac{1}{k+1},$$

which is divergent and this contradicts the boundedness of $\{\|\phi_n\|\}$. So we conclude that $\{\phi_n(f)\}$ is not pointwise convergent on $\mathcal{C}[-\pi,\pi]$; that is, there exists an $f \in \mathcal{C}[-\pi,\pi]$ for which $\{S_n(0; f)\}$ is divergent. □

11.17 EXERCISES

1. Use the Uniform Boundedness Theorem 11.4 to show that the normed linear spaces $(E_0, \|\cdot\|_\infty)$ and $(\ell_1, \|\cdot\|_\infty)$ are not complete.

2. (i) For a normed linear space $(X, \|\cdot\|)$, $\{f_n\}$ is a bounded sequence in $(X^*, \|\cdot\|)$. Prove that the set $\{x \in X : \lim f_n(x) \text{ exists}\}$ is a closed linear subspace of $(X, \|\cdot\|)$.

   (ii) For a Banach space $(X, \|\cdot\|)$, $\{f_n\}$ is a pointwise bounded sequence in $(X^*, \|\cdot\|)$.

   (a) Prove that $Y \equiv \{x \in X : \lim f_n(x) \text{ exists}\}$ is a closed linear subspace $(X, \|\cdot\|)$.

   (b) Prove that if Y is second category in $(X, \|\cdot\|)$ then $Y = X$.

   (c) Prove that the set $\{x \in X : \lim f_n(x) = 0\}$ is a closed linear subspace of $(X, \|\cdot\|)$.

3. (i) For a sequence $\{\alpha_n\}$ of complex numbers, prove that if the series $\Sigma\alpha_n\lambda_n$ converges for all $x \equiv \{\lambda_1, \lambda_2, \ldots, \lambda_n, \ldots\} \in c$ then the series $\Sigma\alpha_n$ is absolutely convergent.

   (ii) For $1 < p < \infty$ and a sequence $\{\alpha_n\}$ of complex numbers, prove that if the series $\Sigma\alpha_n\lambda_n$ converges for all $x \equiv \{\lambda_1, \lambda_2, \ldots, \lambda_n, \ldots\} \in \ell_p$ then $\{\alpha_1, \alpha_2, \ldots, \alpha_n, \ldots\} \in \ell_p$ where $\frac{1}{p} + \frac{1}{q} = 1$.

4. Consider the normed linear space $\mathcal{P}[0,1]$ with norm

$$\| p \|_1 = \int_0^1 | p(t) | \, dt$$

Show that $\beta(p, q) = \int_0^1 pq(t) \, dt$ is a bilinear functional on $\mathcal{P}[0,1] \times \mathcal{P}[0,1]$ which is separately continuous but not jointly continuous.

5. (i) Consider $(X, \|\cdot\|)$ a Banach space and $(Y, \|\cdot\|')$ a normed linear space and $\mathfrak{T}$ a subset of $\mathfrak{B}(X,Y)$. Suppose that for some $x_0 \in X$ the set $\{Tx_0 : T \in \mathfrak{T}\}$ is unbounded in $(Y, \|\cdot\|')$. Prove that the set $\{x \in X : \{Tx : T \in \mathfrak{T}\} \text{ is bounded}\}$ is first category in $(X, \|\cdot\|)$.

   (ii) For any given $f \in \mathcal{C}[-\pi,\pi]$ consider the partial sum of the Fourier series of f

$$S_n(t; f) \equiv \frac{\alpha_0}{2} + \sum_{k=1}^{n} (\alpha_k \cos kt + \beta_k \sin kt)$$

   (a) Prove that for each $t \in [-\pi,\pi]$, the set $\{f \in \mathcal{C}[-\pi,\pi]: \text{the sequence } \{S_n(t; f)\} \text{ is bounded}\}$, is first category in $(\mathcal{C}[-\pi,\pi], \|\cdot\|_\infty)$.

   (b) Prove that there exists a continuous function on $[-\pi,\pi]$ whose Fourier series is divergent at each point of a dense subset of $[-\pi,\pi]$.

6. Consider an algebra A which is also a Banach space $(A, \|\cdot\|)$ where multiplication is a continuous function of x for fixed y and is a continuous function of y for fixed x.

   (i) Prove that there exists a $k > 0$ such that

$$\| xy \| \le k \| x \| \| y \| \quad \text{for all } x,y \in A.$$

   (ii) Prove that there is an equivalent norm for A under which A is a Banach algebra.

   (iii) Suppose that A has an identity e. Using the left regular representation of A in $\mathfrak{B}(A)$ defined by

$$a \mapsto T_a \quad \text{where } T_a x = ax \text{ for all } x \in A$$

   prove that $\|\cdot\|'$ where $\| a \|' = \| T_a \|$ is an equivalent norm for A under which A is a unital Banach algebra.

V. TYPES OF CONTINUOUS LINEAR MAPPINGS

A continuous linear mapping between normed linear spaces generates in a natural way, a continuous linear mapping between the duals of those spaces called its conjugate mapping. In a Hilbert space, because of the close relation between the space and its dual, a continuous linear operator has associated with it and its conjugate another continuous linear operator on the space, called its adjoint.

Projection operators on Banach and Hilbert spaces are an important set of operators especially for the decomposition of the space into simpler component subspaces.

Compact operators on Banach and Hilbert spaces are natural generalisations of finite rank operators and many of their properties are derived from those of finite rank operators.

§12. CONJUGATE MAPPINGS

The process of forming conjugate mappings is a natural extension of the duality between Banach spaces and their duals. Although the technical development of the idea may at first seem contrived, nevertheless it provides a remarkably powerful technique in application.

12.1 **Definition**. Given a continuous linear mapping T of a normed linear space $(X, \|\cdot\|)$ into a normed linear space $(Y, \|\cdot\|')$, the mapping T' of $(Y, \|\cdot\|')^*$ into $(X, \|\cdot\|)^*$ called the *conjugate mapping* of T is defined by

$$T'(g) = g \circ T;$$

that is, $$(T'g)(x) = g(Tx) \quad \text{for all } x \in X.$$

12.2 **Remark**. Since T is a continuous linear mapping of $(X, \|\cdot\|)$ into $(Y, \|\cdot\|')$ and g is a continuous linear functional on $(Y, \|\cdot\|')$, the composite mapping $g \circ T$ is a continuous linear functional on $(X, \|\cdot\|)$; that is, T' as defined does map Y^* into X^*. □

We now determine the particular properties of the conjugate mapping.

12.3 **Theorem**. *Given a continuous linear mapping* T *of a normed linear space* $(X, \|\cdot\|)$ *into a normed linear space* $(Y, \|\cdot\|')$,

(i) *the conjugate mapping* T' *of* $(Y^*, \|\cdot\|')$ *into* $(X^*, \|\cdot\|)$ *is continuous and linear and* $\|T'\| = \|T\|$,

(ii) *the mapping* $T \mapsto T'$ *is an isometric isomorphism of* $\mathfrak{B}(X,Y)$ *into* $\mathfrak{B}(Y^*, X^*)$.

Proof.

(i) Since composition is distributive over addition and is homogeneous over multiplication by a scalar, we have that

$$\begin{aligned} T'(g_1+g_2) &= (g_1+g_2)\circ T = g_1\circ T + g_2\circ T \\ &= T'(g_1)+T'(g_2) \quad \text{for } g_1, g_2 \in Y^*. \end{aligned}$$

and

$$T'(\alpha g) = \alpha g\circ T = \alpha T'(g) \qquad \text{for } \alpha \text{ scalar and } g \in Y^*,$$

so T' is linear.

Since T is a continuous linear mapping we have that

$$\|T'g\| = \|g\circ T\| \le \|T\|\,\|g\| \qquad \text{for all } g \in Y^*$$

so T' is continuous and $\|T'\| \le \|T\|$.

However, given $\varepsilon > 0$ there exists an $x_0 \in X$, $\|x_0\| \le 1$ such that

$$\|Tx_0\|' > \|T\| - \varepsilon.$$

By Corollary 6.3 of the Hahn–Banach Theorem, for $Tx_0 \in Y$ there exists a $g_0 \in Y^*$, $\|g_0\| = 1$ such that $g_0(Tx_0) = \|Tx_0\|$.

Now $\|T'\| \ge \|T'g_0\| \ge |T'g_0(x_0)| = |g_0(Tx_0)| = \|Tx_0\|' > \|T\| - \varepsilon$.

Therefore, $\|T'\| = \|T\|$.

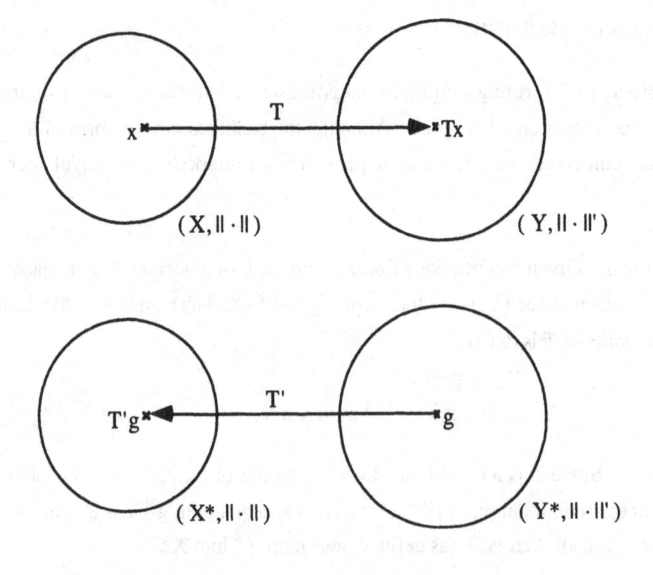

Figure 17 $T': Y^* \to X^*$ generated from $T: X \to Y$.
by $T'g(x) = g(Tx)$ for all $x \in X$.

(ii) Again, since composition is distributive over addition and is homogeneous over multiplication by a scalar, we have

$$(T_1+T_2)'(g) = g \circ (T_1+T_2) = g \circ T_1 + g \circ T_2 = T_1'(g) + T_2'(g)$$
$$= (T_1' + T_2')(g) \quad \text{for all } g \in Y^*$$

and
$$(\alpha T)'(g) = g \circ \alpha T = \alpha(g \circ T)$$
$$= \alpha T'(g) \quad \text{for } \alpha \text{ scalar and all } g \in Y^*,$$

so the mapping $T \mapsto T'$ is linear.

However, since $\| T' \| = \| T \|$, the mapping $T \mapsto T'$ is an isometric isomorphism of $\mathfrak{B}(X,Y)$ into $\mathfrak{B}(Y^*,X^*)$. □

12.4 **Remark**. It is useful to examine the behaviour of conjugates under composition.

(i) Given a continuous linear mapping T from $(X, \|\cdot\|)$ into $(Y, \|\cdot\|')$ and a continuous linear mapping S from $(Y, \|\cdot\|')$ into $(Z, \|\cdot\|'')$, then

$$S \circ T : X \to Z$$

and
$$(S \circ T)' : Z^* \to X^*.$$

Then
$$(S \circ T)'(h) = h \circ (S \circ T) \quad \text{for } h \in Z^*$$
$$= T'(h \circ S) \quad \text{where } h \circ S \in Y^*$$
$$= T' \circ S'(h). \quad \text{for all } h \in Z^*$$

So
$$(S \circ T)' = T' \circ S'.$$

(ii) For the special case where T is a continuous linear operator on $(X, \|\cdot\|)$ then we saw in Section 4.11 that $\mathfrak{B}(X)$ is a normed algebra with identity. Then $\mathfrak{B}(X^*)$ is also a normed algebra with identity. So in this case we need the composition properties of conjugates given in (i). But also for the identity operator I on $(X, \|\cdot\|)$,

$$I'(f) = f \circ I = f \quad \text{for all } f \in X$$

so I' is the identity operator on $(X^*, \|\cdot\|)$.

We notice that, because products are reversed under the mapping $T \mapsto T'$ of $\mathfrak{B}(X)$ into $\mathfrak{B}(X^*)$, the mapping is not an algebra isomorphism, but it does preserve identities. □

12.5 **Example**. Consider finite dimensional linear spaces X_n and Y_m and T a linear mapping of X_n and Y_m. Consider a basis $\{e_1, e_2, \dots, e_n\}$ for X_n and a basis for Y_m with dual basis $\{g_1, g_2, \dots, g_m\}$ for Y_m^*; (see Theorem 4.10.5).

Now T has matrix representation

$$T = [\alpha_{jk}] \quad \left.\begin{matrix} j \in \{1, 2, \dots, m\} \\ k \in \{1, 2, \dots, n\} \end{matrix}\right\}$$

with respect to the bases for X_n and Y_m.

For $x \equiv \lambda_1 e_1 + \lambda_2 e_2 + \dots + \lambda_n e_n$ and $g \equiv \mu_1 g_1 + \mu_2 g_2 + \dots + \mu_m g_m$

we have
$$Tx = \sum_{j=1}^{n} \alpha_{jk} \lambda_k$$

and
$$g(Tx) = \sum_{j=1}^{m} \sum_{k=1}^{n} \alpha_{ij} \lambda_k \bar{\mu}_j$$

and therefore $$T'g = \sum_{j=1}^{n} \alpha_{ij}\, \bar{\mu}_j .$$

We conclude that T' has matrix representation

$$T' = [\alpha_{kj}] \quad \left.\begin{matrix} j \in \{1, 2, \dots, m\} \\ k \in \{1, 2, \dots, n\} \end{matrix}\right\}$$

the transpose of the matrix representation for T. □

12.6 **Example.** Consider the continuous linear operator T on $(c_0, \|\cdot\|_\infty)$ defined for $x \equiv \{\lambda_1, \lambda_2, \dots, \lambda_n, \dots\}$ by

$$Tx = \{\lambda_1, \frac{\lambda_2}{2}, \dots, \frac{\lambda_n}{n}, \dots\}.$$

We note that any continuous linear functional f on $(c_0, \|\cdot\|_\infty)$ is of the form

$$f(x) = \sum \lambda_n \bar{\mu}_n \text{ where } \{\mu_1, \mu_2, \dots, \mu_n, \dots\} \in \ell_1; \text{ (see Example 5.3.1).}$$

So $T'f(x) = f(Tx) = \sum \frac{\lambda_n}{n} \bar{\mu}_n$. Therefore T'f is the continuous linear functional on c_0 generated by $\{\mu_1, \frac{\mu_2}{2}, \dots, \frac{\mu_n}{n}, \dots\} \in \ell_1$. □

The following example seems trivial but we will see that it is quite useful.

12.7 **Example.** Consider any normed linear space $(X, \|\cdot\|)$ with proper linear subspace A. The *inclusion* mapping *in*: $A \to X$ is defined by

$$in(a) = a \quad \text{for all } a \in A.$$

Now *in*': $X^* \to A^*$ is defined by

$$\begin{aligned}(in'f)(a) &= f(in(a)) \\ &= f(a) \quad \text{for all } a \in A\end{aligned}$$

so $$in'(f) = f|_A.$$

Moreover from the Hahn–Banach Theorem 6.2 we have that every continuous linear functional f on A can be extended as a continuous linear functional f on X so $\{f|_A : f \in X^*\} = A^*$. That is, *in*' maps X^* onto A^*. □

We apply this result in Theorem 12.12.

A number of applications of conjugate mappings concern one-to-one and onto relations. There is a duality between these concepts as the following result shows.

12.8 **Theorem.** *Consider a continuous linear mapping* T *from a normed linear space* $(X, \|\cdot\|)$ *into a normed linear space* $(Y, \|\cdot\|')$. *Then* T' *is one-to-one if and only if* T *has dense range.*

Proof. If $T'g = 0$ for some $g \in Y^*$ we have $T'g(x) = 0$ for all $x \in X$.
So $g(Tx) = 0$ for all $x \in X$; that is, $g(T(X)) = 0$.
If $\overline{T(X)} = Y$ then, since g is continuous, $g = 0$; that is, T' is one-to-one.

Conversely, suppose $\overline{T(X)} \neq Y$. Then by Corollary 6.7 to the Hahn–Banach Theorem there exists a nonzero $g_0 \in Y^*$ such that $g_0(\overline{T(X)}) = 0$.
Then $g_0(Tx) = 0$ for all $x \in X$ which implies that $T'g_0 = 0$. But then T' is not one-to-one. □

12.9 **Definition.** Given a continuous linear mapping T from a normed linear space $(X, \|\cdot\|)$ into a normed linear space $(Y, \|\cdot\|')$, the conjugate mapping $T': Y^* \to X^*$ has its own conjugate mapping $T'': X^{**} \to Y^{**}$ called the *second conjugate* of T.

12.10 **Remark.** It is natural to ask about the relation between $T: X \to Y$ and its second conjugate $T'': X^{**} \to Y^{**}$. If we regard X as embedded in X^{**} and Y as embedded in Y^{**} under the natural embedding, then T'' is an extension of T from X to X^{**}.
Consider $T''|_{\hat{X}}$. Then for all $x \in X$

$$\begin{aligned}(T''\hat{x})(g) &= \hat{x}(T'g) = g(Tx)\\ &= \widehat{Tx}(g) \quad \text{for all } g \in Y^*.\end{aligned}$$

So

$$T''(\hat{x}) = \widehat{Tx}\ .$$

Therefore, if we define a mapping $\hat{T}: \hat{X} \to \hat{Y}$ by

$$\hat{T}(\hat{x}) = \widehat{Tx}$$

we see that T'' is an extension of $\hat{T}$ to X^{**}. □

It is of interest to investigate duality relations between a continuous linear mapping and its conjugate when one of these mappings is a topological isomorphism.

12.11 **Theorem.** *Consider a continuous linear mapping* T *of a normed linear space* $(X, \|\cdot\|)$ *into a normed linear space* $(Y, \|\cdot\|')$.

(i) *If* T *is a topological isomorphism of* X *onto* Y *then* T' *is a topological isomorphism of* Y^* *onto* X^* *and* $(T')^{-1} = (T^{-1})'$.

(ii) *If* $(X, \|\cdot\|)$ *is complete and* T' *is a topological isomorphism of* Y^* *onto* X^* *then* T *is a topological isomorphism of* X *onto* Y.

Proof.

(i) If T is a topological isomorphism of X onto Y
then $T \circ T^{-1} = I$ on Y and $T^{-1} \circ T = I$ on X.
Using the properties of conjugates under composition, Remark 12.4, we have

$$(T \circ T^{-1})' = (T^{-1})' \circ T' = I' \text{ on } Y = I \text{ on } Y^*$$

and

$$(T^{-1} \circ T)' = T' \circ (T^{-1})' = I' \text{ on } X = I \text{ on } X^*.$$

So we conclude from algebraic considerations that T' has an inverse mapping $(T')^{-1} = (T^{-1})'$ from Y^* onto X^* and since T^{-1} is continuous then from Theorem 12.3(i), $(T^{-1})' = (T')^{-1}$ is continuous.

(ii) If T' is a topological isomorphism of Y^* onto X^*, then from (i) it follows that T'' is a topological isomorphism of X^{**} onto Y^{**}. Therefore there exist $m, M > 0$ such that

$$m\|F\| \leq \|T''F\|' \leq M\|F\| \quad \text{for all } F \in X^{**}.$$

So restricting T'' to $\hat{X}$ we have from Remark 12.10, that

$$m\|x\| \leq \|Tx\|' \leq M\|x\| \quad \text{for all } x \in X$$

which implies that T is a topological isomorphism of X into Y. We show that $T(X) = Y$. Since $(X, \|\cdot\|)$ is complete then $T(X)$ is complete and so is closed in $(Y, \|\cdot\|')$. But T' is one-to-one so from Theorem 12.8, $T(X) = Y$. □

We now demonstrate how conjugate mapping techniques can produce unexpected powerful results. Some of the most interesting applications are to do with establishing reflexivity properties for Banach spaces.

12.12 **Theorem.** *Given a reflexive Banach space* $(X, \|\cdot\|)$, *a normed linear space* $(Y, \|\cdot\|')$ *topologically isomorphic to* $(X, \|\cdot\|)$ *is also reflexive.*

Proof. Given that T is a topological isomorphism of $(X, \|\cdot\|)$ onto $(Y, \|\cdot\|')$ we have by Theorem 12.11(i) that T' is a topological isomorphism of Y^* onto X^* and T'' is a topological isomorphism of X^{**} onto Y^{**}.
From Remark 12.10 we see that T'' is an extension of $\hat{T}$ which is a topological isomorphism of $\hat{X}$ onto $\hat{Y}$. However, since $(X, \|\cdot\|)$ is reflexive, $X^{**} = \hat{X}$ and so $T'' = \hat{T}$ and we conclude that $Y^{**} = \hat{Y}$; that is, $(Y, \|\cdot\|')$ is reflexive. □

12.13 **Theorem.** *Given a reflexive Banach space* $(X, \|\cdot\|)$, *every closed linear subspace* A *is reflexive.*

Proof. Consider the inclusion mapping $in\colon A \to X$.
Now $\quad in'\colon X^* \to A^*$
and $\quad in''\colon A^{**} \to X^{**} = \hat{X}$ since X is reflexive.
For any $G \in A^{**}$ and $f \in X^*$ we have that

$$in''\,G(f) = \hat{x}(f).$$

so

$$G(in'(f)) = \hat{x}(f).$$

But from Example 12.7 we have that

$$in'(f) = f|_A$$

so

$$G(f|_A) = \hat{x}(f) \quad \text{for all } f \in X^* \qquad (*)$$

Since A is a proper closed linear subspace of X we have from Corollary 6.7 to the Hahn–Banach Theorem that for any $x \in X \setminus A$ there exists an $f \in X^*$ such that $f(A) = 0$ and $f(x) \neq 0$.

So in (*) we have that $\hat{x}$ satisfies $x \in A$ and

$$in'' \text{ maps } G \text{ to } \hat{a} \text{ in } \hat{X}.$$

But $$in'' \text{ maps } \hat{a} \text{ to } \hat{\hat{a}} \text{ in } \hat{X}.$$

Since in' is onto, we have from Theorem 12.8 that in'' is one-to-one.

Therefore, $G = \hat{a}$ and we conclude that $A^{**} = \hat{A}$; that is, A is reflexive. □

Now let us return to the definition of a conjugate mapping. For noncontinuous linear mappings we have difficulty in defining a conjugate as in Definition 12.1. However, in some cases the pursuit of a generalisation is of value.

12.14 **Definition**. Given a linear mapping T of a normed linear space $(X, \|\cdot\|)$ into a normed linear space $(Y, \|\cdot\|')$ which is not necessarily continuous. For $g \in Y^*$, $g \circ T$ is a linear functional on $(X, \|\cdot\|)$ which may not be continuous. For the set

$$D(T') \equiv \{g \in Y^* : g \circ T \text{ is continuous on } X\}$$

we define the mapping T' of $D(T')$ into X^* called the *conjugate mapping* of T by

$$T'(g) = g \circ T;$$

that is, $$T'(g)(x) = g(Tx) \quad \text{for all } x \in X.$$

Of course if T is continuous on X then this definition is that given in Definition 12.1. The usefulness of this generalisation depends on the size of $D(T')$. We show that for linear mappings with closed graph this set is quite substantial.

12.15 **Theorem**. *For a linear mapping* T *of a normed linear space* $(X, \|\cdot\|)$ *into a normed linear space* $(Y, \|\cdot\|')$ *with closed graph,* $D(T')$ *is total in* Y^*.

Proof. We need to show that given a $y_0 \in Y$, $y_0 \neq 0$, there exists a $g_0 \in D(T')$ such that $g_0(y_0) \neq 0$. Since T is linear, $(0, y_0) \notin G_T$. Since G_T is a closed linear subspace in $(X \times X, \|\cdot\|_\pi)$ we have from Corollary 6.7 to the Hahn–Banach Theorem that there exists a nonzero continuous linear functional h on $(X \times X, \|\cdot\|_\pi)$ such that

$$h(0,y_0) \neq 0 \text{ and } h(x,Tx) = 0 \quad \text{for all } x \in X.$$

Define $g_0 \in Y^*$ by

$$g_0(y) = h(0,y) \text{ so that } g_0(y_0) \neq 0.$$

If we define $f \in X^*$ by

$$f(x) = h(x,0)$$

then $$0 = h(x,Tx) = f(x) + g_0(Tx) \quad \text{for all } x \in X.$$

So $$g_0 \circ T = -f \in X^*,$$

which implies that $g_0 \in D(T')$. □

It is instructive to see how the size of $D(T')$ is related to the continuity of T.

12.16 **Theorem**. *For the linear mapping* T *of a normed linear space* (X, ||·||) *into a normed linear space* (Y, ||·||'), D(T') = Y *if and only if* T *is continuous.*

Proof. If T is continuous then clearly D(T') = Y*.

Conversely, if D(T') = Y* then $\sup\{|g(Tx)| : \|x\| \le 1\} = \|g \circ T\|$ for each $g \in Y^*$. That is, the set $\{\|Tx\| : \|x\| \le 1\}$ is weakly bounded so by Theorem 11.12(i) from the Uniform Boundedness Theorem, the set $\{\|Tx\| : \|x\| \le 1\}$ is bounded which implies that T is continuous. □

No matter what the size of D(T'), the conjugate T' has regular properties of its own.

12.17 **Theorem**. *For a linear mapping* T *of a normed linear space* (X, ||·||) *into a normed linear space* (Y, ||·||'), *the conjugate mapping* T' *has a closed graph in* $(Y^* \times X^*, \|\cdot\|_\pi)$.

Proof. Consider a sequence $\{g_n\}$ in D(T') which is convergent to $g \in Y^*$ and where the sequence $\{T'g_n\}$ is convergent to some $f \in X^*$. Then for $x \in X$, $\{g_n(Tx)\}$ is convergent to g(Tx) but $g_n(Tx) = T'g_n(x) \to f(x)$ as $n \to \infty$ for all $x \in X$.
Therefore, g(Tx) = f(x) for all $x \in X$. Since f is continuous we have that $g \in D(T')$ and T'g = f so T' has closed graph. □

From Theorem 12.17 we can deduce a further property.

12.18 **Corollary**. *For a linear mapping T of a normed linear space* (X, ||·||) *into a Banach space* (Y, ||·||'), D(T') *is closed.*

Proof. Consider $g \in \overline{D(T')}$. For any sequence $\{g_n\}$ in D(T') convergent to g we have

$$\|T'g_m - T'g_n\| \le \|T'\| \, \|g_m - g_n\| \quad \text{for all } m,n \in \mathbb{N}$$

so $\{T'g_n\}$ is a Cauchy sequence in (Y, ||·||'). But (Y, ||·||') is complete so there exists a $y \in Y$ such that $\{T'g_n\}$ is convergent to y. But T' has closed graph so y = T'g and therefore D(T') is closed. □

12.19. **EXERCISES**

1. A continuous linear mapping T of a normed linear space $(X, \|\cdot\|)$ into a normed linear space $(Y, \|\cdot\|')$ is said to be *norm increasing* if

$$\| Tx \|' \geq \| x \| \quad \text{for all } x \in X.$$

Prove that when T is norm increasing and $(X, \|\cdot\|)$ is complete then T is onto if and only if T' is one-to-one.

2. Consider a continuous linear mapping T of a normed linear space $(X, \|\cdot\|)$ into a normed linear space $(Y, \|\cdot\|')$. Prove that T is an isometric isomorphism of $(X, \|\cdot\|)$ onto $(Y, \|\cdot\|')$ if and only if T' is an isometric isomorphism of $(Y^*, \|\cdot\|')$ onto $(X^*, \|\cdot\|)$.

3. Given a normed linear space $(X, \|\cdot\|)$ and a linear subspace A dense in X, prove that A^* is isometrically isomorphic to X^* under the conjugate of the inclusion mapping *in*: $A \to X$ where *in*(a) = a; (see Theorem 4.8).

4. Given Banach spaces $(X, \|\cdot\|)$ and $(Y, \|\cdot\|')$, prove that if $(X, \|\cdot\|)$ is reflexive and there exists a continuous linear mapping T of $(X, \|\cdot\|)$ onto $(Y, \|\cdot\|')$ then $(Y, \|\cdot\|')$ is reflexive.

5. Given a Banach space $(X, \|\cdot\|)$ with dual space X^* and second dual X^{**}, denote the natural embedding of X into X^{**} by Q and the natural embedding of X^* into X^{***} by Q_1 ; that is,

$$(Qx)(f) = f(x) \quad \text{for all } f \in X^*$$

and

$$(Q_1 f)(F) = F(f) \quad \text{for all } F \in X^{**}.$$

(i) Prove that $Q'Q_1$ is the identity mapping on X^*.

(ii) Hence prove that if X is reflexive then X^* is reflexive.

6. Consider a closed linear subspace M of a normed linear space $(X, \|\cdot\|)$ and the inclusion mapping *in*: $M \to X$ where *in*(m) = m.

Consider the conjugate mapping *in*': $X^* \mapsto M^*$ and prove that

(i) ker *in*' = $M^\perp$ and

(ii) M^* is isometrically isomorphic to $X^*/M^\perp$.

7. Consider a closed linear subspace M of a normed linear space $(X, \|\cdot\|)$ and the quotient space $(X/M, \|\cdot\|)$.

Prove that if $(X, \|\cdot\|)$ is reflexive then $(X/M, \|\cdot\|)$ is reflexive.

(See Theorem 6.11(ii).)

8. For a closed linear subspace M of a normed linear space $(X, \|\cdot\|)$ we write

$$M^{\perp\perp} \equiv \{F \in X^{**} : F(f) = 0 \text{ for all } f \in M^{\perp}\}.$$

Prove that

(i) $M^{\perp\perp}$ is isometrically isomorphic to M^{**},

(ii) if $(X, \|\cdot\|)$ is reflexive then $\hat{M} = M^{\perp\perp}$,

(iii) if X/M is reflexive then for every $F \in X^{**}$ there exists an $x \in X$ such that

$$F(f) = \hat{x}(f) \text{ for all } f \in M^{\perp}$$

and if also M is reflexive then $(X, \|\cdot\|)$ is reflexive.

9. For Banach spaces $(X, \|\cdot\|)$ and $(Y, \|\cdot\|')$ prove that the mapping $T \mapsto T'$ of $\mathfrak{B}(X, Y)$ into $\mathfrak{B}(X^*, Y^*)$ is onto if and only if $(Y, \|\cdot\|')$ is reflexive.

10. Consider normed linear spaces $(X, \|\cdot\|)$ and $(Y, \|\cdot\|')$ and a linear mapping T of X into Y. Prove that T is continuous if and only if for every sequence $\{x_n\}$ in X where $f(x_n) \to 0$ as $n \to \infty$ for every $f \in X^*$, we have $g(Tx_n) \to 0$ as $n \to \infty$ for every $g \in Y^*$.

11. Using the Uniform Boundedness Theorem 11.4 prove the following special case of the Closed Graph Theorem 10.14.
If T *is a linear mapping of a Banach space* $(X, \|\cdot\|)$ *into a reflexive Banach space* $(Y, \|\cdot\|')$ *and* T *has a closed graph, then* T *is continuous.*

§13. ADJOINT OPERATORS ON HILBERT SPACE

Given a continuous linear operator T on a Hilbert space H, we are able to associate with T another continuous linear operator on H derived from its conjugate T' on H* and the mapping relating H and H* given in the Riesz Representation Theorem 5.2.1 and discussed in Remark 5.2.4.

13.1 **Notation**. We recall that the mapping $z \to f_z$ of H into H* defined by

$$f_z(x) = (x, z) \qquad \text{for all } x \in H$$

is one-to-one and onto and is conjugate linear and norm-preserving. We denote this mapping by ϕ. Now ϕ is an isometry onto so is invertible and ϕ^{-1} is also an isometry onto.

13.2 **Definition**. for a continuous linear operator T on a Hilbert space H we define the adjoint operator T* on H by

$$T^* = \phi^{-1} \circ T' \circ \phi.$$

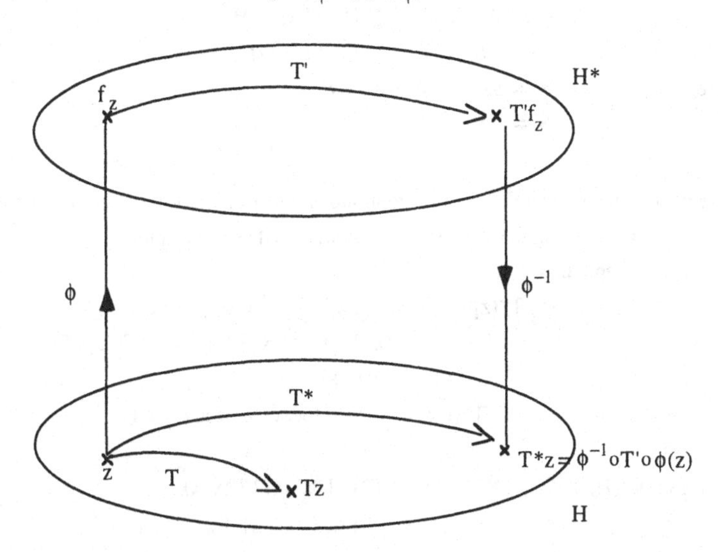

Figure 18. $T^* = \phi^{-1} \circ T' \circ \phi$.

13.3 **Remark**. Since ϕ, ϕ^{-1} and T' are additive so too is $\phi^{-1} \circ T' \circ T$. Since ϕ and ϕ^{-1} are both conjugate homogeneous and T is homogeneous, then $\phi^{-1} \circ T' \circ \phi$ is homogeneous. Therefore $T^* = \phi^{-1} \circ T' \circ \phi$ is linear.
Both ϕ and ϕ^{-1} as isometries are continuous and T' is continuous so $\phi^{-1} \circ T' \circ \phi$ is continuous.
We conclude that $T^* = \phi^{-1} \circ T' \circ \phi$ is a continuous linear operator on H. □

The relation between a continuous linear operator T and its adjoint T^* can be expressed more directly through the inner product on H.

13.4 **Theorem.** *Given a continuous linear operator* T *on a Hilbert space* H,

$$(Tx, z) = (x, T^*z) \quad \textit{for all } x,z \in H.$$

Moreover, this statement defines the adjoint operator T^* *on* H.

Proof. Given $z \in H$, we have from the definition of the conjugate T' on H^*,

$$T' f_z(x) = f_z(Tx) = (Tx, z) \quad \text{for all } x \in H$$

and from the definition of T^*,

$$T' f_z(x) = f_{T^*z}(x) = (x, T^*z) \quad \text{for all } x \in H.$$

So $\quad (Tx, z) = (x, T^*z) \quad$ for all $x,z \in H$.

If S is an operator on H such that

$$(Tx, z) = (x, Sz) \quad \text{for all } x,z \in H$$

then $\quad (x, T^*z) = (x, Sz) \quad$ for all $x,z \in H$.

So given $z \in H$, $\quad (x, T^*z-Sz) = 0 \quad$ for all $x \in H$.

Therefore, $\quad T^*z = Sz \quad$ for all $z \in H$;

that is, $\quad S = T^*$. □

13.5 **Remark.** We showed that T^* is a continuous linear operator on H but this could be established directly using the statement in Theorem 13.4 as a definition.

For any $z_1, z_2 \in H$ and all $x \in H$

$$\begin{aligned}(x, T^*(z_1+z_2)) &= (Tx, z_1+z_2) = (Tx, z_1) + (Tx, z_2)\\ &= (x, T^* z_1) + (x, T^* z_2) = (x, T^* z_1+T^* z_2)\end{aligned}$$

so $\quad T^* (z_1+z_2) = T^* z_1+T^* z_2$.

Similarly, we show that $\quad T^*(\alpha x) = \alpha T^*x \quad$ for scalar α and $x \in H$.

So T^* is linear.

But also for all $x \in H$,

$$\begin{aligned}\| T^*x \|^2 &= (T^*x, T^*x) = (TT^*x, x)\\ &\le \| TT^*x \| \| x \| \le \| T \| \| T^*x \| \| x \|\end{aligned}$$

since T is continuous.

Therefore, $\quad \| T^*x \| \le \| T \| \| x \| \quad$ for all $x \in H$

and we conclude that T^* is continuous and

$$\| T^* \| \le \| T \|.$$

□

The taking of adjoints for continuous linear operators on Hilbert space generalises the following matrix operation in the finite dimensional case.

13.6 **Example.** Consider a linear operator T on a finite dimensional inner product space H_n. We examine the form of the adjoint operator T^* on H_n.

With respect to an orthonormal basis $\{e_1, e_2, \ldots, e_n\}$ for H_n, the linear operator T has matrix representation

$$T = [\alpha_{jk}], \qquad j,k \in \{1,2, \ldots, n\}.$$

For $x \equiv \lambda_1 e_1 + \lambda_2 e_2 + \ldots + \lambda_n e_n$ and $z \equiv \mu_1 e_1 + \mu_2 e_2 + \ldots + \mu_n e_n$,

we have
$$Tx = \sum_{k=1}^{n} \alpha_{jk} \lambda_k$$

and
$$(Tx, z) = \sum_{j,k=1}^{n} \alpha_{jk} \lambda_k \bar{\mu}_j \quad \text{so} \quad (x, T^*z) = \sum_{j,k=1}^{n} \lambda_k \alpha_{jk} \bar{\mu}_j$$

Therefore, T^* has matrix representation

$$T^* = [\bar{\alpha}_{kj}], \qquad j, k \in \{1, 2, \ldots, n\},$$

the conjugate transpose of the matrix representation of T. □

We now examine properties of the operator $T \mapsto T^*$ on $\mathfrak{B}(H)$. These properties could be deduced from the definition using the conjugate mapping, but we use the direct method depending on the definition of the adjoint by the inner product given in Theorem 13.4.

13.7 **Theorem**. *Given a continuous linear operator* T *on a Hilbert space* H, *the mapping* $T \mapsto T^*$ *on* $\mathfrak{B}(H)$ *has the properties*

(i) $(T_1+T_2)^* = T_1^*+T_2^*$

(ii) $(\alpha T)^* = \bar{\alpha}\, T^*$ *for scalar* α

(iii) $(T_1 T_2)^* = T_2^* T_1^*$

(iv) $T^{**} = T$

(v) $\| T^* \| = \| T \|$

(vi) $\| T^*T \| = \| T \|^2$.

Proof. The proofs of properties (i)–(iv) all follow the same pattern. We prove property (iv).

Given $z \in H$,

$$(x, T^{**}z) = (T^*x, z) = \overline{(z, T^*x)} = \overline{(Tz, x)} = (x, Tz) \quad \text{for all } x \in H.$$

So $\qquad T^{**}z = Tz \qquad$ for all $z \in H$;

that is, $\qquad T^{**} = T$.

(v) We have already shown in Remark 13.5 that $\| T^* \| \leq \| T \|$.

From (iv), $\qquad \| T \| = \| T^{**} \| \leq \| T^* \|$ and so $\| T^* \| = \| T \|$.

(vi) From Theorem 4.11.3 we have that $\mathfrak{B}(H)$ is a normed algebra and so

$$\| T^*T \| \leq \| T^* \| \, \| T \| = \| T \|^2 \quad \text{by (v)}.$$

But also
$$\| Tx \|^2 = (Tx, Tx) = (T^*Tx, x)$$
$$\leq \| T^*Tx \| \, \| x \| \leq \| T^*T \| \, \| x \|^2 \qquad \text{for all } x \in H.$$

Therefore, $\qquad \| T \|^2 \leq \| T^*T \|$

and we conclude that $\qquad \| T^*T \| = \| T \|^2$. □

The adjoint mapping on a Hilbert space H is an element of the operator algebra $\mathfrak{B}(H)$. This suggests that this operation can be generalised to normed algebras.

13.8 **Definitions.** Given an algebra A over the complex numbers $\mathbb{C}$, an operator $x \mapsto x^*$ on A which satisfies the properties

(i) $(x+y)^* = x^* + y^*$

(ii) $(\alpha x)^* = \bar{\alpha} x^*$ for scalar α

(iii) $(xy)^* = y^* x^*$

(iv) $x^{**} = x$,

is called an *involution* on A. An algebra A with an involution is called a *$*$-algebra*. With a normed algebra $(A, \|\cdot\|)$ we need the involution to be related to the norm. A Banach $*$-algebra $(A, \|\cdot\|)$ where the involution and the norm are related by

(vi) $\| x^*x \| = \| x \|^2$ for all $x \in A$

is called a B**algebra*.

13.9 **Remarks.**

(i) The submultiplicative property of the norm (Definition 4.11.1) enables us to deduce from (vi) and (iv) of Definitions 3.8 that

(v) $\| x^* \| = \| x \|$.

(ii) This abstract structure, a B* algebra is a generalisation of $\mathfrak{B}(H)$ for a Hilbert space H. It can be seen from Theorem 13.7 that $\mathfrak{B}(H)$ is the prototype of B* algebras.

(iii) A major inquiry in the development of Banach algebra theory was to determine whether, given a B* algebra A, there exists a Hilbert space H so that A can be represented as a subalgebra of $\mathfrak{B}(H)$; (by "represented" we mean that there exists an isometric algebra isomorphism which preserves the $*$ operation). The problem was first solved in the affirmative by I.M. Gelfand and M.A. Naimark, *Math. Sbornik* **12** (1943), 411–504. □

13.10 **Self-adjoint operators**

For a complex Hilbert space H, the involution on $\mathfrak{B}(H)$ reveals important structural properties of $\mathfrak{B}(H)$.

13.10.1 **Definition.** A continuous linear operator T on a Hilbert space H is said to be *self-adjoint* if $T^* = T$.

13.10.2 **Example.** In Example 13.6 we had a linear operator T on a finite dimensional inner product space H_n with orthonormal basis $\{e_1, e_2, \ldots, e_n\}$ with matrix representation

$$T = [\alpha_{jk}] \qquad j,k \in \{1,2, \ldots, n\}.$$

It is clear that T is self-adjoint if and only if

$$\alpha_{jk} = \bar{\alpha}_{kj} \qquad j, k \in \{1, 2, \ldots, n\}.$$

So the self-adjoint operators are those with hermitian symmetric matrix representation. We know that in matrix theory this set of matrices is of crucial significance. □

13.10.3 **Remarks.**

(i) The set of self-adjoint operators is always nontrivial; it always contains the zero operator and the identity operator but for any continuous linear operator T on H we have from properties (i), (iii) and (iv) of Theorem 13.7 that $T + T^*$, TT^* and T^*T are all self-adjoint operators on H. Properties (i) and (ii) of Theorem 13.7 show that the set of self-adjoint operators is a real linear subspace of $\mathfrak{B}(H)$.

(ii) The set of self–adjoint operators on H is also closed in $\mathfrak{B}(H)$:

For a sequence $\{T_n\}$ of self-adjoint operators convergent to a continuous linear operator T in $(\mathfrak{B}(H), \|\cdot\|)$ we have

$$\|T-T^*\| \le \|T-T_n\| + \|T_n-T_n^*\| + \|T_n^*-T^*\| = \|T-T_n\| + \|(T_n-T)^*\|$$

by property (i) of Theorem 13.7

$$= 2\|T-T_n\|$$ by property (v) of Theorem 13.7

and we conclude that $T^* = T$.

(iii) But further, given a continuous linear operator T on H, using properties (i), (ii) and (v) of Theorem 13.7 we see that the operators

$$\frac{T+T^*}{2} \text{ and } \frac{T-T^*}{2i}$$

are self-adjoint. But then

$$T = \left(\frac{T+T^*}{2}\right) + i\left(\frac{T-T^*}{2i}\right)$$

that is, every continuous linear operator T on H can be expressed in terms of self-adjoint operators in this way and it is clear that such a representation is unique.

In this respect, the set of self-adjoint operators in $\mathfrak{B}(H)$ acts like the real numbers $\mathbb{R}$ in the complex numbers $\mathbb{C}$. □

In discussing properties of a continuous linear operator on a Hilbert space over the complex numbers, the following set in the complex plane gives a picture which determines many properties of the operator.

13.10.4 **Definitions.** For a linear operator T on an inner product space X, the *numerical range* of T is the set of complex numbers

$$W(T) \equiv \{(Tx, x) : \|x\| = 1\}.$$

When this set is bounded, the *numerical radius* of T is the real number

$$w(T) \equiv \sup\{\,|(Tx, x)| : \|x\| = 1\}.$$

13.10.5 Remarks.

(i) It is clear that if T is continuous then

$$|(Tx, x)| \leq \|T\| \qquad \text{for all } \|x\| = 1$$

so W(T) is a bounded set in the complex plane contained in $B[0; \|T\|]$.

Further $$w(T) \leq \|T\|.$$

(ii) In a finite dimensional inner product space H_n with orthonormal basis $\{e_1, e_2, \ldots, e_n\}$ and linear operator T with matrix representation

$$T = [\alpha_{jk}] \qquad j,k \in \{1,2, \ldots, n\}$$

we have, for $x \equiv \lambda_1 e_1 + \lambda_2 e_2 + \ldots + \lambda_n e_n$ that

$$(Tx, x) = \sum_{j,k=1}^{n} \alpha_{jk} \lambda_k \bar{\lambda}_j$$

so W(T) is a quadratic form associated with the matrix.

In the earliest work done on Hilbert space, interest was centred on such a quadratic form. It then came to be realised that the significance of the object lies in its abstract formulation.

(iii) The Toeplitz–Hausdorff Theorem states that the numerical range of a linear operator on a Hilbert space is always a convex subset of the complex plane. (See a proof in F.F. Bonsall and J. Duncan, *Numerical Ranges II*, Cambridge University Press (1973), pp 5–6.) However, we will not need to use this fact.

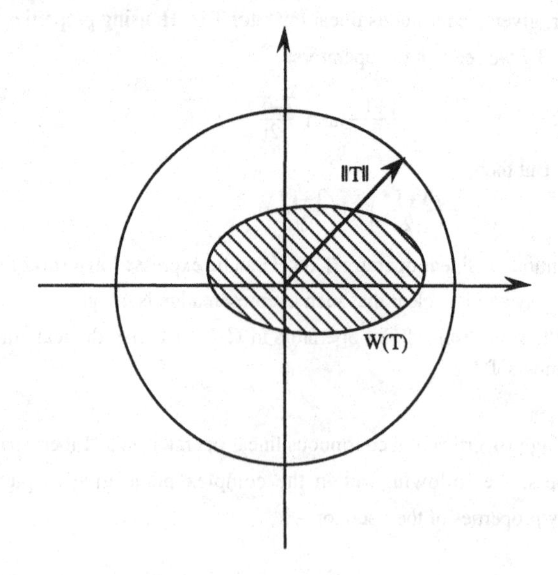

Figure 19. The numerical range of a continuous linear operator T, a subset of the complex plane

We will see later in Section 17 that the concept of numerical range of a linear operator has been usefully generalised to normed linear spaces. □

Self-adjoint operators can be neatly characterised in terms of their numerical range. But to establish this characterisation we first develop the following numerical range properties.

13.10.6 **Lemma.** *For a linear operator* T *on a complex inner product space* X,

$$4(Tx, y) = (T(x+y), x+y) - (T(x-y), x-y) + i(T(x+iy), x+iy) - i(T(x-iy), x-iy)$$

for all $x,y \in X$, ***the general polarisation formula***,

and so

(i) $T = 0$ *if and only if* $W(T) = \{0\}$,

(ii) *when* T *is continuous,* $\frac{1}{2}\| T \| \leq w(T) \leq \| T \|$.

Proof. The polarisation formula follows by straightforward calculation.

(i) If $W(T) = \{0\}$ then from the general polarisation formula

$$(Tx, y) = 0 \quad \text{for all } x,y \in X$$

from which it follows that $T = 0$.

The converse is obvious.

(ii) From the polarisation formula

$$4\,|(Tx, y)| \leq w(T)\,(\| x+y \|^2 + \| x-y \|^2 + \| x+iy \|^2 + \| x-iy \|^2)$$

and by the parallelogram law

$$|(Tx, y)| \leq w(T)\,(\| x \|^2 + \| y \|^2).$$

So $\qquad |(Tx, y)| \leq 2\,w(T)$ for all $\| x \|, \| y \| \leq 1$.

But $\qquad \| T \| = \sup\{\| Tx \| : \| x \| \leq 1\} = \sup\{(Tx, \frac{Tx}{\|Tx\|}) : \| x \| \leq 1\}$

$$\leq \sup\{|(Tx, y)| : \| x \|, \| y \| \leq 1\}.$$

So $\qquad \| T \| \leq 2\,w(T)$

and we have $\qquad \frac{1}{2}\| T \| \leq w(T) \leq \| T \|$. □

13.10.7 **Remark.** It should be noted that although this lemma holds for complex inner product spaces it does not hold in general for real inner product spaces. Consider Euclidean space $(\mathbb{R}^2, \|\cdot\|_2)$ and linear operator T with matrix representation

$$T = \begin{bmatrix} 0 & -1 \\ 1 & 0 \end{bmatrix}.$$

For any $x \equiv (\lambda_1, \lambda_2)$ we have

$$(Tx, x) = (\lambda_1 \;\; \lambda_2)\begin{bmatrix} 0 & -1 \\ 1 & 0 \end{bmatrix}\begin{pmatrix} \lambda_1 \\ \lambda_2 \end{pmatrix} = (\lambda_2 \;\; -\lambda_1)\begin{pmatrix} \lambda_1 \\ \lambda_2 \end{pmatrix} = 0$$

But clearly $T \neq 0$. □

13.10.8 **Theorem.** *A continuous linear operator* T *on a complex Hilbert space* H *is self-adjoint if and only if* W(T) *is real.*

Proof. If T is self-adjoint then, for any $x \in H$

$$(Tx, x) = (x, Tx) = \overline{(Tx, x)}$$

and so W(T) is real.

Conversely, if W(T) is real then, for any $x \in H$

$$(Tx, x) = \overline{(Tx, x)} = \overline{(x, T^*x)} = (T^*x, x)$$

and so $((T^*-T)x, x) = 0$.

This implies that $W(T^*-T) = \{0\}$ and from Lemma 13.10.6(i) we have $T^* = T$. □

Perhaps the most important property of the numerical range is that it helps us to locate the elements in $\mathfrak{B}(H)$ which have continuous inverses.

13.10.9 **Theorem.** *Given a continuous linear operator* T *on a Hilbert space* H, *for any* $\lambda \notin \overline{W(T)}$ *the continuous linear operator* $T - \lambda I$ *has a continuous inverse on* H.

Proof. For $\lambda \notin \overline{W(T)}$ we write $d \equiv d(\lambda, \overline{W(T)}) > 0$.

For $x \in H$, $\| x \| = 1$, $\| (T-\lambda I) x \| \geq | ((T-\lambda I)x, x) | = | (Tx, x) - \lambda | \geq d$ (*)

so $$\| (T-\lambda I)x \| \geq d \| x \| \quad \text{for all } x \in H,$$

which implies that $T-\lambda I$ is a topological isomorphism.

Now $(T-\lambda I)(H)$ is a closed linear subspace of H. Suppose that $(T-\lambda I)(H)$ is a proper subspace of H. Then by Corollary 2.2.21 there exists a $z \in H$, $\| z \| = 1$ such that z is orthogonal to $(T-\lambda I)(H)$, so $((T-\lambda I)z, z) = 0$ which contradicts (*).

So $T-\lambda I$ is a topological isomorphism of H onto H. □

Using the characterisation for self-adjoint operators given in Theorem 13.10.8 we can make the following deduction.

13.10.10 **Corollary.** *Given a self-adjoint operator* T *on a complex Hilbert space* H, *for any complex number* $\lambda \equiv \alpha + i\beta$ *where* $\beta \neq 0$, *the continuous linear operator* $T-\lambda I$ *has a continuous inverse on* H.

Theorem 13.10.8 also enables us to define other sets of self-adjoint operators using the order relation of the real numbers.

13.10.11 **Definition.** A self-adjoint operator T on a Hilbert space H is said to be a *positive operator* if $W(T) \geq 0$; that is, T is a positive operator if $\lambda \geq 0$ for all $\lambda \in W(T)$.

13.10.12 **Remarks.**

(i) The set of positive operators is always nontrivial; it contains the zero and identity operators and for any continuous linear operator T on H we have that T^*T and TT^* are positive operators since

$$(T^*Tx, x) = (x, T^*Tx) = (Tx, Tx) = \| Tx \|^2 \geq 0 \text{ for all } x \in H.$$

(ii) The set of positive operators is a *positive cone* in $\mathcal{B}(H)$; that is, for a positive operator T and $\alpha \geq 0$ we have that αT is also a positive operator and for positive operators T_1 and T_2 we have that T_1+T_2 is also a positive operator. This positive cone is also closed in $\mathcal{B}(H)$: For a sequence $\{T_n\}$ of positive operators convergent to a self-adjoint operator T in $(\mathcal{B}(H), \|\cdot\|)$ we have

$$| (T_nx, x) - (Tx, x) | \leq \| T_n - T \| \| x \|^2 \quad \text{for all } x \in H$$

so $(Tx,x) \geq 0$ for all $x \in H$.

(iii) The positive cone of positive operators induces a partial order relation on the set of self-adjoint operators. For self-adjoint operators T_1 and T_2 we say that

$$T_1 \leq T_2 \quad \text{if} \quad T_2 - T_1 \geq 0.$$

From Lemma 13.10.6(i) it follows that if $T_1 \leq T_2$ and $T_2 \leq T_1$ then $T_1 = T_2$. □

As in Corollary 13.10.10 we have the following property for positive operators.

13.10.13 **Corollary**. *Given a positive operator* T *on a complex Hilbert space* H, *for any complex number* $\lambda \equiv \alpha + i\beta$ *where* $\alpha < 0$ *or* $\beta \neq 0$, *the continuous linear operator* $T - \lambda I$ *has a continuous inverse on* H.

The following particular case is important.

13.10.14 **Corollary**. *For any continuous linear operator* T *on a Hilbert space* H, *the continuous linear operators* $I+T^*T$ *and* $I+TT^*$ *are topological isomorphisms of* H *onto* H.

13.10.15 **Remark**. In the attempt to establish the Gelfand-Naimark Representation Theorem for B* algebras mentioned in Remarks 13.9(iii), a key step was to use the B* algebra structure to prove that $e+x^*x$ and $e+xx^*$ are regular for every element x of a unital B* algebra. We notice that the proof of Corollary 13.10.14 uses the underlying Hilbert space structure on which the operators act. The Gelfand–Naimark Theorem had to achieve this result without the assumption of any such underlying structure. □

13.11 Normal and unitary operators

For a complex Hilbert space, further structural properties of $\mathcal{B}(H)$ are revealed by drawing attention to other special sets of continuous linear operators on H.

13.11.1 **Definitions**. A continuous linear operator T on a Hilbert space H is said to be a *normal operator* if $T^*T = TT^*$, and is said to be a *unitary operator* if $T^*T = TT^* = I$.

We will first explore the properties of normal operators.

13.11.2 **Remarks.**

(i) Clearly self-adjoint operators are normal so the set of normal operators contains the closed real linear subspace of self-adjoint operators in $\mathfrak{B}(H)$. But also given any normal operator T on H and any scalar λ, then λT is also normal. Further, the set of normal operators is closed in $\mathfrak{B}(H)$: For a sequence $\{T_n\}$ of normal operators convergent to a continuous linear operator T in $(\mathfrak{B}(H), \|\cdot\|)$ we have

$$\|T^*T-TT^*\| \le \|T^*T-T^*T_n\| + \|T^*T_n-T_n^*T_n\| + \|T_n^*T_n-T_nT_n^*\| + \|T_nT_n^*-T_nT^*\| + \|T_nT^*-TT^*\|$$
$$\le 2\|T^*\|\,\|T-T_n\| + 2\|(T-T_n)^*\|\,\|T_n\|$$

and so $T^*T = TT^*$.

(ii) Given a complex finite dimensional inner product space H_n with an orthonormal basis and a linear operator T on H_n we saw in Example 13.6 that the matrix representation of T^* is the conjugate transpose of the matrix representation of T. The linear operator T commutes with its adjoint T^* if and only if the matrix representations do the same. □

Normal operators have the following characterisation.

13.11.3 **Theorem.** *A continuous linear operator* T *on a complex Hilbert space* H *is normal if and only if* $\|Tx\| = \|T^*x\|$ *for all* $x \in H$.

Proof. Now $\|Tx\|^2 = (Tx, Tx) = (x, T^*Tx)$ and $\|T^*x\|^2 = (T^*x, T^*x) = (x, TT^*x)$ so $\|Tx\| = \|T^*x\|$ for all $x \in H$ if and only if $(x, (T^*T-TT^*)x) = 0$ for all $x \in H$; that is, by Lemma 13.10.6(i) if and only if $T^*T = TT^*$. □

The limit developed in Proposition 4.11.12 is very significant for normal operators.

13.11.4 **Theorem.** *A normal operator* T *on a complex Hilbert space* H *has the properties*

(i) $\|T^2\| = \|T\|^2$ *and*

(ii) $\lim_{n\to\infty} \|T^n\|^{1/n} = \|T\|$.

Proof.

(i) From Theorem 13.11.3 we have that

$$\|T^2x\| = \|T^*Tx\| \quad \text{for all } x \in H$$

so

$$\|T^2\| = \|T^*T\|.$$

But from Theorem 13.7(vi) we have that

$$\|T^*T\| = \|T\|^2.$$

Therefore

$$\|T^2\| = \|T\|^2.$$

(ii) Since powers of T are also normal operators we have that

$$\|T^{2^n}\| = \|T\|^{2^n} \quad \text{for all } n \in \mathbb{N}.$$

We have from Proposition 4.11.12 that $\lim_{n\to\infty} \| T^n \|^{1/n}$ exists.

So $\lim_{n\to\infty} \| T^n \|^{1/n} = \lim_{n\to\infty} \| T^{2^n} \|^{1/2^n} = \| T \|$. □

We now discuss the properties of unitary operators.

13.11.5 **Remark.**

(i) The algebraic formulation of the definition tells us that T is unitary if and only if T has a continuous inverse and $T^{-1} = T^*$.

(ii) Such operators generalise rotation operators in Euclidean space. For example, in $(\mathbb{R}^2, \|\cdot\|_2)$ the rotation operator T_θ which rotates by an angle θ about the origin with matrix representation,

$$T_\theta = \begin{bmatrix} \cos\theta & -\sin\theta \\ \sin\theta & \cos\theta \end{bmatrix}$$

has

$$T_\theta^* = \begin{bmatrix} \cos\theta & \sin\theta \\ -\sin\theta & \cos\theta \end{bmatrix}$$

and clearly $T_\theta^* = T_\theta^{-1}$.

Equivalently, in $(\mathbb{C}, |\cdot|)$

$$T_\theta z = e^{i\theta} z$$

and

$$T_\theta^* z = e^{-i\theta} = T_\theta^{-1} z \quad \text{for all } z \in \mathbb{C}.$$

(iii) We noted in Remarks 13.10.3(iii) that the set of self-adjoint operators acts like the real numbers in the complex numbers. We see now that the set of unitary operators acts like the set of complex numbers with modulus 1. □

There are useful equivalent formulations for unitary operators.

13.11.6 **Theorem.** *For a continuous linear operator* T *on a Hilbert space* H *the following conditions are equivalent.*

(i) $T^*T = I$,

(ii) $(Tx, Ty) = (x, y)$ *for all* $x,y \in H$,

(iii) $\| Tx \| = \| x \|$ *for all* $x \in H$.

Proof.

(i) ⇒ (ii) If $T^*T = I$ then

$$(Tx, Ty) = (x, T^*Ty) = (x, y) \quad \text{for all } x,y \in H.$$

(ii) ⇒ (iii) If $(Tx, Ty) = (x, y)$ for all $x,y \in H$ then $(Tx, Tx) = (x, x)$

so $\| Tx \| = \| x \|$ for all $x \in H$.

(iii) $\Rightarrow$ (i) If $\| Tx \| = \| x \|$ for all $x \in H$

then $(x, x) = \| x \|^2 = \| Tx \|^2 = (Tx, Tx) = (x, T^*Tx)$

so $((T^*T-I)x, x) = 0$ for all $x \in H$.

By Lemma 13.10.6(i), $T^*T = I$. □

13.11.7 **Remarks.**

(i) Property (iii) says that T is an isometric isomorphism of H into H and (ii) says that T is an inner product isomorphism. So (ii) $\Leftrightarrow$ (iii) says that a continuous linear operator T on H is an inner product isomorphism if and only if T is an isometric isomorphism; (see Exercise 2.4.13).

(ii) If H is finite dimensional then since T is one-to-one it is also onto and this is enough to give us that T is unitary. However, in general there exist continuous linear operators which satisfy the conditions of Theorem 13.11.6 but are not onto and so are not unitary. For example, the shift operator T on $(\ell_2, \|\cdot\|_2)$ defined by

$$T(\{\lambda_1, \lambda_2, \ldots, \lambda_n, \ldots\}) = \{0, \lambda_1, \lambda_2, \ldots, \lambda_n, \ldots\}$$

is an isometric isomorphism of $(\ell_2, \|\cdot\|_2)$ onto a proper closed linear subspace of $(\ell_2, \|\cdot\|_2)$. □

For a characterisation of unitary operators in general we need the following result.

13.11.8 **Theorem.** *A continuous linear operator* T *on a Hilbert space* H *is unitary if and only if* T *is an isometric isomorphism of* H *onto* H.

Proof. Taking into account the algebraic formulation of the definition of a unitary operator we need only prove that if T is an isometric isomorphism of H onto H then $TT^* = I$. Now T^{-1} exists on H and from Theorem 13.11.6 we have

$$T^{-1} = (T^*T)\, T^{-1} = T^*(TT^{-1}) = T^*.$$

So $TT^* = TT^{-1} = I$ and we conclude that T is unitary. □

Along the same lines as Corollary 13.10.10 for self-adjoint operators and Corollary 13.10.13 for positive operators we have the following result for unitary operators.

13.11.9 **Theorem.** *Given a unitary operator* T *on a complex Hilbert space* H, *for all* $\lambda \in \mathbb{C}, |\lambda| \neq 1$, *the operator* $T-\lambda I$ *has a continuous inverse on* H.

Proof. Since both T and T^* are unitary and, from Theorem 13.11.3 are isometric isomorphisms, $\| T \| = \| T^* \| = 1$. So from Theorem 13.10.9 we see that both $T-\lambda I$ and $T^*-\lambda I$ have continuous inverses on H for $|\lambda| > 1$. Also both T and T^* have continuous inverses on H. Now consider $0 < |\lambda| < 1$. Then $T^* - \frac{1}{\lambda} I$ has a continuous inverse on H. But $T-\lambda I = -\lambda(T^* - \frac{1}{\lambda} I)T$ so $T-\lambda I$ has a continuous inverse $-\frac{1}{\lambda} T^*(T^* - \frac{1}{\lambda} I)^{-1}$ on H. □

13.12. EXERCISES

1. Consider a Hilbert space H.
   (i) Prove that T is a topological isomorphism of H onto H if and only if T^* is a topological isomorphism of H onto H.
   (ii) Prove that if T is a topological isomorphism of H onto H then
   $$(T^{-1})^* = (T^*)^{-1}.$$

2. (i) In 2-dimensional Unitary space $(\mathbb{C}^2, \|\cdot\|_2)$ consider the operator T with two distinct eigenvalues α_1 and α_2 with corresponding eigenvectors x_1 and x_2 with $\| x_1 \|_2 = \| x_2 \|_2 = 1$.
   Prove that W(T) is a closed elliptical disc (with interior) with foci at α_1 and α_2 and major axis $\dfrac{|\alpha_1-\alpha_2|}{\sqrt{1-(x_1, x_2)}}$ and minor axis $\dfrac{|(x_1, x_2)|\,|\alpha_1-\alpha_2|}{\sqrt{1-(x_1, x_2)}}$.
   (ii) In 3-dimensional Unitary space $(\mathbb{C}^3, \|\cdot\|_2)$ consider the operator T with matrix representation
   $$T = \begin{bmatrix} 0 & 0 & \lambda \\ 1 & 0 & 0 \\ 0 & 1 & 0 \end{bmatrix} \quad \text{where } |\lambda| = 1.$$
   Prove that W(T) is an equilateral triangle (with interior) whose vertices are the three cube roots of λ.

3. Consider a linear operator T on a complex inner product space X.
   Using the general polarisation formula, or otherwise, prove that T is continuous if and only if W(T) is a bounded set in the complex plane.
   Does this characterisation for the continuity of T hold for real inner product spaces?
   Given a complex Hilbert space H and $\mathfrak{B}(H)$ the Banach algebra of continuous linear operators on H prove that the numerical radius is an equivalent linear space norm but show that it is not an equivalent algebra norm for $\mathfrak{B}(H)$.

4. Given a continuous linear operator T on a Hilbert space H where there exists an $x_0 \in H$, $\| x_0 \| = 1$ such that $\| T \| = (Tx_0, x_0)$.
   Prove that x_0 is an eigenvector of T with eigenvalue $\| T \|$.

5. Consider a self-adjoint operator T on a Hilbert space H.
   Prove that $\| T \| = \sup \{(Tx, x) : \| x \| = 1\}$.
   (Hint: Consider the expression
   $$4 \| Tx \|^2 \left(T(\alpha x + \frac{Tx}{\alpha}),\ \alpha x + \frac{Tx}{\alpha}\right) - \left(T(\alpha x - \frac{Tx}{\alpha}),\ \alpha x - \frac{Tx}{\alpha}\right) \text{ where } \alpha = \| Tx \|^{1/2}.)$$

6. (i) Consider the positive operator T on a Hilbert space H.

(a) Prove that $\langle .,. \rangle$ defined by

$$\langle x, y \rangle = (Tx, y) \quad \text{for all } x,y \in H$$

is a positive hermitian form on H.

(b) Prove that $|(Tx, y)|^2 \le (Tx, x)(Ty, y)$ for all $x,y \in H$, and deduce that $\| Tx \|^2 \le (Tx, x) \| T \|$ for all $x \in H$.

(ii) Prove that for any sequence $\{T_n\}$ of positive operators on H, $\| T_n x \| \to 0$ as $n \to \infty$ for each $x \in H$ if and only if $(T_n x, x) \to 0$ as $n \to \infty$ for each $x \in H$.

7. (i) For a sequence of continuous linear operators $\{T_n\}$ on a Hilbert space H we are given that $\{(T_n x, y)\}$ is convergent to (Tx, y) for all $x,y \in H$.

(a) Prove that T so defined is a continuous linear operator on H; (see also Exercise 10.15.8(i)).

(b) Prove that if the operators T_n are self-adjoint then so too is T.

(ii) Consider a sequence of self-adjoint operators $\{T_n\}$ on H such that $T_n \le T_{n+1}$ for all $n \in \mathbb{N}$, where $\{(T_n x, x)\}$ is convergent for each $x \in H$ and where there exists an $K > 0$ such that $|(T_n x, x)| \le K$ for all $\| x \| = 1$. Prove that

(a) there exists a continuous linear operator T on H such that $\{T_n x\}$ is convergent to Tx for each $x \in H$ and

(b) T is a self–adjoint operator on H.

(Hint: Use inequality Exercise 13.12.6(i)(b) to show that $\| T_m x - T_n x \|^4 \le ((T_m x, x) - (T_n x, x)) K \| x \|^2$ for all $m > n$ and $x \in H$.)

8. A continuous linear operator T on a Hilbert space H is called a *contraction* if $\| T \| \le 1$.

(i) Prove that T is a contraction if and only if $I - T^*T \ge 0$.

(ii) Given a continuous linear operator S which is invertible on H, prove that TS^{-1} is a contraction if and only if $T^*T \le S^*S$.

9. Given a continuous linear operator T on a complex Hilbert space write

$$H \equiv \tfrac{1}{2}(T + T^*) \text{ and } K \equiv \tfrac{1}{2}(T - T^*).$$

(i) Prove that T is a normal operator if and only if $HK = KH$.

(ii) Given that T is normal, prove that T has a continuous inverse H if and only if H^2+K^2 has a continuous inverse on H and in this case $T^{-1} = T^*(H^2+K^2)^{-1}$.

(iii) Prove that T is unitary if and only if $H^2+K^2 = I$.

10. Consider a normal operator T on a complex Hilbert space H.

(i) Prove that

(a) $\ker T^2 = \ker T$ and

(b) $\overline{T^2(H)} = \overline{T(H)}$.

(ii) Prove that T(H) is dense in H if and only if T is one-to-one.

11. Consider a continuous linear operator T on a complex Banach space X.

(i) Prove that

$$V(T) \subseteq \bigcap_{\alpha \in \mathbb{C}} \{\lambda \in \mathbb{C} : |\lambda - \alpha| \leq \| T - \alpha I \| \}.$$

(ii) Prove that

$$\sup \operatorname{Re} V(T) = \lim_{\alpha \to 0+} \frac{\| I + \alpha T \| - 1}{\alpha} \quad \text{for } \alpha \in \mathbb{R}.$$

(iii) Prove that $V(T) \subseteq \mathbb{R}$ if and only if

$$\lim_{\alpha \to 0} \frac{\| I + i\alpha T \| - 1}{\alpha} = 0 \quad \text{for } \alpha \in \mathbb{R}.$$

12. Consider a self-adjoint operator T on a complex Hilbert space H.

(i) Prove that the continuous linear operators T+iI and T−iI have continuous inverses on H.

(ii) The *Cayley transform* of T is a continuous linear operator U on H defined by

$$U = (T - iI)(T + iI)^{-1}.$$

Prove that U is a unitary operator.

(iii) Prove that T can be recovered from the Cayley transform U to give

$$T = i\,(I + U)(I - U)^{-1}.$$

(Note: The Cayley transform generalises the Möbius transform for complex numbers

$$w = \frac{z - i}{z + i}$$

which maps the real numbers into the unit circle. To recover we have

$$z = i\left(\frac{1 + w}{1 - w}\right)$$

and (iii) gives the analogous result for $\mathfrak{B}(H)$.)

13. Given a complex Hilbert space H, a subalgebra A of $\mathfrak{B}(H)$ is said to be *self-adjoint* if it contains the identity operator and the adjoint of each of its elements. If A is also a closed subalgebra it is called a *C* algebra.* Given an arbitrary subset M of $\mathfrak{B}(H)$, the *commutant* of M denoted by M^c is the set of all continuous linear operators on H which commute with every operator in M. Prove that

(i) M^c is a unital Banach subalgebra of $\mathfrak{B}(H)$,

(ii) if M is self-adjoint then M^c is a C* algebra.

§14. PROJECTION OPERATORS

We now examine the properties of the special set of projection operators on a Banach space and on a Hilbert space. Such operators play a significant role in the decomposition of a space into component subspaces.

14.1 **Definition**. A linear operator P on a linear space X is called a *projection operator* (an *algebraic projection*) if it is idempotent; that is, $P^2 = P$.

For linear spaces there is an important relation between projection operators and direct sum decomposition of the linear space, expressed in the following result.

14.2 **Lemma**. *Consider a linear space* X.

(i) *If* P *is a projection on* X *then* $X = P(X) \oplus \ker P$.

(ii) *If* $X = M \oplus N$ *where* M *and* N *are linear subspaces of* X *then the linear operator* P *on* X *defined by*

$$Px = m \quad where\ x = m+n, \quad m \in M\ and\ n \in N$$

is a projection operator on X.

Proof.

(i) For any $x \in X$, we have $x = Px + (I-P)x$.

Now $Px \in P(X)$ and $P(I-P)x = (P-P^2)x = 0$, so $(I-P)x \in \ker P$.

Therefore, $X = P(X) + \ker P$.

For $Px \in P(X) \cap \ker P$ we have $Px = P(Px) = 0$ since $Px \in \ker P$, so $Px = 0$.

We conclude that $X = P(X) \oplus \ker P$.

(ii) For $X = M \oplus N$ with operator P defined on X by

$$Px = m \quad \text{where } x = m+n, \quad m \in M \text{ and } n \in N$$

the unique decomposition implied by the direct sum gives us that P is linear. But also

$$P^2x = m, \text{ so } P^2 = P;$$

that is, P is a projection operator. □

14.3 **Corollary**. *On a linear space* X, *a linear operator* P *is a projection operator if and only if* I–P *is a projection operator.*

Proof. $(I-P)^2 = 1 - 2P + P^2 = I - P$ if and only if $P^2 = P$. □

In a normed linear space the natural projection operators are continuous and the natural decomposition of the space is by closed linear subspaces.

14.4 **Theorem.**

(i) *Given a normed linear space* $(X, \|\cdot\|)$, *if* P *is a continuous projection operator on* X, *then* $X = P(X) \oplus \ker P$ *where* P(X) *and* ker P *are closed linear subspaces of* X.

(ii) *Given a Banach space* $(X, \|\cdot\|)$, *if* $X = M \oplus N$ *where* M *and* N *are closed linear subspaces of* X *then the linear operator* P *defined on* X *by*

$$Px = m \quad \textit{where } x = m+n, \quad m \in M \textit{ and } n \in N,$$

is a continuous projection on X.

Proof.

(i) Since P is continuous, ker P is closed. But also I–P is continuous so $P(X) = \ker(I-P)$ is also closed.

(ii) Renorm X by $\|x\|' = \|m\| + \|n\|$.

Since $(X, \|\cdot\|)$ is a Banach space and M and N are closed linear subspaces then $(X, \|\cdot\|')$ is also a Banach space; (see Exercise 1.26.3(iii)).

Since $\quad \|Px\| = \|m\| \leq \|m\| + \|n\| = \|x\|'$ for all $x \in X$,

then P is a continuous linear mapping of $(X, \|\cdot\|')$ into $(X, \|\cdot\|)$.

However, $\quad \|x\| = \|m\| + \|n\| = \|x\|'$ for all $x \in X$

so by Corollary 10.10 to the Open Mapping Theorem we have that $\|\cdot\|'$ and $\|\cdot\|$ are equivalent norms for X. Therefore, P is a continuous projection operator on $(X, \|\cdot\|)$. □

14.5 **Remark**. Given a linear space X and a linear subspace M of X there always exists at least one linear subspace N complementary to M in X; that is, such that $X = M \oplus N$. Given a normed linear space $(X, \|\cdot\|)$ and a closed linear subspace M of X, if there exists a closed linear subspace N of X such that $X = M \oplus N$ we say that M is *complemented* in X. In terms of projection operators, Theorem 14.4 implies that a closed linear subspace M is complemented in X if and only if there exists a continuous projection operator P from X onto M. F.J. Murray, *Trans. Amer. Math. Soc.* **41**, (1937), 138–152 has given an example of a closed linear subspace M in $(\ell_p, \|\cdot\|_p)$ where $1 < p < \infty$, $p \neq 2$, for which there does not exist any closed linear subspace complementary to M; that is, where M is not the range of any continuous projection operator. □

In Hilbert space we have an orthogonality relation which gives a naturally preferred way of forming direct sum decompositions of the space and the preferred continuous projection operators on the space.

We first investigate orthogonality of subspaces in Hilbert space.

14.6 **Definition**. Given sets A and B in an inner product space X, we say that A is *orthogonal* to B if every element of A is orthogonal to every element of B; that is,

$$(x, y) = 0 \qquad \text{for all } x \in A \text{ and } y \in B.$$

For a set A, the *orthogonal complement* of A, denoted by $A^{\perp}$, is defined by

$$A^{\perp} \equiv \{x \in X : (x, y) = 0 \text{ for all } y \in A\}.$$

14.7 **Lemma**. *For any set* A *in an inner product space* X, *the orthogonal complement* $A^{\perp}$ *is a closed linear subspace of* X.

Proof. Now $A^{\perp} = \bigcap\{\ker f_y : y \in A\}$ where, given $y \in A$, f_y is the continuous linear functional on X defined by $f_y(x) = (x, y)$ for all $x \in X$; (see Example 4.10.6).
So $\ker f_y$ is a closed linear subspace of X for each $y \in A$ and therefore $\bigcap\{\ker f_y : y \in A\}$ is a closed linear subspace of X. □

The following property is worth noting.

14.8 **Corollary**. *For any set* A *in an inner product space* X,

$$(\overline{\text{sp } A})^{\perp} = A^{\perp}.$$

Proof. Clearly $(\overline{\text{sp } A})^{\perp} \subseteq A^{\perp}$.
If $y \perp A$ then $y \perp \text{sp } A$.
If x is a cluster point of sp A then there exists a sequence $\{x_n\}$ in sp A such that $\{x_n\}$ is convergent to x. For $y \in X$,

$$|(y, x) - (y, x_n)| \leq \|y\| \|x - x_n\|.$$

So if $y \in A^{\perp}$ then $(y, x_n) = 0$ for all $n \in \mathbb{N}$ and $(y, x) = 0$; that is, $y \in (\overline{\text{sp } A})^{\perp}$. □

The following theorem justifies the term "orthogonal complement".

14.9 **Theorem**. *Given any closed linear subspace* M *of a Hilbert space* H, *then*

$$H = M \oplus M^{\perp}.$$

Proof. Clearly $M \cap M^{\perp} = \{0\}$ so $M + M^{\perp} = M \oplus M^{\perp}$.
We show firstly that $M \oplus M^{\perp}$ is closed.
If x is a cluster point of $M \oplus M^{\perp}$ then there exists a sequence $\{x_n\}$ in $M \oplus M^{\perp}$ such that $\{x_n\}$ is convergent to x in H.
For each $n \in \mathbb{N}$ we write $x_n = u_n + v_n$ where $u_n \in M$ and $v_n \in M^{\perp}$.
Since $(u_n, v_m) = 0$ for all $m, n \in \mathbb{N}$

$$\|x_n - x_m\|^2 = \|u_n - u_m\|^2 + \|v_n - v_m\|^2.$$

So $\{u_n\}$ is a Cauchy sequence in M and $\{v_n\}$ is a Cauchy sequence in $M^{\perp}$. Since both M and $M^{\perp}$ are closed and H is complete there exist $u \in M$ and $v \in M^{\perp}$ such that $\{u_n\}$ is convergent to u and $\{v_n\}$ is convergent to v.

But $$\| x_n - (u+v) \| \le \| u_n - u \| + \| v_n - v \|$$
so $x = u+v \in M \oplus M^\perp$ which implies that $M \oplus M^\perp$ is closed.
Suppose that $M \oplus M^\perp$ is a proper closed linear subspace of H. Then from Corollary 2.2.21, there exists a $z \neq 0$ such that z is orthogonal to $M \oplus M^\perp$. But then z is orthogonal to M so $z \in M^\perp$, and z is orthogonal to $M^\perp$. Therefore, $(z, z) = 0$ which contradicts $z \neq 0$. We conclude that $H = M \oplus M^\perp$. □

This result gives us another useful elementary property.

14.10 **Corollary**. *Given any closed linear subspace* M *of a Hilbert space* H,
$$M^{\perp\perp} = M.$$

Proof. Now $H = M \oplus M^\perp$ and $H = M^\perp \oplus M^{\perp\perp}$. But $M \subseteq M^{\perp\perp}$.
Consider $x \in H$ where $x = m + n$, $m \in M$ and $n \in M^\perp$
$= y + n'$, $y \in M^{\perp\perp}$ and $n' \in M^\perp$.
Then $y - m = n - n' \in M^\perp$. But $y - m \in M^{\perp\perp}$ so $y = m$ and we conclude that $M^{\perp\perp} = M$. □

We use orthogonal complements to show the close relation between the kernel and range of a continuous linear operator and its adjoint.

14.11 **Theorem**. *For a continuous linear operator* T *on a Hilbert space* H
(i) $\overline{T(H)}^\perp = \ker T^*$,
(ii) $\overline{T(H)} = (\ker T^*)^\perp$.

Proof.
(i) Now $y \in T(H)^\perp$ if and only if $(Tx, y) = 0$ for all $x \in H$,
that is, $(x, T^*y) = 0$ for all $x \in H$ which is $T^*y = 0$,
that is, if and only if $y \in \ker T^*$.
But $\overline{T(H)}^\perp = T(H)^\perp$ by Corollary 14.8.
(ii) It follows from Corollary 14.10, that
$$\overline{T(H)} = \overline{T(H)}^{\perp\perp} = (\ker T^*)^\perp \text{ from (i).}$$ □

14.12 **Remark**. Theorem 14.9 implies that in Hilbert space, direct sum decompositions by closed linear subspaces exist; that is, given a closed linear subspace M of a Hilbert space H, there exists a closed linear subspace complementary to M, in particular $M^\perp$. So from Theorem 14.4(ii) we see that given any closed linear subspace M of a Hilbert space H there exists a continuous projection operator on H with range M, in particular the projection operator P with kernel $M^\perp$. □

So in Hilbert space we are naturally directed to examine those continuous projection operators whose range and kernel are orthogonal subspaces.

14.13 **Definition**. On an inner product space X, a projection operator P where P(X) is orthogonal to ker P is called an *orthogonal projection.*

Such operators have the following characterisation.

14.14 **Theorem**. *On an inner product space* X*, a projection operator* P *on* X *is an orthogonal projection if and only if*

$$(Px, y) = (x, Py) \quad \textit{for all } x,y \in X.$$

Proof. If P(X) and ker P are orthogonal then for any $x = m + n$ and $y = m' + n'$ where $m, m' \in P(X)$ and $n, n' \in \ker P$,

$$(Px, y) = (m, m' + n') = (m, m')$$

and $$(x, Py) = (m + n, m') = (m, m')\ .$$

Conversely, if $x \in P(X)$ and $y \in \ker P$ then $Px = x$ and $Py = 0$ so that

$$(x, y) = (Px, y) = (x, Py) = 0$$

and so P(X) and ker P are orthogonal. □

It is interesting to note that orthogonal projections are automatically continuous.

14.15 **Theorem**. *An orthogonal projection* P *on a inner product space* X *is continuous and if* $P \neq 0$ *then* $\| P \| = 1$.

Proof. For $x = m + n$ where $m \in P(X)$ and $n \in \ker P$ we have $Px = m$ and $(Px, n) = 0$.

So $$\| x \|^2 = \| Px \|^2 + \| n \|^2$$

and $$\| Px \| \leq \| x \| \qquad \text{for all } x \in X;$$

that is, P is continuous.

But also $\| P \| \leq 1$. Since $P^2 = P$ then $\| P \| \leq \| P \|^2$ and so $\| P \| \geq 1$ if $P \neq 0$.

Therefore, $\| P \| = 1$. □

14.16 **Corollary**. *On a Hilbert space* H*, an orthogonal projection* P *is a positive operator.*

Proof. Clearly Theorem 14.14 implies that P is self–adjoint. Then for all $x \in H$

$$(Px, x) = (P^2x, x) = (Px, Px) = \| Px \|^2 \geq 0.$$ □

We have the following important characterisation of the kernel of an orthogonal projection on a Hilbert space.

14.17 **Theorem**. *If* P *is an orthogonal projection on a Hilbert space* H, *then*

$$\ker P = P(H)^{\perp} \text{ and } P(H) = \ker P^{\perp}.$$

Proof. From Theorem 14.4(i), P(H) is closed. From Corollary 14.16, P is self-adjoint. So our result follows from Theorem 14.11(i) and Corollary 14.10. □

We now investigate the structure of the set of orthogonal projections on a Hilbert space H as a subset of $\mathfrak{B}(H)$. We do this because we will ultimately show that certain classes of operators on a Hilbert space can be represented by orthogonal projections.

14.18 **Theorem**. *Consider* P *and* Q *orthogonal projections on an inner product space* X. *Then* PQ *is an orthogonal projection if and only if* PQ = QP *and in this case*

$$PQ(X) = P(X) \cap Q(X).$$

Proof. If PQ is an orthogonal projection on X then from Theorem 14.14,

$$(PQx, y) = (x, PQy) = (Px, Qy) = (QPx, y) \text{ for all } x,y \in X,$$

so PQ = QP.

Conversely, if PQ = QP then $(PQ)^2 = PQPQ = P^2Q^2 = PQ$.

But also, $(PQx, y) = (Qx, Py) = (x, QPy) = (x, PQy)$ for all $x,y \in X$,

so we conclude from Theorem 14.14 that PQ is an orthogonal projection.

If $x \in P(X) \cap Q(X)$ then $Px = x$ and $Qx = x$ so $PQx = x$ and

$$P(X) \cap Q(X) \subseteq PQ(X).$$

If $x \in PQ(X)$ then $x = PQx = P(Qx)$ so $x \in P(X)$. But also $x \in QP(X)$ so $x \in Q(X)$. Then

$$PQ(X) \subseteq P(X) \cap Q(X).$$

Therefore,

$$PQ(X) = P(X) \cap Q(X).$$

□

14.19 **Definition**. Orthogonal projections P and Q on an inner product space X are said to be *orthogonal* to each other if PQ = 0.

The following theorem gives some explanation of this property.

14.20 **Theorem**. *Consider orthogonal projections* P *and* Q *on a Hilbert space* H. *Then* P *and* Q *are orthogonal to each other if and only if* P(H) *is orthogonal to* Q(H).

Proof. If P(H) is orthogonal to Q(H) then $Q(H) \subseteq P(H)^{\perp}$ so from Theorem 14.14 $Q(H) \subseteq \ker P$ and so $PQ(H) = \{0\}$; that is, PQ = 0.

Conversely, if PQ = 0 then for all $x \in Q(H)$ we have $Px = P(Qx) = 0$.

So $Q(H) \subseteq \ker P = P(H)^{\perp}$ which implies that P(H) is orthogonal to Q(H). □

14.21 **Theorem.** *Consider* P *and* Q *orthogonal projections on a Hilbert space* H. *Then* P+Q *is an orthogonal projection on* H *if and only if* P *and* Q *are orthogonal to each other and then* (P+Q)(H) = P(H) + Q(H).

Proof. If P and Q are orthogonal projections which are orthogonal to each other, then from Theorem 14.20, $PQ = 0 = QP$.
Then $(P+Q)^* = P^* + Q^* = P + Q$ but also $(P+Q)^2 = P^2 + PQ + QP + Q^2 = P + Q$. So from Theorem 14.14, P+Q is an orthogonal projection.

Conversely, if P+Q is an orthogonal projection, for $x \in P(H)$ we have $Px = x$ so

$$\|x\|^2 = \|Px\|^2 \le \|Px\|^2 + \|Qx\|^2 = (Px, x) + (Qx, x) = ((P+Q)x, x)$$
$$= \|(P+Q)x\| \, \|x\| \le \|x\|^2 \text{ since } \|P+Q\| \le 1.$$

Therefore, $\|Qx\| = 0$ and so $x \in \ker Q$ and $P(H) \subseteq \ker Q = Q(H)^\perp$ by Theorem 14.17, and P(H) is orthogonal to Q(H).
But this argument also shows that for $x \in P(H)$ we have $x \in \ker Q$ and so $x \in (P+Q)(H)$.
Therefore $P(H) \subseteq (P+Q)(H)$.
The argument is symmetric with respect to P and Q so $Q(H) \subseteq (P+Q)(H)$. Therefore,

$$P(H) + Q(H) \subseteq (P+Q)(H).$$

However, for any $x \in (P+Q)(H)$ we have $x = (P+Q)x \in P(H) + Q(H)$ so

$$(P+Q)(H) \subseteq P(H) + Q(H)$$

and we conclude that $\quad (P+Q)(H) = P(H) + Q(H).$ □

Projection operators play a role in simplifying the action of a linear operator by studying its restriction to special component subspaces.

14.22 **Definitions.** Given a linear operator T on a linear space X we say that a linear subspace M of X is *invariant* under T if $T(M) \subseteq M$.
Given linear subspaces M and N of X such that $X = M \oplus N$ we say that a linear operator T on X is *reduced* by M and N if both M and N are invariant under T.

14.23 **Remarks.** For a linear operator T on a linear space X with an invariant subspace M, we can regard the restriction of T to M as a linear operator on M.
When a linear operator T on a linear space X is reduced by linear subspaces M and N we can study T on X by T restricted to M and T restricted to N and such restrictions may be considerably simpler than the original T. For example, when X is n-dimensional then we will have our greatest simplification if there exists a set of n one-dimensional subspaces which reduce the linear operator T. This is equivalent to X having a basis $\{e_1, e_2, \ldots, e_n\}$ each element of which is an eigenvector of T. □

We can characterise the reduction of a linear operator in terms of associated projection operators.

14.24 **Theorem**. *Consider a linear space* X *and linear subspaces* M *and* N *where* $X = M \oplus N$ *and denote by* P *and* Q *the projection operators associated with* M *and* N. *Then a linear operator* T *on* X *is reduced by* M *and* N *if and only if*

$$PTx = TPx \quad \textit{and} \quad QTx = TQx \quad \textit{for all } x \in X.$$

Proof: Write $x = m + n$ where $m \in M$ and $n \in N$.
Suppose that T is reduced by M and N.
Then $PTx = PTm + PTn = Tm = TPx$ since $Tm \in M = P(X)$,
and similarly $QTx = TQx$ since $Tn \in N = Q(X)$.
Conversely, $Tm = TPx = PTx \in M$ and $Tn = TQx = QTx \in N$. □

Of special interest is the case of closed invariant subspaces for continuous linear operators on a Hilbert space.

14.25 **Theorem**. *Consider a continuous linear operator* T *on a Hilbert space* H. *A closed linear subspace* M *of* H *is invariant under* T *if and only if* $M^\perp$ *is invariant under* T^*.

Proof: Given that $T(M) \subseteq M$ then for all $x \in M$, $(Tx, y) = 0$ for all $y \in M^\perp$.
Then $(x, T^*y) = 0$ so $T^*y \in M^\perp$. □

This result has the following implication.

14.26 **Corollary**. *Given a continuous linear operator* T *on a Hilbert space* H, *if the closed linear subspace* M *of* H *is invariant under* T *then* T *is reduced by* M *and* $M^\perp$.

14.27 **Remarks**. The following famous problem has been the object of considerable research over many years.

The invariant subspace problem.
Does a continuous linear operator T *on a Banach space* X *have a nontrivial invariant closed linear subspace*?

Every continuous linear operator T on a Banach space X has trivial invariant subspaces $\{0\}$ and X. On a finite dimensional complex linear space every linear operator has a nonzero eigenvector so the problem really concerns infinite dimensional Banach spaces. The general problem was solved in the negative by Per Enflo, *Acta Math.* **158** (1987), 213–313 and independently by C.J. Read, *Bull. London Math. Soc.* **16** (1984), 337–401, who in *Bull. London Math. Soc.* **17** (1985), 305–317 gave a counterexample in the Banach space $(\ell_1, \|\cdot\|_1)$. The solution is still not known for continuous linear operators in Hilbert space. However, for certain classes of linear operators on a complex Banach space it is known that the solution is positive and we will have more to say about this in Section 18. □

14.28 EXERCISES

1. (i) Consider a finite dimensional subspace M_n of an inner product space X. Prove that there exists an orthogonal projection of X onto M_n.
   (ii) Consider a finite dimensional subspace M_n of a normed linear space $(X, \|\cdot\|)$. Prove that there exists a continuous projection of X onto M_n.

2. (i) Given a proper closed linear subspace M of a Hilbert space H, consider the mapping P of H into M defined for each $x \in H$ by
$$\| x-Px \| = d(x, M).$$
   Prove that P is linear, continuous and an orthogonal projection operator on H.
   (ii) Given a proper closed linear subspace M of a rotund reflexive Banach space X, consider the mapping P of X into M defined for each $x \in X$ by
$$\| x-Px \| = d(x, M).$$
   Prove that P is linear, continuous and a projection operator on X with $\| P \| = 1$.

3. Consider a Banach space $(X, \|\cdot\|)$ with linear subspaces $M_1, M_2, \ldots, M_n$ such that
$$X = M_1 \oplus M_2 \oplus \ldots \oplus M_n$$
   with corresponding projection operations $P_1, P_2, \ldots, P_n$ from X onto $M_1, M_2, \ldots, M_n$.
   Prove that the subspaces $M_1, M_2, \ldots, M_n$ are closed if and only if the projections $P_1, P_2, \ldots, P_n$ are continuous.

4. Consider a closed linear subspace M of a Banach space $(X, \|\cdot\|)$ such that X/M has finite dimension. Prove that every projection of X onto M is continuous.

5. Consider a Banach space $(X, \|\cdot\|)$ with closed linearly independent subspaces M and N. Prove that $M \oplus N$ is closed if and only if there exists a $d > 0$ such that $\| m-n \| \geq d$ for all $m \in M$ and $n \in N$ and $\| m \| = \| n \| = 1$.

6. Consider a normed linear space $(X, \|\cdot\|)$ with a closed linear subpace M. Prove that M is complemented in X if
   (i) M is finite dimensional, or
   (ii) X/M is finite dimensional.

7. Consider a continuous projection operator P on a Hilbert space H. Prove that
   (i) if $\| P \| \leq 1$ then P is an orthogonal projection
   (ii) if P is a normal operator then P is an orthogonal projection.

8. Consider a continuous projection operator P on a normed linear space $(X, \|\cdot\|)$.

   (i) Prove that the conjugate operator P' on X* is also a continuous projection operator.

   (ii) Prove that $P'(X^*) = (\ker P)^{\perp} \equiv \{f \in X^* : f(\ker P) = 0\}$ and $\ker P' = P(X)^{\perp} \equiv \{f \in X^* : f(\ker P) = 0\}$.

   (iii) Prove that $P'(X^*)$ is finite dimensional if and only if $P(X)$ is finite dimensional and $\dim P'(X^*) = \dim P(X)$.

9. Consider a continuous linear operator T on a Banach space $(X, \|\cdot\|)$. Prove that if T is reduced by the closed subspaces M and N then the conjugate operator T' is reduced by the subspaces

$$M^{\perp} \equiv \{f \in X^* : f(M) = 0\}\ ,\ N^{\perp} \equiv \{f \in X^* : f(N) = 0\}\ .$$

10. (i) Consider a linear operator T on a linear space X reduced by linear subspaces M and N. Prove that

    (a) T^{-1} exists if and only if $T|_M^{-1}$ and $T|_N^{-1}$ exist,

    (b) if T^{-1} exists then T^{-1} is reduced by M and N and

$$T^{-1}|_M = T|_M^{-1} \text{ and } T^{-1}|_N = T|_N^{-1}.$$

   (ii) Consider a continuous linear operator T on a Banach space $(X, \|\cdot\|)$ reduced by closed linear subspaces M and N. Prove that T^{-1} exists and is continuous if and only if $T|_M^{-1}$ and $T|_N^{-1}$ both exist and are continuous.

11. Consider a normal operator T on a Hilbert space H and M a closed linear subspace of H where M and $M^{\perp}$ are invariant under T.

    Prove that

   (i) $\|T\| = \max\{\|T|_M\|, \|T|_{M^{\perp}}\|\}$,

   (ii) $T|_M$ is a normal operator on M,

   (iii) $T|_M^* = T^*|_M$.

12. Consider a sequence $\{P_n\}$ of continuous projection operators on a Banach space X, pointwise convergent to an operator P on X.

   (i) Prove that P is a continuous projection operator on X.

   (ii) Prove that if X is a Hilbert space and $\{P_n\}$ is a sequence of orthogonal projection operators then P is also an orthogonal projection operator on X.

13. Consider a sequence $\{P_n\}$ of orthogonal projection operators on a Hilbert space H where $P_j P_k = 0$ for $j, k \in \mathbb{N}, j \neq k$.

    Prove that the operator P on H defined by $P = \sum_{n=1}^{\infty} P_n$ is an orthogonal projection on H and P(H) is the closure of the span of $\bigcup_{n=1}^{\infty} P_n(H)$.

§15. COMPACT OPERATORS

When we generalise the study of linear mappings on finite dimensional to infinite dimensional normed linear spaces it is reasonable that we select for special attention those continuous linear mappings which retain some of the special properties of those on finite dimensional spaces. An important set of such linear mappings is the set of compact mappings.

15.1 **Definition**. A linear mapping T of a normed linear space $(X, \|\cdot\|)$ into a normed linear space $(Y, \|\cdot\|')$ is said to be *compact* if for each bounded sequence $\{x_n\}$ in $(X, \|\cdot\|)$ the sequence $\{Tx_n\}$ has a subsequence convergent in $(Y, \|\cdot\|')$.

The reason for the term "compact" mapping is evident from the following characterisation.

15.2 **Theorem**. *A linear mapping* T *of a normed linear space* $(X, \|\cdot\|)$ *into a normed linear space* $(Y, \|\cdot\|')$ *is compact if and only if for each bounded set* B *in* $(X, \|\cdot\|)$, *the set* $\overline{T(B)}$ *is compact in* $(Y, \|\cdot\|')$.

Proof. Suppose that T is compact and B is a bounded set in X. Consider a sequence $\{y_n\}$ in $\overline{T(B)}$. For each $n \in \mathbb{N}$, choose $x_n \in B$ such that $\| y_n - Tx_n \| < \frac{1}{n}$.

Now the sequence $\{Tx_n\}$ has a subsequence $\{Tx_{n_k}\}$ convergent to say $y \in \overline{T(B)}$ and then $\{y_{n_k}\}$ is also convergent to y. So $\overline{T(B)}$ is compact.

Conversely, for any bounded sequence $\{x_n\}$ in X, the set $\overline{\{Tx_n : n \in \mathbb{N}\}}$ is compact so $\{Tx_n\}$ has a convergent subsequence and therefore T is compact. □

The following elementary lemma is useful

15.3 **Lemma.** *A linear mapping* T *of a normed space* $(X, \|\cdot\|)$ *into a normed linear space* $(Y, \|\cdot\|')$ *is compact if and only if for each sequence* $\{x_n\}$ *in the closed unit ball of* $(X, \|\cdot\|)$ *the sequence* $\{Tx_n\}$ *has a subsequence convergent in* $(Y, \|\cdot\|')$.

Proof. Consider a sequence $\{x_n\}$ in $(X, \|\cdot\|)$ where there exists an $r > 0$ such that $\| x_n \| \leq r$ for all $n \in \mathbb{N}$. Then $\| \frac{x_n}{r} \| \leq 1$ for all $n \in \mathbb{N}$. If $\{T(\frac{x_n}{r})\}$ has a subsequence $\{T(\frac{x_{n_k}}{r})\}$ convergent to y in $(Y, \|\cdot\|')$ then subsequence $\{T(x_{n_k})\}$ is convergent to ry in $(Y, \|\cdot\|')$. We conclude that T is compact.

The converse is obvious. □

Compact mappings are automatically continuous.

15.4 **Theorem**. *A compact mapping* T *of a normed linear space* $(X, \|\cdot\|)$ *into a normed linear space* $(Y, \|\cdot\|')$ *is continuous.*

Proof. If T is not continuous then for each $n \in \mathbb{N}$ there exists an $x_n \in X$ such that $\| Tx_n \| > n \| x_n \|$. But then the sequence $\left\{\frac{x_n}{\|x_n\|}\right\}$ is bounded but the sequence $\left\{T\left(\frac{x_n}{\|x_n\|}\right)\right\}$ cannot have a convergent subsequence so T is not compact. □

15.5 **Remark**. Not all continuous linear mappings are compact. Consider the identity operator on any normed linear space and the closed unit ball of the space. If the identity operator is compact then the closed unit ball is compact and by Riesz Theorem 2.1.9, this implies that the space is finite dimensional. So the identity operator on any infinite dimensional normed linear space is continuous but not compact. □

The compactness property generalises the finite dimensional situation.

15.6 **Theorem**. *A linear mapping* T *of a finite dimensional normed linear space* $(X_n, \|\cdot\|)$ *into a normed linear space* $(Y, \|\cdot\|')$ *is compact.*

Proof. Since T is automatically continuous by Corollary 2.1.10, for a bounded subset B of $(X_n, \|\cdot\|)$, T(B) is bounded in $(Y, \|\cdot\|')$. But T(B) is a bounded subset of the finite dimensional subspace $(T(X_n), \|\cdot\|'_{T(X_n)})$. Therefore, $\overline{T(B)}$ is compact by Corollary 2.1.7. □

However, Theorem 15.6 serves to motivate the following idea.

15.7 **Definition**. A linear mapping T of a normed linear space $(X, \|\cdot\|)$ into a normed linear space $(Y, \|\cdot\|')$ is said to be of *finite rank* if T(X) is finite dimensional.

15.8 **Remarks**.

(i) Not all finite rank mappings are continuous. Consider a discontinuous linear functional f on an infinite dimensional normed linear space $(X, \|\cdot\|)$ and the associated one dimensional linear operator T on X defined for a given $z \in X$ by $Tx = f(x)\, z$. Then T is finite rank but not continuous.

(ii) However, a continuous finite rank mapping is always compact. The proof follows as in Theorem 15.6. □

It is interesting to see how the compactness criterion actually gives us information about the range of a compact mapping.

15.9 **Theorem**. *Consider a compact mapping* T *of a normed linear space* $(X, \|\cdot\|)$ *into a normed linear space* $(Y, \|\cdot\|')$.

(i) T(X) *is separable.*

(ii) *If* T(X) *is of second category in itself then* T *is of finite rank.*

Proof.

(i) For the closed unit ball B of $(X, \|\cdot\|)$, $T(X) = \bigcup_{n=1}^{\infty} nT(B)$.

But $\overline{T(B)}$ is compact so it is separable. Then $\bigcup_{n=1}^{\infty} n\overline{T(B)}$ is also separable and we conclude that T(X) is separable.

(ii) If also $(T(X), \|\cdot\|'_{T(X)})$ is second category then by Lemma 10.5, for the closed unit ball B of $(X, \|\cdot\|)$, $\overline{T(B)}$ is also a neighbourhood of the origin in $(T(X), \|\cdot\|'_{T(X)})$. Since $(T(X), \|\cdot\|'_{T(X)})$ has a compact neighbourhood of the origin we conclude from Riesz Theorem 2.1.9 that T(X) is finite dimensional.

One of the main applications of compact operators is in the theory of integral equations. It is with an eye to this setting that we consider an example of a compact operator on the Banach space $(\mathcal{C}[a,b], \|\cdot\|_\infty)$.

For a discussion of compactness in this space we need the following concept.

15.10 **Definition**. A subset A of $(\mathcal{C}[a,b], \|\cdot\|_\infty)$ is said to be *equicontinuous* on [a,b] if given $\varepsilon > 0$ and $x_0 \in [a,b]$ there exists a $\delta(\varepsilon, x_0) > 0$ such that for all $f \in A$,

$$|f(x) - f(x_0)| < \varepsilon \quad \text{for all } x \in [a,b] \text{ where } |x - x_0| < \delta.$$

We need to use the following characterisation of compactness in $(\mathcal{C}(X), \|\cdot\|_\infty)$, where (X, d) is a compact metric space, (see AMS §9).

15.11 **The Ascoli–Arzelà Theorem.**

Given a compact metric space (X, d), *a subset* A *of* $(\mathcal{C}(X) \|\cdot\|_\infty)$ *has* $\bar{A}$ *compact if and only if* A *is bounded and equicontinuous.*

15.12 **Example**. Consider the Fredholm operator K defined on $\mathcal{C}[a,b]$ by

$$(Kf)(x) = \int_a^b k(x,t)\, f(t)\, dt$$

with kernel k a continuous complex function on the square region $S \equiv \{(x,t) : a \le x \le b, a \le t \le b\}$. Now K is a continuous linear operator on $(\mathcal{C}[a,b], \|\cdot\|_\infty)$, (see AMS §7), but further K is a compact operator.

Proof. We use the characterisation given in Theorem 15.2.
Consider a bounded set B in $(\mathcal{C}[a,b], \|\cdot\|_\infty)$. Then there exists an $M > 0$ such that

$$\|f\|_\infty \le M \qquad \text{for all } f \in B.$$

We need to show that $\overline{K(B)}$ is compact. By the Ascoli–Arzelà Theorem 15.11 we need only show that K(B) is bounded and equicontinuous. Since K is a continuous linear operator it follows that K(B) is bounded.
Now k is uniformly continuous on the square region S. So given $\varepsilon > 0$ there exists a $\delta > 0$ such that

$$\left| k(x,t) - k(x_2,t) \right| < \varepsilon \quad \text{for all } t \in [a,b] \text{ and } x_1, x_2 \in [a,b] \text{ and } |x_1 - x_2| < \delta.$$

Therefore, for any $f \in B$

$$\begin{aligned} |(Kf)(x_1) - (Kf)(x_2)| &\le \int_a^b |k(x_1,t) - k(x_2,t)|\,|f(t)|\,dt \\ &\le (b-a)\,\|f\|_\infty \sup\{|k(x_1,t) - k(x_2,t)| : a \le t \le b\} \\ &< \varepsilon(b-a)M \qquad \text{for } |x_1 - x_2| < \delta \text{ and all } f \in B; \end{aligned}$$

that is, K(B) is equicontinuous. □

We now investigate the structure of the set of compact mappings among the continuous linear mappings.

15.13 **Notation.** Given normed linear spaces $(X, \|\cdot\|)$ and $(Y, \|\cdot\|')$, we denote by $\mathcal{K}(X,Y)$ the set of compact mappings of X into Y and for the case where Y = X, by $\mathcal{K}(X)$ the set of compact operators on X.

15.14 **Theorem.** *Given normed linear spaces* $(X, \|\cdot\|)$ *and* $(Y, \|\cdot\|')$, *then* $\mathcal{K}(X,Y)$ *is*

(i) *a linear subspace of* $\mathcal{B}(X,Y)$ *and*
(ii) *a closed linear subspace of* $\mathcal{B}(X,Y)$ *if* $(Y, \|\cdot\|')$ *is complete.*

Proof.
(i) Given compact mappings S and T consider $\{x_n\}$ a bounded sequence in $(X, \|\cdot\|)$. Then there exists a subsequence $\{x_{n_k}\}$ such that $\{Sx_{n_k}\}$ is convergent and a further subsequence $\{x_{n_{k_\ell}}\}$ such that $\{Tx_{n_{k_\ell}}\}$ is convergent.
Therefore $\{(\alpha S + \beta T)x_{n_{k_\ell}}\}$ is convergent for any scalars α and β. So $\mathcal{K}(X, Y)$ is a linear subspace of $\mathcal{B}(X, Y)$.
(ii) Consider a sequence $\{T_n\}$ of compact mappings convergent to a continuous linear mapping T in $(\mathcal{B}(X,Y), \|\cdot\|)$ and a bounded sequence $\{x_n\}$ in $(X, \|\cdot\|)$.
Since T_1 is compact there exists a subsequence $\{x_{n,1}\}$ of $\{x_n\}$ such that $\{T_1 x_{n,1}\}$ is convergent in $(Y, \|\cdot\|')$. Since T_2 is compact there exists a subsequence $\{x_{n,2}\}$ of $\{x_{n,1}\}$ such that $\{T_2 x_{n,2}\}$ is convergent in $(Y, \|\cdot\|')$. Continuing this process we consider the diagonal sequence

$$\{x_{n,n}\} = \{x_{1,1}, x_{2,2}, \dots, x_{n,n}, \dots\}.$$

This is a subsequence of each subsequence $\{x_{n,1}\}, \{x_{n,2}\}, \dots, \{x_{n,k}\}, \dots$ by construction.

Given $\varepsilon > 0$ there exists a $\nu \in \mathbb{N}$ such that

$$\| T_n - T \| < \varepsilon \quad \text{for all } n \geq \nu.$$

So $$\| Tx_{m,m} - Tx_{n,n} \| \leq \| Tx_{m,m} - T_\nu x_{m,m} \| + \| T_\nu x_{m,m} - T_\nu x_{n,n} \| + \| T_\nu x_{n,n} - Tx_{n,n} \|$$
$$\leq \| T - T_\nu \| \, \| x_{m,m} \| + \| T_\nu x_{m,m} - T_\nu x_{n,n} \| + \| T - T_\nu \| \, \| x_{n,n} \|.$$

But $\{T_\nu x_{n,n}\}$ is convergent so we conclude that $\{Tx_{n,n}\}$ is Cauchy in $(Y, \|\cdot\|')$. Since $(Y, \|\cdot\|')$, is complete then $\{Tx_{n,n}\}$ is convergent; that is, T is compact.

So we conclude that $\mathcal{K}(X,Y)$ is closed in $(\mathcal{B}(X,Y), \|\cdot\|)$. □

Theorem 15.14(ii) provides a useful method for proving that a linear operator is compact.

15.15 **Example**. Given Hilbert sequence space $(\ell_2, \|\cdot\|_2)$, the diagonal operator T defined for $x \equiv \{\lambda_1, \lambda_2, \dots, \lambda_n, \dots\}$ by $Tx = \{\alpha_1\lambda_1, \alpha_2\lambda_2, \dots, \alpha_n\lambda_n, \dots\}$ where $\alpha_n \to 0$ as $n \to \infty$, is compact.

Proof. For each $n \in \mathbb{N}$, consider the finite rank operator F_n where

$$F_n x = \{\alpha_1\lambda_1, \alpha_2\lambda_2, \dots, \alpha_n\lambda_n, 0, \dots\}.$$

Clearly for each $n \in \mathbb{N}$, F_n is linear and continuous. Since $\alpha_n \to 0$ as $n \to \infty$, given $\varepsilon > 0$ there exists a $\nu \in \mathbb{N}$ such that $|\alpha_n| < \varepsilon$ when $n > \nu$. Then

$$\| (T - F_n) x \|_2 = \sqrt{\sum_{k=n+1}^{\infty} |\alpha_k|^2 |\lambda_k|^2} < \varepsilon \sqrt{\sum_{k=n+1}^{\infty} |\lambda_k|^2}.$$

So $$\| T - F_n \| < \varepsilon \quad \text{for all } n > \nu;$$

that is, T is the limit of continuous finite rank operators F_n which are compact.
By Theorem 15.14(ii) we conclude that T is compact. □

The following structural property of $\mathcal{K}(X)$ is of interest.

15.16 **Theorem**. *Given a compact operator* T *and a continuous linear operator* S *on a normed linear space* $(X, \|\cdot\|)$, *then* ST *and* TS *are both compact operators on* $(X, \|\cdot\|)$.

Proof. Consider a bounded sequence $\{x_n\}$ in $(X, \|\cdot\|)$. Then since S is continuous the sequence $\{Sx_n\}$ is also bounded. Since T is compact the sequence $\{T(Sx_n)\}$ has a convergent subsequence; that is, TS is a compact operator.

But also the sequence $\{Tx_n\}$ has a convergent subsequence $\{Tx_{n_k}\}$ and since S is continuous the sequence $\{S(Tx_{n_k})\}$ is also convergent; that is, ST is a compact operator. □

This result has a consequence for the continuity of inverses.

15.17 **Corollary.** *Given a compact operator* T *on an infinite dimensional normed linear space* $(X, \|\cdot\|)$, *if* T^{-1} *exists on* $(X, \|\cdot\|)$ *then* T^{-1} *is not continuous.*

Proof. If T^{-1} is continuous on X then by Theorem 15.16 we have that $I = T^{-1}T$ is also compact. But from Remark 15.5 we see that I is not compact so T^{-1} is not continuous. □

We now show how the compactness property is inherited by conjugate mappings.

15.18 **Schauder's Theorem.**
A compact mapping T *of a normed linear space* $(X, \|\cdot\|)$ *into a normed linear space* $(Y, \|\cdot\|')$ *has compact conjugate mapping* T' *of* $(Y^*, \|\cdot\|')$ *into* $(X^*, \|\cdot\|)$.

Proof. Since T is compact, for the closed unit ball B in $(X, \|\cdot\|)$, $\overline{T(B)}$ is compact in $(Y, \|\cdot\|')$. Consider $\{f_n\}$ a sequence in the closed unit ball of $(Y^*, \|\cdot\|')$. In $(\mathfrak{C}(\overline{T(B)}), \|\cdot\|_\infty)$ the set $A \equiv \{f_n|_{\overline{T(B)}} : n \in \mathbb{N}\}$ is bounded and since

$$|f_n(y) - f_n(y_1)| \le \|y - y_1\|' \quad \text{for all } n \in \mathbb{N}$$

then A is also equicontinuous.
Therefore by the Ascoli–Arzelà Theorem 15.11, $\bar{A}$ is compact in $(\mathfrak{C}(\overline{T(B)}), \|\cdot\|_\infty)$.
So $\{f_n\}$ has a subsequence $\{f_{n_k}\}$ where
$\{f_{n_k}|_{\overline{T(B)}} : n \in \mathbb{N}\}$ is convergetn in $(\mathfrak{C}(\overline{T(B)}), \|\cdot\|_\infty)$. But

$$\|T' f_{n_k} - T' f_{n_j}\| = \sup\{|(f_{n_k} - f_{n_j})(Tx)| : x \in B\}.$$

So $\{T' f_{n_k}\}$ is a Cauchy sequence in $(X^*, \|\cdot\|)$. But $(X^*, \|\cdot\|)$ is complete so $\{T' f_{n_k}\}$ is convergent in $(X^*, \|\cdot\|)$. Then by Lemma 15.3, T' is compact. □

For Hilbert space, the following theorem reveals the special structure of compact operators as limits of finite rank operators.

15.19 **Theorem.** *Given a continuous linear operator* T *on a Hilbert space* H, *the following conditions are equivalent.*

(i) T *is compact.*

(ii) *For any orthonormal set* $\{e_\alpha\}$ *in* H *and* $\varepsilon > 0$ *the set*

$$\{\alpha : |(Te_\alpha, e_\alpha)| \ge \varepsilon\} \textit{ is finite.}$$

(iii) *There exists a sequence* $\{F_n\}$ *of continuous finite rank operators on* H *such that*

$$\|T - F_n\| \to 0 \textit{ as } n \to \infty.$$

Proof.

(i) $\Rightarrow$ (ii) For T compact suppose that (ii) does not hold. Then for some orthonormal set $\{e_\alpha\}$ and some $r > 0$, the set $\{\alpha : |(Te_\alpha, e_\alpha)| \geq r\}$ is infinite. So it has a countably infinite subset which gives rise to an orthonormal sequence $\{e_n\}$ such that

$$|(Te_n, e_n)| \geq r \qquad \text{for all } n \in \mathbb{N}.$$

Since T is compact there is a subsequence $\{e_{n_k}\}$ such that $\{Te_{n_k}\}$ converges to an $x \in H$. Discarding a finite number of terms of this sequence we may suppose that

$$\|Te_{n_k} - x\| < \tfrac{r}{2} \qquad \text{for all } k \in \mathbb{N}.$$

Then $\quad |(Te_{n_k}, e_{n_k}) - (x, e_{n_k})| = |(Te_{n_k} - x, e_{n_k})| \leq \|Te_{n_k} - x\| < \tfrac{r}{2}$

and so $$|(x, e_{n_k})| > \tfrac{r}{2} \qquad \text{for all } k \in \mathbb{N}.$$

But this contradicts Bessel's inequality, Theorem 3.13.

(ii) $\Rightarrow$ (iii) Given $n \in \mathbb{N}$, consider the family $\mathfrak{F}$ of all orthonormal sets $\{e_\alpha\}$ in H where $|(Te_\alpha, e_\alpha)| > \frac{1}{n}$ for all α.

By (ii) each such orthonormal set is finite. Consider $\mathfrak{F}$ partially ordered by set inclusion. Now the union of any totally ordered subfamily of orthonormal sets from $\mathfrak{F}$ is itself a member of $\mathfrak{F}$ and is an upper bound for this subfamily. It follows by Zorn's Lemma that $\mathfrak{F}$ has a maximal member $\{e_\beta\}$. As a member of $\mathfrak{F}$, $\{e_\beta\}$ is finite.
Consider the finite dimensional linear subspace $M \equiv \text{sp}\,\{e_\beta\}$. Then

$$|(Tx, x)| < \tfrac{1}{n} \qquad \text{for all } x \in M^\perp \text{ and } \|x\| = 1$$

for otherwise, $\mathfrak{F}$ contains $\{e_\beta\} \cup \{x\}$ which contradicts the maximality of $\{e_\beta\}$.
Consider the projection P_n of H onto M and write $x = (I-P_n)z$ where $z \in H$.

Then $$|(T(I-P_n)z, (I-P_n)z)| < \tfrac{1}{n};$$

that is, $$|((I-P_n)\,T(I-P_n)z, z)| < \tfrac{1}{n} \qquad \text{for all } z \in H \text{ and } \|z\| \leq 1.$$

Therefore, by Lemma 13.10.6(ii), $\|(I-P_n)T(I-P_n)\| < \frac{2}{n}$.

The continuous linear operator $F_n = P_nT + TP_n - P_nT\,P_n$ has finite rank and $\|T - F_n\| < \frac{2}{n}$.

(iii) $\Rightarrow$ (i) Since continuous finite rank operators are compact, the result follows from Theorem 15.14(ii). □

For separable Banach spaces with a Schauder basis we have a comparable result. But to establish this we need an important properties of Schauder bases.

15.20 **Lemma.** *Consider a separable Banach space* $(X, \|\cdot\|)$ *with a Schauder basis* $\{e_n\}$. *There exists a* $K \geq 1$ *such that for any* $x \equiv \sum_{k=1}^{\infty} \lambda_k e_k$,

$$\left\|\sum_{k=1}^{n} \lambda_k e_k\right\| \leq K \left\|\sum_{k=1}^{\infty} \lambda_k e_k\right\| \quad \textit{for all } n \in \mathbb{N}.$$

Proof. Consider the Banach space X with norm

$$\| x \|' = \sup \{ \| \sum_{k=1}^{n} \lambda_k e_k \| : n \in \mathbb{N} \}, \qquad \text{(see Exercise 1.26.17).}$$

Now $$\| x \| = \| \sum_{k=1}^{\infty} \lambda_k e_k \| \leq \sup \{ \| \sum_{k=1}^{n} \lambda_k e_k \| : n \in \mathbb{N} \} = \| x \|'$$

By Corollary 10.10 to the Open Mapping Theorem, norms $\|\cdot\|$ and $\|\cdot\|'$ are equivalent, so there exists a $K > 0$ such that

$$\sup \{ \| \sum_{k=1}^{n} \lambda_k e_k \| : n \in \mathbb{N} \} \leq K \| \sum_{k=1}^{\infty} \lambda_k e_k \| \text{ for all } x = \sum_{k=1}^{\infty} \lambda_k e_k.$$

Obviously, $K \geq 1$. □

15.21 **Lemma**. *For every compact subset* A *in a separable Banach space* $(X, \|\cdot\|)$ *with a Schauder basis* $\{e_n\}$,

$$\lim_{n \to \infty} \sup_{x \in A} \| x - P_n x \| = 0$$

where for $x \equiv \sum_{k=1}^{\infty} \lambda_k e_k$ *we define for each* $n \in \mathbb{N}$, *the projection operator* P_n *on* X *by*

$$P_n x = \sum_{k=1}^{n} \lambda_k e_k.$$

Proof. Since A is compact, given $\varepsilon > 0$ there exists an ε-net $\{y_1, y_2, \ldots, y_m\}$ in X such that

$$A \subseteq \bigcup_{k=1}^{n} B(y_k; \varepsilon)\,; \quad \text{(see AMS §8).}$$

For $x \in A$ there exists some y_{k_0} such that $\| x - y_{k_0} \| < \varepsilon$.

But also there exists a $\nu \in \mathbb{N}$ such that, for all $k \in \{1, 2, \ldots, m\}$

$$\| y_k - P_n y_k \| < \varepsilon \qquad \text{for all } n > \nu.$$

Therefore,

$$\begin{aligned} \| x - P_n x \| &\leq \| x - y_{k_0} \| + \| y_{k_0} - P_n y_{k_0} \| + \| P_n y_{k_0} - P_n x \| \\ &\leq \| x - y_{k_0} \| (1 + \| P_n \|) + \| y_{k_0} - P_n y_{k_0} \| \\ &< \varepsilon (1 + \| P_n \|) + \varepsilon \qquad \text{for all } n > \nu. \end{aligned}$$

From Lemma 15.20 we have that

$$\| P_n x \| = \| \sum_{k=1}^{n} \lambda_k e_k \| \leq K \| \sum_{k=1}^{\infty} \lambda_k e_k \| = K \| x \| \qquad \text{for all } x \in X,$$

so $$\| P_n \| \leq K \qquad \text{for all } n \in \mathbb{N}.$$

Then $$\sup_{x \in A} \| x - P_n x \| < (2+K)\varepsilon \qquad \text{for all } n > \nu.$$ □

15.22 **Theorem**. *Consider a continuous linear operator T on a separable Banach space* $(X, \|\cdot\|)$ *with a Schauder basis* $\{e_n\}$. *Then* T *is compact if and only if there exists a sequence* $\{F_n\}$ *of continuous finite rank operators on* X *such that* $\| T - F_n \| \to 0$ *as* $n \to \infty$.

Proof. For each $n \in \mathbb{N}$ consider the natural projection P_n on X defined for each $x = \sum_{k=1}^{\infty} \lambda_k e_k$ by $P_n(x) = \sum_{k=1}^{n} \lambda_k e_k$ where $\| P_n \| \leq K$.

Suppose that T is compact. Then for the closed unit ball B of $(X, \|\cdot\|)$, $\overline{T(B)}$ is compact. Therefore by Lemma 15.21, given $\varepsilon > 0$ there exists a $\nu \in \mathbb{N}$ such that

$$\| Tx - P_nTx \| \leq (2+K)\varepsilon \quad \text{for all } x \in B \text{ and } n > \nu.$$

So

$$\| T - P_nT \| < (2+K)\varepsilon \quad \text{for all } n > \nu.$$

But P_nT is a continuous finite rank operator on $(X, \|\cdot\|)$.

The converse follows again from Theorem 15.14(ii). □

15.23 **Remark**. Theorems 15.19 and 15.21 state that on a Hilbert space or on a separable Banach space with Schauder basis every compact operator can be represented as the limit of continuous finite rank operators. In 1932, Stefan Banach asked whether such a representation holds for compact linear operators on any Banach space. However, Per Enflo, *Acta Math.* **130** (1973), 309–317, constructed a separable reflexive Banach space having a compact operator which is not the limit of continuous finite rank operators. So it was deduced that the space constructed by Enflo does not have a Schauder basis. This provided a negative solution to the Basis Problem. □

15.24 EXERCISES

1. (i) Consider a compact mapping T of a Banach space $(X, \|\cdot\|)$ into a Banach space $(Y, \|\cdot\|')$. Prove that
   (a) T(X) is closed in $(Y, \|\cdot\|')$ if and only if T(X) is finite dimensional,
   (b) every closed linear subspace of T(X) is finite dimensional.

   (ii) Prove that on a Banach space $(X, \|\cdot\|)$ every compact projection operator P has P(X) finite dimensional.

   (iii) Prove that on a Hilbert space H every compact self-adjoint operator T has T(X) finite dimensional.

2. (i) Given a bounded sequence of complex numbers $\{\alpha_n\}$ prove that the diagonal operator T on $(m, \|\cdot\|_\infty)$ defined for $x \equiv \{\lambda_1, \lambda_2, \ldots, \lambda_n, \ldots\}$ by $Tx = \{\alpha_1\lambda_1, \alpha_2\lambda_2, \ldots, \alpha_n\lambda_n, \ldots\}$ is compact if and only if $\alpha_n \to 0$ as $n \to \infty$.

   (ii) Consider a continuous linear operator S on $(m, \|\cdot\|_\infty)$ and an operator P on $(\mathfrak{B}(m), \|\cdot\|_\infty)$ defined for $x \equiv \{\lambda_1, \lambda_2, \ldots, \lambda_n, \ldots\} \in m$ by

$$P(S)(x) = \{S(e_1)\,\lambda_1, S(e_2)\,\lambda_2, \ldots, S(e_n)\,\lambda_n, \ldots\}$$

   where $\{e_1, e_2, \ldots, e_n, \ldots\}$ is the usual basis for m.

(a) Prove that P is a continuous projection operator on $(\mathfrak{B}(m), \|\cdot\|_\infty)$.

(b) Prove that if S is a compact operator on $(m, \|\cdot\|_\infty)$ then P(S) is a compact operator on $(m, \|\cdot\|_\infty)$.

3. Consider continuous linear operators S and T on a Banach space $(X, \|\cdot\|)$. We know that ST = I does not necessarily imply that TS = I. However, if T is a compact operator prove that
   (i) $S(I - T) = I$ if and only if $(I - T)S = I$, and
   (ii) if S has inverse $I - T$ then $I - (I-T)^{-1}$ is compact.

4. Consider a compact mapping T of a normed linear space $(X, \|\cdot\|)$ into a normed linear space $(Y, \|\cdot\|')$. Prove that if a sequence $\{x_n\}$ in X has the property that $\{f(x_n)\} \to 0$ as $n \to \infty$ for all $f \in X^*$ then $\| Tx_n \| \to 0$ as $n \to \infty$.

5. (i) Consider a compact mapping T of an infinite dimensional Banach space $(X, \|\cdot\|)$ into a Banach space $(Y, \|\cdot\|')$. Prove that $\overline{\{Tx : \| x \| = 1\}}$ contains 0 in Y.
   (ii) Consider a separable Hilbert space H with an orthonormal basis $\{e_n\}$.
      (a) Prove that if T is a compact mapping of H into a normed linear space then $Te_n \to 0$ as $n \to \infty$.
      (b) Prove that if T is a continuous linear mapping of H into a Banach space where $\sum \| Te_n \|^2 < \infty$ then T is compact.

6. Prove the converse of Schauder's Theorem 15.18
   A continuous linear mapping T *from a Banach space* $(X, \|\cdot\|)$ *into a Banach space* $(Y, \|\cdot\|')$ *is compact if its conjugate mapping* T' *from* $(Y^*, \|\cdot\|')$ *into* $(X^*, \|\cdot\|')$ *is compact.*

7. Consider continuous linear operators T and S on a Banach space $(X, \|\cdot\|)$ where T has a continuous inverse on X and $T - S$ is compact. Prove that
   (i) S(X) is closed and has finite codimension,
   (ii) either S has a continuous inverse on X or is not one-to-one.

8. Consider a continuous linear operator T on a complex Hilbert space H.
   (i) Prove that T is compact if and only if T*T is compact.
   (ii) Suppose that $\lim_{n\to\infty} \| T^n \|^{1/n} = 0$. Prove that T is compact if and only if T T* is compact.
   (Hint: Use Theorem 15.19(ii).)

9. (i) Consider an incomplete normed linear space $(X, \|\cdot\|)$ and a compact operator T on $(X, \|\cdot\|)$. Prove that the unique continuous linear extension $\tilde{T}$ on the completion $(\tilde{X}, \|\cdot\|)$ is also compact.

(ii) Consider the Fredholm operator K defined on $\mathfrak{C}[0, 1]$ by

$$(Kf)(x) = \int_0^1 k(x,t)\, f(t)\, dt$$

with kernel k a continuous complex function on the square region $S \equiv \{(x, t) : 0 \le x \le 1, 0 \le t \le 1\}$. Prove that

(a) K is a compact operator on $(\mathfrak{C}[0, 1], \|\cdot\|_2)$,

(b) $\tilde{K}$ is a compact operator on the Hilbert space $(\mathfrak{L}[0, 1], \|\cdot\|_2)$.

10. Consider a compact operator T on a Banach space $(X, \|\cdot\|)$ and M a closed invariant subspace of T. Prove that the mapping $x + M \mapsto Tx + M$ is a compact operator on the quotient space $(X/M, \|\cdot\|)$.

VI. SPECTRAL THEORY

In Section 4 we noted that, given a normed linear space the algebra of continuous linear operators on the space is a noncommutative unital normed algebra and as such, interest lies in studying the set of regular elements in this algebra.

Spectral theory develops this line of investigation by determining, for a continuous linear operator T on a complex normed linear space X, properties of the set of scalars

$$\{\lambda \in \mathbb{C} : T - \lambda I \text{ is singular in } \mathfrak{B}(X)\}.$$

This study enables us to decompose certain classes of operators on Hilbert space.

The decomposition theorems are called Spectral Theorems and we develop such a theorem first for compact normal operators and then extend it for compact operators.

§16. THE SPECTRUM

As many of the properties of an operator algebra can be derived directly from its being a unital normed algebra, it is more straightforward to begin by developing our theory for elements of a unital normed algebra.

The spectrum of an element of a unital normed algebra is a useful representation in the complex plane of the singular elements associated with any given element of the algebra.

16.1 **Definitions**. Given an element x in a complex unital normed algebra $(A, \|\cdot\|)$, the *spectrum* of x is the set of complex numbers

$$\sigma(x) \equiv \{\lambda \in \mathbb{C} : x - \lambda e \text{ is singular}\},$$

and the *resolvent set* of x is the complement of $\sigma(x)$

$$\rho(x) \equiv \{\lambda \in \mathbb{C} : x - \lambda e \text{ is regular}\}.$$

For any element of a complex unital normed algebra it is important to locate its spectrum as a subset of the complex plane and the following properties give us our first general information in this regard.

To avoid unnecessary complications we will assume completeness; that is, mostly we will work in a complex unital Banach algebra, and in particular we will consider the algebra of operators on a complex Banach space.

16.2 **Theorem**. *For any element* x *in a complex unital Banach algebra* $(A, \|\cdot\|)$, $\sigma(x)$ *is compact.*

Proof. Consider the mapping ϕ from $\mathbb{C}$ into A defined by $\phi(\lambda) = x - \lambda e$.
Now $|\phi(\lambda) - \phi(\mu)| = |\lambda - \mu|$ so ϕ is continuous.
From Corollary 4.11.17 we see that the set of singular elements is closed. As the spectrum $\sigma(x)$ is the inverse image under ϕ of the singular elements in A, we deduce that $\sigma(x)$ is closed.
When $|\lambda| > \|x\|$ then $\|x/\lambda\| < 1$ and from Corollary 4.11.11 we deduce that $e - x/\lambda$ is regular. Then $x - \lambda e$ is regular so we conclude that $\sigma(x) \subseteq \{\lambda \in \mathbb{C} : |\lambda| \leq \|x\|\}$. Then as a bounded closed subset of the complex plane, $\sigma(x)$ is compact. □

Because of the spectral property given in Theorem 16.2, it is useful to define the following constant.

16.3 **Definition**. Given an element x in a complex unital Banach algebra $(A, \|\cdot\|)$, the *spectral radius* of x is defined as

$$v(x) \equiv \sup\{|\lambda| : \lambda \in \sigma(x)\}.$$

16.4 **Remark**. The spectral radius is the radius of the smallest closed disc centred at 0 and containing the spectrum. From Theorem 16.2 we have $\sigma(x) \leq \{\lambda : |\lambda| \leq \|x\|\}$
and so

$$v(x) \leq \|x\| \quad \text{for all } x \in A. \qquad \square$$

We now establish deeper properties for the spectrum. We show that in a complex unital Banach algebra the spectrum is nonempty and we have a formula for the spectral radius in terms of the norm although the spectrum is a purely algebraic notion. To do so we need to use complex analysis of analytic functions.

16.5 **Definition**. Given an element x in a complex unital normed algebra $(A, \|\cdot\|)$, the *resolvent operator* R maps $\rho(x)$ into A and is defined by

$$R(\lambda) = (x - \lambda e)^{-1}.$$

16.6 **Lemma**. *Given an element* x *in a complex unital normed algebra* $(A, \|\cdot\|)$,
(i) $R(\lambda) - R(\mu) = (\lambda - \mu)\, R(\lambda)\, R(\mu)$ *for all* $\lambda, \mu \in \rho(x)$ *and* $\|R(\lambda)\| \to 0$ as $\lambda \to \infty$, *and*
(ii) *for any* $f \in A^*$, $f \circ R$ *is analytic on* $\rho(x)$.
(iii) *When* A *is complete, for all* $|\lambda| > v(x)$,

$$f \circ R(\lambda) = -\frac{1}{\lambda}\left(f(e) + \sum_{n=1}^{\infty} \frac{f(x^n)}{\lambda^n}\right).$$

Proof.

(i) Since $x^{-1} - y^{-1} = x^{-1}(y - x)y^{-1}$ for all regular elements $x, y \in A$, we have

$$R(\lambda) - R(\mu) = (\lambda-\mu) R(\lambda) R(\mu)$$

so R is continuous on $\rho(x)$.

But also from Theorem 4.11.18,

$$(e - x/\lambda)^{-1} \to e \text{ as } \lambda \to \infty$$

so
$$\| R(\lambda) \| = \| (x-\lambda e)^{-1} \| = \frac{1}{|\lambda|} \| (e - x/\lambda)^{-1} \| \to 0 \text{ as } \lambda \to \infty.$$

(ii) For any $f \in A^*$ and $\lambda, \mu \in \rho(x)$, $\lambda \neq \mu$

$$\frac{f \circ R(\lambda) - f \circ R(\mu)}{\lambda-\mu} = f(R(\lambda) R(\mu))$$

and so
$$\lim_{\lambda\to\mu} \frac{f \circ R(\lambda) - f \circ R(\mu)}{\lambda-\mu} = f(R^2(\mu))$$

and then $f \circ R$ is analytic on $\rho(x)$.

(iii) For $|\lambda| > \| x \|$ we have from Theorem 4.11.10 that

$$R(\lambda) \equiv (x-\lambda e)^{-1} = -\frac{1}{\lambda}\left(e + \sum_{n=1}^{\infty} \frac{x^n}{\lambda^n}\right).$$

For any $f \in A^*$ we have

$$f \circ R(\lambda) = -\frac{1}{\lambda}\left(f(e) + \sum_{n=1}^{\infty} \frac{f(x^n)}{\lambda^n}\right).$$

But $f \circ R$ is analytic on $\rho(x) \supseteq \{\lambda \in \mathbb{C} : |\lambda| > \nu(x)\}$ so this is the Laurent series expansion for $f \circ R$ on $\{\lambda \in \mathbb{C} : |\lambda| > \nu(x)\}$. □

Lemma 16.6 enables us to establish the following spectral properties.

16.7 **Theorem**. *Given an element* x *in a complex unital normed algebra* $(A, \|\cdot\|)$,

(i) $\sigma(x) \neq \varnothing$ *and*

(ii) *when* A *is complete*

$$\nu(x) = \lim_{n\to\infty} \| x^n \|^{1/n} = \inf\{\| x^n \|^{1/n} : n \in \mathbb{N}\}.$$

Proof.

(i) Suppose that $\sigma(x) = \varnothing$, then $\rho(x) = \mathbb{C}$ and for any $f \in A^*$, $f \circ R$ is an entire function. But also since $f \in A^*$, $|f \circ R(\lambda)| \leq \| f \| \| R(\lambda) \|$ so from Lemma 16.6(i),

$$f \circ R(\lambda) \to 0 \text{ as } \lambda \to \infty$$

By Liouville's Theorem we deduce that $f \circ R(\lambda) = 0$ for all $\lambda \in \mathbb{C}$.

But this holds for any $f \in A^*$ and since, by Remark 6.6, A^* is total on A, we conclude that $R(\lambda) = 0$ for all $\lambda \in \mathbb{C}$. That is, the zero element is the inverse of $x - \lambda e$ for all $\lambda \in \mathbb{C}$. But no element has zero as an inverse so we conclude that $\sigma(x) \neq \varnothing$.

(ii) From Lemma 16.6(iii) we have that for any $f \in A^*$ and any real $\alpha > \nu(x)$, the series $\sum_{n=1}^{\infty} \frac{f(x^n)}{\alpha^n}$ is convergent and so the sequence $\left\{\frac{f(x^n)}{\alpha^n}\right\}$ is bounded. But this holds for all $f \in A^*$, so by the Theorem 11.12(i), we conclude that there exists a $K > 0$ such that

$$\left\| \frac{x^n}{\alpha^n} \right\| < K \quad \text{for all } n \in \mathbb{N}.$$

Then $\| x^n \|^{1/n} < \alpha K^{1/n}$ for all $n \in \mathbb{N}$ and so $\limsup_{n\to\infty} \| x^n \|^{1/n} < \alpha$. We deduce that

$$\limsup_{n\to\infty} \| x^n \|^{1/n} \le \nu(x).$$

Now for $\nu(x) > 0$ there exists a $\lambda \in \mathbb{C}$ such that $|\lambda| < \nu(x)$ where $x - \lambda e$ is singular. If also

$$\limsup_{n\to\infty} \| x^n \|^{1/n} < |\lambda|$$

then by Theorem 4.11.10, $x - \lambda e$ is regular. So we conclude that

$$\nu(x) = \limsup_{n\to\infty} \| x^n \|^{1/n}.$$

From Proposition 4.11.12 we have

$$\nu(x) = \lim_{n\to\infty} \| x^n \|^{1/n} = \inf\{\| x^n \|^{1/n} : n \in \mathbb{N}\}. \qquad \square$$

16.8 **Remark.** We note that we invoked the completeness condition when we used series arguments like those of Theorem 4.11.10 and Corollary 4.11.11. Completeness guarantees the boundedness of the spectrum and a meaningful definition of spectral radius. $\square$

We now present a useful algebraic property of the spectrum.

16.9 **The Spectral Mapping Theorem for polynomials.**
Consider a complex unital normed algebra $(A, \|\cdot\|)$ *and a polynomial* p *with complex coefficients. For every* $x \in A$,

$$\sigma(p(x)) = p(\sigma(x)) \equiv \{p(\lambda) : \lambda \in \sigma(x)\}.$$

Proof. Clearly $p(x) \in A$. For any $\lambda \in \mathbb{C}$ consider the polynomial $p(t) - \lambda$. Now if p is of degree n then by the Fundamental Theorem of Algebra, $p(t) - \lambda$ has n roots, $\lambda_1, \lambda_2, \ldots, \lambda_n$ and we can write

$$p(x) - \lambda e = \alpha \prod_{i=1}^{n} (x - \lambda_i e) \quad \text{for some } \alpha \in \mathbb{C}.$$

If $\lambda \in \sigma(p(x))$ then there exists some $i_0 \in \{1, 2, \ldots, n\}$ such that $x - \lambda_{i_0} e$ is singular, that is $\lambda_{i_0} \in \sigma(x)$. Then $\lambda = p(\lambda_{i_0}) \in p(\sigma(x))$ and so $\sigma(p(x)) \subseteq p(\sigma(x))$.

Conversely, if $\lambda \in p(\sigma(x))$ then there exists $i_0 \in \{1, 2, \ldots, n\}$ such that $\lambda_{i_0} \in \sigma(x)$ and $\lambda = p(\lambda_{i_0})$. However, if $\lambda \notin \sigma(p(x))$ then every factor $x - \lambda_i e$, of $p(x) - \lambda e$ is regular so $\lambda_i \notin \sigma(x)$ for all $i \in \{1, 2, \ldots, n\}$. So we conclude that $p(\sigma(x)) \subseteq \sigma(p(x))$. $\square$

16.10 EXERCISES

1. An element z of a complex unital Banach algebra $(A, \|\cdot\|)$ is called a *topological divisor of zero* if there exists a sequence $\{z_n\}$ in A where $\| z_n \| = 1$ for all $n \in \mathbb{N}$ and either $z_n z \to 0$ or $z z_n \to 0$ as $n \to \infty$. Prove that
   (i) every divisor of zero is a topological divisor of zero,
   (ii) the set of topological divisors of zero is closed in A,
   (iii) every topological divisor of zero is a singular element,
   (iv) the boundary of the set of singular elements is contained in the set of topological divisors of zero.

2. Consider a complex unital Banach algebra A and a complex unital Banach subalgebra A'. Prove that for any element $x \in A'$,
   (i) $\sigma_{A'}(x) \subseteq \sigma_A(x)$ and
   (ii) $\partial\sigma_A(x) \subseteq \partial\sigma_{A'}(x)$.

3. Consider a complex unital Banach algebra $(A, \|\cdot\|)$. Prove that each of the following conditions implies that $(A, \|\cdot\|)$ is isometrically isomorphic to $\mathbb{C}$.
   (i) A is a division algebra; that is, every nonzero element of A has an inverse.
   (ii) Zero is the only topological divisor of zero.
   (iii) There exists an $M > 0$ such that $\| xy \| \geq M \| x \| \| y \|$ for all $x, y \in A$.
   (iv) For every regular element $x \in A$, $\| x^{-1} \| = \dfrac{1}{\| x \|}$.

4. Consider a complex unital normed algebra A. Prove that
   (i) if $x \in A$ is regular then $x^{-1} - e/\lambda = (x^{-1}/\lambda)(\lambda e - x)$ and deduce that $\sigma(x^{-1}) = \{ 1/\lambda : \lambda \in \sigma(x) \}$,
   (ii) if we are given that $e - xy$ is regular for $x, y \in A$ then $e - yx$ is regular where $(e-yx)^{-1} = e + y(e-xy)^{-1} x$ and deduce that for any $x, y \in A$
   $$\sigma(xy) \setminus \{0\} = \sigma(yx) \setminus \{0\}.$$

5. Consider the complex unital commutative Banach algebra $(\mathfrak{C}[0,1], \|\cdot\|_\infty)$ and $f \in \mathfrak{C}[0,1]$. Prove that
   (i) f is either regular or is a topological divisor of zero,
   (ii) $\lambda \in \sigma(f)$ if and only if $\overline{\lambda} \in \sigma(\overline{f})$,
   (iii) $f = \overline{f}$ if and only if $\sigma(f) \subseteq \mathbb{R}$.

6. Consider a complex unital Banach algebra A. Prove that for any given $x \in A$ the spectral radius $\nu(x)$ has the following properties
   (i) $\nu(\lambda x) = |\lambda| \nu(x)$,
   (ii) $\nu(xy) = \nu(yx)$,
   (iii) if $xy = yx$ then $\nu(xy) \leq \nu(x)\,\nu(y)$ and $\nu(x+y) \leq \nu(x) + \nu(y)$.

§17. THE SPECTRUM OF A CONTINUOUS LINEAR OPERATOR

We now consider the particular case of the algebra of continuous linear operators on a normed linear space. Although Section 16 gives us general structural properties of these as elements of the unital normed algebra of operators, here it is important to determine whether a particular continuous linear operator has an inverse on the space and whether the inverse is also continuous.

17.1 The spectrum in linear spaces

We begin with purely algebraic considerations in $\mathcal{L}(X)$, the algebra of linear operators on a linear space X. It is clear that a linear operator T on a linear space X has an inverse T^{-1} on X if and only if T is one-to-one and onto. So we have the following definitions.

17.1.1 **Definitions.** Given a linear operator T on a linear space X, an *eigenvalue* of T is a scalar λ where there exists an $x \neq 0$ such that $Tx = \lambda x$. Given an eigenvalue λ of T, the elements $x \in X$ such that $Tx = \lambda x$ are called *eigenvectors* of T and the linear space $\{x \in X : Tx = \lambda x\}$ is called the *eigenspace* associated with the eigenvalue λ. The set of eigenvalues of T is called the *point spectrum* of T and is denoted by $P\sigma(T)$. Clearly $\lambda \in P\sigma(T)$ if and only if $T - \lambda I$ is not one-to-one.

There is a fundamental relation between the eigenvalues and their associated eigenspaces.

17.1.2 **Theorem.** *Given a linear operator* T *on a linear space* X *with distinct eigenvalues* $\lambda_1, \lambda_2, \ldots, \lambda_n$, *any set* $\{x_1, x_2, \ldots, x_n\}$ *of corresponding eigenvectors is linearly independent.*

Proof. We use proof by induction.
Now $\{x_1\}$ is linearly independent. Suppose that $\{x_1, x_2, \ldots, x_k\}$ is linearly independent and consider $\alpha_1 x_1 + \alpha_2 x_2 + \ldots + \alpha_k x_k + \alpha_{k+1} x_{k+1} = 0$.
Then
$$T(\alpha_1 x_1 + \alpha_2 x_2 + \ldots + \alpha_{k+1} x_{k+1}) = \alpha_1 Tx_1 + \alpha_2 Tx_2 + \ldots + \alpha_{k+1} Tx_{k+1}$$
$$= \alpha_1 \lambda_1 x_1 + \alpha_2 \lambda_2 x_2 + \ldots + \alpha_{k+1} \lambda_{k+1} x_{k+1} = 0.$$
Then
$$\alpha_1(\lambda_1 - \lambda_{k+1})x_1 + \alpha_2(\lambda_2 - \lambda_{k+1})x_2 + \ldots + \alpha_k(\lambda_k - \lambda_{k+1})x_k = 0$$
and since $\{x_1, x_2, \ldots, x_k\}$ is linearly independent
$$\alpha_1(\lambda_1 - \lambda_{k+1}) = \alpha_2(\lambda_2 - \lambda_{k+1}) = \ldots = \alpha_k(\lambda_k - \lambda_{k+1}) = 0.$$
Since the eigenvalues are distinct, we have $\alpha_1 = \alpha_2 = \ldots = \alpha_k = 0$.
Since $x_{k+1} \neq 0$ we conclude that $\alpha_1 = \alpha_2 = \ldots = \alpha_k = \alpha_{k+1} = 0$
and so $\{x_1, x_2, \ldots, x_{k+1}\}$ is linearly independent. □

The following is an elementary consequence.

17.1.3 **Corollary**. *Given a linear operator* T *on a linear space* X, *for eigenspaces* M_1 *and* M_2 *associated with eigenvalues* λ_1 *and* λ_2, $M_1 \cap M_2 = \{0\}$.

So we have some insight into the structure of the point spectrum of a linear operator on a finite dimensional linear space.

17.1.4 **Corollary**. *For a linear operator* T *on an* n*-dimensional linear space* X_n, *the point spectrum* $P\sigma(T)$ *contains no more than* n *elements.*

Proof. X_n contains the direct sum of its eigenspaces so T cannot have more than n eigenvalues. □

For finite dimensional linear spaces, the failure of a linear operator to have an inverse can be reduced to its failing to be one-to-one.

17.1.5 **Theorem**. *A linear operator* T *on an* n*-dimensional linear space* X_n *is one-to-one if and only if it is onto.*

Proof. Consider $\{e_1, e_2, \dots, e_n\}$ a basis for X_n.
Suppose that T is one-to-one and consider the set $\{Te_1, Te_2, \dots, Te_n\}$. If for scalars $\{\alpha_1, \alpha_2, \dots, \alpha_n\}$ we have $\sum_{k=1}^{n} \alpha_k Te_k = 0$ then $T\left(\sum_{k=1}^{n} \alpha_k e_k\right) = 0$ so $\sum_{k=1}^{n} \alpha_k e_k = 0$ since T is one-to-one. But $\{e_1, e_2, \dots, e_n\}$ is a basis for X_n so $\alpha_1 = \alpha_2 = \dots = \alpha_n = 0$. Then $\{Te_1, Te_2, \dots, Te_n\}$ is linearly independent and so is a basis for X_n. We conclude that T is onto.

Conversely, if T is onto then for each $e_k \in X_n$ where $k \in \{1, 2, \dots, n\}$ there exists an $x_k \in X_n$ such that $e_k = Tx_k$. Then for such a set $\{x_1, x_2, \dots, x_n\}$, if $\sum_{k=1}^{n} \alpha_k x_k = 0$ then $\sum_{k=1}^{n} \alpha_k Tx_k = 0$ and $\sum_{k=1}^{n} \alpha_k e_k = 0$ so $\alpha_1 = \alpha_2 = \dots = \alpha_n = 0$ since $\{e_1, e_2, \dots, e_n\}$ is a basis for X_n. Then $\{x_1, x_2, \dots, x_n\}$ is linearly independent and so is a basis for X_n.
If $Tx = 0$ then $x = \sum_{k=1}^{n} \beta_k x_k$ for scalars β_k where $k \in \{1, 2, \dots, n\}$. So

$$0 = \sum_{k=1}^{n} \beta_k Tx_k = \sum_{k=1}^{n} \beta_k e_k$$

and then $\beta_1 = \beta_2 = \dots = \beta_n = 0$ since $\{e_1, e_2, \dots, e_n\}$ is a basis for X_n. Therefore $x = 0$ and we conclude that T is one-to-one. □

17.2 The spectrum in normed linear spaces

Given a continuous linear operator T on a normed linear space $(X, \|\cdot\|)$, T is a regular element of $\mathfrak{B}(X)$ if and only if T^{-1} exists and is continuous on X. When X is finite dimensional we have seen in Theorem 17.1.5 that T^{-1} exists on X if and only if T is one-to-one and the continuity of T and T^{-1} is automatic by Corollary 2.1.10. When X is infinite dimensional the situation is a little more complicated.

In general, a continuous linear operator T on a Banach space $(X, \|\cdot\|)$ is not invertible in $\mathfrak{B}(X)$ if and only if

(i) T is not one-to-one, or

(ii) T is one-to-one but T(X) is not dense in $(X, \|\cdot\|)$, or

(iii) T is one-to-one and T(X) is dense in $(X, \|\cdot\|)$ but T^{-1} is not continuous on T(X).

So we are led to the following decomposition of the spectrum of T.

17.2.1 **Definitions.** The spectrum of a continuous linear operator T on a complex Banach space $(X, \|\cdot\|)$,

$$\sigma(T) \equiv \{\lambda \in \mathbb{C} : \lambda I - T \text{ is singular in } \mathfrak{B}(X)\}$$

can be separated into three disjoint component sets:

the *point spectrum* consists of the eigenvalues of T

$$P\sigma(T) \equiv \{\lambda \in \mathbb{C} : \lambda I - T \text{ is not one-to-one}\},$$

the *residual spectrum* is the set

$$R\sigma(T) \equiv \{\lambda \in \mathbb{C} : \lambda I - T \text{ is one-to-one but } (\lambda I - T)(X) \text{ is not dense}\}$$

and the *continuous spectrum* is the set

$$C\sigma(T) \equiv \{\lambda \in \mathbb{C} : \lambda I - T \text{ is one-to-one and } (\lambda I - T)(X) \text{ is dense but } (\lambda I - T)^{-1} \text{ is not continuous on } (\lambda I - T)(X)\}.$$

So $$\sigma(T) = P\sigma(T) \cup R\sigma(T) \cup C\sigma(T).$$

17.2.2 **Remarks.**

(i) Theorem 17.1.5 shows that for a continuous linear operator T on a complex finite dimensional normed linear space, $\sigma(T) = P\sigma(T)$.

(ii) We recall from Corollary 1.24.10 that for a continuous linear operator T on a Banach space $(X, \|\cdot\|)$, if $\lambda I - T$ is one-to-one and $(\lambda I - T)(X)$ is dense and $(\lambda I - T)^{-1}$ is continuous on $(\lambda I - T)(X)$ then $\lambda I - T$ is onto and so $\lambda I - T$ is regular in $\mathfrak{B}(X)$. □

17.3 The spectrum and numerical range

It is important to be able to determine the spectrum of a continuous linear operator T on a complex Banach space $(X, \|\cdot\|)$, but it is not always a simple set to specify. From Section 16 we saw that $\sigma(T)$ is a compact set contained in the closed disc of radius $\|T\|$.

In Hilbert space H the closure of the numerical range of T is a smaller set contained in the closed disc of radius $\|T\|$ which also contains the spectrum. In Theorem 13.10.9 we showed that for any $\lambda \notin \overline{W(T)}$, $T - \lambda I$ is regular in $\mathfrak{B}(H)$. This result has the following expression in terms of the spectrum of T.

17.3.1 **Theorem.** *For a continuous linear operator* T *on a complex Hilbert space,*

$$\sigma(T) \subseteq \overline{W(T)}.$$

Theorem 13.10.8 then has an immediate implication for the spectrum of self-adjoint operators.

17.3.2 **Corollary.** *For a self-adjoint operator* T *on a complex Hilbert space,*

$$\sigma(T) \subseteq [-\| T \|, \| T \|].$$

For the spectrum of positive operators we have the following result.

17.3.3 **Corollary.** *For a positive operator* T *on a complex Hilbert space,*

$$\sigma(T) \subseteq [0, \| T \|].$$

For the spectrum of normal operators, Theorems 13.11.4 and 16.2 have the following implication.

17.3.4 **Corollary.** *For a normal operator* T *on a complex Hilbert space,*

$$v(T) = w(T) = \| T \|$$

and there exists a $\lambda \in \sigma(T)$ *such that* $|\lambda| = \| T \|$.

For the spectrum of unitary operators, numerical range considerations do not give us further insight. Theorem 13.11.9 has the following expression in terms of the spectrum.

17.3.5 **Corollary.** *For a unitary operator* T *on a complex Hilbert space*

$$\sigma(T) \subseteq \{\lambda \in \mathbb{C} : |\lambda| = 1\}.$$

In the 1960s there was a successful generalisation of the concept of numerical range of a continuous linear operator on a normed linear space. This development was largely due to F.F. Bonsall and an exposition of the basic theory can be found in F.F. Bonsall and J. Duncan, *Numerical Ranges I and II*, Cambridge University Press (1971) and (1973).

17.3.6 **Definition.** Given a complex normed linear space $(X, \|\cdot\|)$, for each $x \in X$, $\| x \| = 1$, consider the set $D(x) \equiv \{f \in X^* : \| f \| = 1, f(x) = 1\}$. Corollary 6.3 to the Hahn–Banach Theorem guarantees that for each $x \in X$, $\| x \| = 1$, we have that $D(x)$ is nonempty. For a continuous linear operator T on X we call the set of complex numbers

$$V(T) \equiv \{f(Tx) : x \in X, \ \| x \| = 1, f \in D(x)\}$$

the *numerical range* of T.

Clearly, when X is an inner product space, $V(T) = W(T)$.

The result which gave the stamp of success to this work is the generalisation of Theorem 17.3.1.

17.3.7. **Theorem.** *For a continuous linear operator* T *on a complex Banach space* $(X, \|\cdot\|)$,

$$\sigma(T) \subseteq \overline{V(T)}.$$

Proof. For $\lambda \notin \overline{V(T)}$ we write $d \equiv d(\lambda, \overline{V(T)}) > 0$. For $x \in X$, $\| x \| = 1$ and $f \in D(x)$,

$$\| (T-\lambda I)x \| \geq | f((T-\lambda I)x) | \geq d$$

so

$$\| (T-\lambda I)x \| \geq d \| x \| \quad \text{for all } x \in X$$

which implies that $T-\lambda I$ is a topological isomorphism. Now $(T-\lambda I)(X)$ is a closed linear subspace X. Suppose that $(T-\lambda I)(X)$ is a proper subspace. Then by Corollary 8.7 there exists an $x_0 \notin (T-\lambda I)(X)$, $\| x_0 \| = 1$ and an $f_0 \in D(x_0)$ such that

$$| f_0(x) | \leq \frac{d}{2\|T-\lambda I\|} \quad \text{for all } x \in (T-\lambda I)(X) \text{ and } \| x \| \leq 1.$$

But then

$$\frac{d}{2} \geq | f_0((T-\lambda I)x_0) | = | f_0(Tx_0) - \lambda | \geq d(\lambda, V(T)) \geq d,$$

a contradiction. So we conclude that $T-\lambda I$ is onto and $\lambda \notin \sigma(T)$. □

17.4 **The spectrum of a conjugate operator**

In Theorem 12.11 we studied the relation between the regularity of a continuous linear operator on a normed linear space and its conjugate on the dual space. This result has the following implications for the spectrum.

17.4.1 **Theorem.** *For a continuous linear operator* T *on a complex normed linear space* $(X, \|\cdot\|)$ *with conjugate operator* T' *on* $(X^*, \|\cdot\|)$,

(i) $\sigma(T') \subseteq \sigma(T)$ *and*

(ii) *if* $(X, \|\cdot\|)$ *is complete then* $\sigma(T') = \sigma(T)$.

We are particularly interested in the relation on Hilbert space where for λ scalar, $T-\lambda I$ is regular if and only if $T^* - \overline{\lambda} I$ is regular.

17.4.2 **Corollary**. *For a continuous linear operator* T *on a complex Hilbert space* H *with adjoint* T* *on* H, $\quad \sigma(T^*) = \{\overline{\lambda} : \lambda \in \sigma(T)\}$.

The point spectrum of a normal operator on Hilbert space has particularly satisfying properties which suggest a decomposition of the space.

17.4.3 **Theorem.** *Consider a normal operator* T *on a complex Hilbert space* H.

(i) λ *is an eigenvalue of* T *if and only if* $\overline{\lambda}$ *is an eigenvalue of* T*, *and* λ *and* $\overline{\lambda}$ *have the same eigenspace.*

(ii) *If* λ *and* μ *are distinct eigenvalues for* T *then the corresponding eigenspaces* M_λ *and* M_μ *are orthogonal.*

Proof.

(i) Since $T-\lambda I$ is also normal, by Theorem 13.11.3 we have that

$$\| (T-\lambda I)x \| = \| (T^*-\overline{\lambda} I)x \| \quad \text{for all } x \in H,$$

and our result follows.

(ii) For $x \in M_\lambda$ and $y \in M_\mu$ we have

$$\lambda(y, x) = (y, \bar{\lambda}x) = (y, T^*x) \quad \text{by (i)}$$
$$= (Ty, x) = (\mu y, x) = \mu(y, x).$$

Since $\lambda \neq \mu$ then $(y, x) = 0$. □

17.5 EXERCISES

1. (i) Given a continuous linear operator T on a complex Banach space with conjugate operator T', prove that
   (a) $R\sigma(T) \subseteq P\sigma(T')$,
   (b) $P\sigma(T) \subseteq P\sigma(T') \cup R\sigma(T')$,
   (c) $C\sigma(T') \subseteq C\sigma(T)$.

   (ii) Given a continuous linear operator T on a complex Hilbert space with adjoint T*, prove that
   (a) $\lambda \in R\sigma(T)$ if and only if $\bar{\lambda} \in P\sigma(T^*)$,
   (b) if $\lambda \in P\sigma(T)$ then $\bar{\lambda} \in P\sigma(T^*) \cup R\sigma(T^*)$,
   (c) $\lambda \in C\sigma(T^*)$ if and only if $\bar{\lambda} \in C\sigma(T)$.

2. (i) Consider the *right shift* operator $S_{\mathbb{R}}$ on $(\ell_2, \|\cdot\|_2)$ defined for $x \equiv \{\lambda_1, \lambda_2, \dots, \lambda_n, \dots\}$ by
$$S_{\mathbb{R}}(x) = \{0, \lambda_1, \lambda_2, \dots, \lambda_n, \dots\}.$$
   Prove that
   (a) $P\sigma(S_{\mathbb{R}}) = \emptyset$,
   (b) $R\sigma(S_{\mathbb{R}}) = \{\lambda \in \mathbb{C} : |\lambda| < 1\}$,
   (c) $C\sigma(S_{\mathbb{R}}) = \{\lambda \in \mathbb{C} : |\lambda| = 1\}$.

   (ii) The *left shift* operator $S_{\mathbb{L}}$ on $(\ell_2, \|\cdot\|_2)$ is defined for $x \equiv \{\lambda_1, \lambda_2, \dots, \lambda_n, \dots\}$ by
$$S_{\mathbb{L}}(x) = \{\lambda_2, \lambda_3, \dots, \lambda_n, \dots\}.$$
   Prove that
   (a) $S_{\mathbb{R}}^* = S_{\mathbb{L}}$,
   (b) $P\sigma(S_{\mathbb{L}}) = R\sigma(S_{\mathbb{R}})$,
   (c) $R\sigma(S_{\mathbb{L}}) = P\sigma(S_{\mathbb{R}})$,
   (d) $C\sigma(S_{\mathbb{L}}) = C\sigma(S_{\mathbb{R}})$.

3. Consider the diagonal operator T on $(\ell_2, \|\cdot\|_2)$ defined by the bounded sequence of scalars $\{\alpha_1, \alpha_2, \dots, \alpha_n, \dots\}$ where for $x \equiv \{\lambda_1, \lambda_2, \dots, \lambda_n, \dots\} \in \ell_2$,
$$Tx = \{\alpha_1\lambda_1, \alpha_2\lambda_2, \dots, \alpha_n\lambda_n, \dots\}.$$
   (i) Prove that
   (a) $P\sigma(T) = \{\alpha_n : n \in \mathbb{N}\}$,
   (b) $\sigma(T) = \overline{\{\alpha_n : n \in \mathbb{N}\}}$.

(ii) Show that every nonempty compact subset of the complex plane is the spectrum of some diagonal operator; (See Exercise 4.12.2.)

4. Consider the complex Banach space $(\mathcal{C}[0,1], \|\cdot\|_\infty)$.

(i) For the integral operator I defined by

$$I(f)(x) = \int_0^x f(t)\, dt$$

prove that

(a) $\mathbb{C} \setminus \{0\} = \rho(I)$,

(b) $\{0\} = R\sigma(I)$.

(ii) For the integral operator I on the subspace $\mathcal{C}_0[0,1]$ determine $\sigma(T)$.

(iii) For the multiplication operator M defined by

$$M(f)(x) = x\, f(x)$$

determine $P\sigma(M)$, $R\sigma(M)$ and $C\sigma(M)$.

5. Given a continuous linear operator T on a complex Banach space $(X, \|\cdot\|)$, $\lambda \in \mathbb{C}$ is called an *approximate eigenvalue* of T if given $\varepsilon > 0$ there exists an $x \in X$, $\|x\| = 1$ such that

$$\|(T-\lambda I)(x)\| < \varepsilon.$$

The *approximate spectrum* of T, denoted $AP\sigma(T)$ is the set of all approximate eigenvalues of T.

(i) Prove that $AP\sigma(T)$ is closed.

(ii) Prove that $\lambda \in AP\sigma(T)$ if and only if $\lambda I - T$ does not have a continuous inverse on $(\lambda I - T)(X)$.

(iii) Prove that $P\sigma(T) \cup C\sigma(T) \subseteq AP\sigma(T) \subseteq \sigma(T)$.

(iv) For T' the conjugate operator, prove that

(a) $AP\sigma(T) = \{\lambda \in \mathbb{C} : T' - \lambda I \text{ is not onto}\}$,

(b) $AP\sigma(T') = \{\lambda \in \mathbb{C} : T - \lambda I \text{ is not onto}\}$.

6. Consider a normal operator T on a complex Hilbert space H. Prove that

(i) $R\sigma(T) = \emptyset$,

(ii) if H is separable then $P\sigma(T)$ is countable.

7. Consider a continuous linear operator T on a complex Banach space $(X, \|\cdot\|)$ and a closed linear subspace M of X. Prove that

(i) $P\sigma(T) \supseteq P\sigma(T|_M)$,

(ii) $R\sigma(T) \subseteq R\sigma(T|_M)$,

(iii) $C\sigma(T|_M) \subseteq C\sigma(T) \cup P\sigma(T)$,

(iv) $\rho(T) \subseteq \rho(T|_M) \cup R\sigma(T|_M)$.

8. Consider a continuous linear operator T on a complex Banach space X.

(i) The operator T is said to be *dissipative* if $\sup \operatorname{Re} V(T) \le 0$. Prove that for such an operator $\|(I-rT)x\| \ge \|x\|$ for all $x \in X$ and $r \ge 0$.

(ii) The operator T is said to be *hermitian* if $V(T) \subseteq \mathbb{R}$. Prove that T is hermitian if and only if both iT and $-iT$ are dissipative.

(iii) Prove that for an hermitian operator T,

$$\|(I \pm irT)x\| \ge \|x\| \qquad \text{for all } x \in X \text{ and } r > 0.$$

9. Consider a continuous linear operator T on a complex Banach space X.

(i) Prove that if $\lambda \in V(T)$ and $|\lambda| = \|T\|$ and X is rotund then λ is an eigenvalue of T.

(ii) Given that T is a dissipative operator, prove that

$$\|T^2x\| \ge r(\|rx+Tx\| - r\|x\|) \quad \text{for all } x \in X \text{ and } r \ge 0.$$

(iii) Given that T is a continuous linear operator, prove that if $\lambda \in \partial\, \overline{\text{co}}\, V(T)$ and $(\lambda I-T)^2x = 0$ for some $x \ne 0$ then $(\lambda I-T)x = 0$.
Deduce that if T is an hermitian operator and $\lambda \in \sigma(T)$ and $(\lambda I-T)^2x = 0$ for some $x \ne 0$ then $(\lambda I-T)x = 0$.

§18. THE SPECTRUM OF A COMPACT OPERATOR

Compact operators have properties which give their spectrum a very simple structure. The key to revealing this structure is the following property: Given a compact operator T on a normed linear space, the linear operator I – T is one-to-one if and only if I – T is onto.

We develop the proof of this property through stages.

18.1 **Theorem.** *Given a compact operator* T *on a normed linear space* $(X, \|\cdot\|)$, *if* $I-T$ *is one-to-one then* $I-T$ *has a continuous inverse on* $(I-T)(X)$.

Proof. Suppose that $I-T$ does not have a continuous inverse. Then for each $n \in \mathbb{N}$ there exists $x_n \in X$ such that

$$\|(I-T)x_n\| < \frac{1}{n}\|x_n\|.$$

Now $\{\frac{x_n}{\|x_n\|}\}$ is a bounded sequence and since T is compact there is a subsequence $\{\frac{x_{n_k}}{\|x_{n_k}\|}\}$ such that $\{T(\frac{x_{n_k}}{\|x_{n_k}\|})\}$ is convergent to some $y \in Y$. But

$$\|(I-T)(\frac{x_{n_k}}{\|x_{n_k}\|})\| < \frac{1}{n_k}$$

so $\{\frac{x_{n_k}}{\|x_{n_k}\|}\}$ is also convergent to y. Then $Ty = y$. However, $\|y\| = 1$ and then $I-T$ is not one-to-one. □

18.2 **Lemma**. *Given a compact operator* T *on a normed linear space* $(X, \|\cdot\|)$, *if* $I-T$ *is one-to-one on* X *then* $(I-T)(X)$ *is a closed linear subspace of* $(X, \|\cdot\|)$.

Proof. Consider a cluster point y of $(I-T)(X)$. Then there exists a sequence $\{y_n\}$ in $(I-T)(X)\backslash\{y\}$ such that $\{y_n\}$ is convergent to y. Write $y_n = (I-T)(x_n)$. Then from Theorem 18.1 there exists an $m > 0$ such that

$$m\|x\| \le \|(I-T)x\| \quad \text{for all } x \in X$$

and

$$m\|x_n\| \le \|(I-T)x_n\| = \|y_n\| \quad \text{for all } n \in \mathbb{N}.$$

Since $\{y_n\}$ is convergent, $\{y_n\}$ is bounded and so $\{x_n\}$ is bounded. Since T is compact, $\{x_n\}$ has a subsequence $\{x_{n_k}\}$ such that $\{Tx_{n_k}\}$ is convergent. As $x_n = y_n + Tx_n$, then $\{x_{n_k}\}$ is also convergent to say, x. Then $y_{n_k} = (I-T)(x_{n_k})$ is convergent to $y = (I-T)(x)$; this tells us that $y \in (I-T)(X)$ and $(I-T)(X)$ is closed. □

18.3 **Remark**. If we had assumed that the normed linear space $(X, \|\cdot\|)$ is complete then it would have followed directly from Theorem 18.1 and Theorem 1.24.9 that $(I-T)(X)$ is closed in $(X, \|\cdot\|)$. □

18.4 **Lemma**. *Given a compact operator* T *on a normed linear space* $(X, \|\cdot\|)$, *for each* $n \in \mathbb{N}$, $(I-T)^n = I - S_n$ *where* S_n *is a compact operator on* X.

Proof. $(I-T)^n = \sum_{r=0}^{n} (-1)^r \binom{n}{r} T^r = I - S_n$ where $S_n \equiv -\left(\sum_{r=1}^{n} (-1)^r \binom{n}{r} T^r\right)$.

By Theorem 15.16, S_n is compact. □

We are now in a position to prove the first part of our key property.

18.5 **Theorem**. *Given a compact operator* T *on a normed linear space* $(X, \|\cdot\|)$, *if* $I - T$ *is one-to-one then* $I-T$ *is onto.*

Proof. We write, $\mathcal{R}_n \equiv (I-T)^n(X)$ for each $n \in \mathbb{N}$. Then clearly
$X \supseteq \mathcal{R}_1 \supseteq \mathcal{R}_2 \supseteq \ldots \supseteq \mathcal{R}_n \supseteq \ldots$. Suppose that $\mathcal{R}_n \neq \mathcal{R}_{n+1}$ for each $n \in \mathbb{N}$. By Lemmas 18.2 and 18.4, $\mathcal{R}_{n+1}$ is closed in $\mathcal{R}_n$, so by Riesz Lemma 2.1.8 for each $n \in \mathbb{N}$ there exists an $x_n \in \mathcal{R}_n$, $\| x_n \| = 1$ such that $d(x_n, \mathcal{R}_{n+1}) \geq \frac{1}{2}$.
Now for $n > m$ we have

$$\| Tx_m - Tx_n \| = \| x_m - (I-T)x_m - x_n + (I-T)x_n \| \geq \frac{1}{2}$$

since $(I-T)x_m + x_n - (I-T)x_n \in \mathcal{R}_{m+1}$. But then although $\{x_n\}$ is bounded, $\{Tx_n\}$ cannot have a convergent subsequence and this contradicts T being compact.
So there exists some $\mu \in \mathbb{N}$ such that $\mathcal{R}_\mu = \mathcal{R}_{\mu+1}$ and then $\mathcal{R}_{n+1} = \mathcal{R}_n$ for all $n \geq \mu$.
Since $I - T$ is one-to-one it follows that $(I-T)^n$ is one-to-one for all $n \in \mathbb{N}$.
Given $x \in X$, $(I-T)^\mu x = (I-T)^{2\mu} y$ for some $y \in X$ so $(I-T)^\mu (x - (I-T)^\mu y) = 0$.
Since $(I-T)^\mu$ is one-to-one, $x = (I-T)^\mu y$. That is, $\mathcal{R}_\mu = X$.
But this implies that $(I-T)(X) = X$. □

Theorems 18.1 and 18.5 have the following implications for the spectrum of a compact operator.

18.6 **Theorem**. *Consider a compact operator* T *on a complex normed linear space* $(X, \|\cdot\|)$.
(i) $0 \in \sigma(T)$.
(ii) *For* $\lambda \neq 0$, $T - \lambda I$ *is regular in* $\mathcal{B}(X)$ *if and only if* $T - \lambda I$ *is one-to-one.*

Proof.
(i) follows from Corollary 15.17.
(ii) follows from $T - \lambda I = -\lambda(I - T/\lambda)$ and T/λ is a compact operator. □

The following structural property is immediate.

18.7 **Corollary**. *For a compact operator* T *on a complex normed linear space* $(X, \|\cdot\|)$,

$$\sigma(T) = \{0\} \cup P\sigma(T).$$

A description of this spectral structure can now be given.

18.8 **Theorem**. *Given a compact operator* T *on a complex normed linear space* $(X, \|\cdot\|)$, $P\sigma(T)$ *is countable and has* 0 *as its only possible cluster point.*

Proof. We prove that for any given $\varepsilon > 0$, the set

$$P_\varepsilon \equiv \{\lambda \in P\sigma(T) : |\lambda| \geq \varepsilon\}$$

is finite. Then $P\sigma(T) \setminus \{0\} = \bigcup_{n=1}^{\infty} P_{1/n}$ and so $P\sigma(T)$ is countable.

Suppose on the contrary, that there exists an $r > 0$ such that P_r is infinite.

Then there exists a sequence $\{\lambda_n\}$ in P_r where $\lambda_n \neq \lambda_m$ for all $n \neq m$.

Consider a sequence $\{x_n\}$ of eigenvectors where for each $n \in \mathbb{N}$, $Tx_n = \lambda_n x_n$.

From Theorem 17.1.2, $\{x_n : n \in \mathbb{N}\}$ is a linearly independent set.

Consider $M_n \equiv \{x_1, x_2, \ldots, x_n\}$.

Then for each $n \in \mathbb{N}$, since M_n is finite dimensional it is closed and from the linear independence of the set $\{x_n : n \in \mathbb{N}\}$, $M_{n-1} \subsetneqq M_n$.

From Riesz Lemma 2.1.8 we have for each $n \in \mathbb{N}$, there exists a $y_n \in M_n$, $\|y_n\| = 1$ such that $d(y_n, M_{n-1}) \geq \frac{1}{2}$.

Now for $m < n$, $\|Ty_n - Ty_m\| = \|\lambda_n y_n - (\lambda_n y_n - Ty_n + Ty_m)\|$.

But any $x \in M_n$ has the form $x = \sum_{k=1}^{n} \alpha_k x_k$.

So $(T-\lambda_n I)(x) = \sum_{k=1}^{n} \alpha_k(\lambda_k - \lambda_n)x_k$ and $(T-\lambda_n I)(M_n) \subseteq M_{n-1}$.

Therefore, $(\lambda_n I - T)(y_n) + Ty_m \in M_{n-1}$

and so
$$\|Ty_n - Ty_m\| \geq \frac{|\lambda_n|}{2} \geq \frac{r}{2}.$$

But then, although $\{y_n\}$ is a bounded sequence, $\{Ty_n\}$ cannot have a convergent subsequence and this contradicts T being compact.

So we conclude that $\sigma(T)$ is countable. □

18.9 **Remark**. Quite independently of our general theory we have shown that the spectrum of a compact operator on a complex normed linear space is nonempty and compact. □

We examine the relation between the regularity of a compact operator and its conjugate. We begin with a result similar to Theorem 12.11 but without using completeness.

18.10 **Theorem**. *Consider a compact operator* T *on a normed linear space* $(X, \|\cdot\|)$. *Then* $I - T$ *is a topological isomorphism on* X *if and only if* $I - T'$ *is a topological isomorphism on* X^*.

Proof. We need only show the result comparable to Theorem 12.11 (ii).
If $I - T'$ is a topological isomorphism on X^* then by Theorem 12.8 its conjugate $I - T''$ is one-to-one on X^{**} so $I - T$ is one-to-one on X. By Theorems 18.1 and 18.5, $I - T$ is a topological isomorphism on X. □

As in Theorem 17.4.1 we can relate the spectra of a compact operator and its conjugate. The result now holds without completeness.

18.11 **Corollary**. *For a compact operator* T *on a complex normed linear space* $(X, \|\cdot\|)$ *with conjugate operator* T' , $\sigma(T) = \sigma(T')$.

The converse of Theorem 18.5 is established using conjugate operators.

18.12 **Theorem.** *Given a compact operator* T *on a normed linear space* $(X, \|\cdot\|)$, *if* $I - T$ *is onto then* $I - T$ *is one-to-one.*

Proof. Since $I - T$ is onto, by Theorem 12.8, its conjugate $I - T'$ is one-to-one on X^*. Since T is compact, by Schauder's Theorem 15.18, its conjugate T' is also compact. It follows from Theorems 18.1 and 18.5 that $I - T'$ is a topological isomorphism on X^* and so our result follows from Theorem 18.10. □

We now determine the special properties of the eigenspaces corresponding to the eigenvalues of a compact operator.

18.13 **Theorem**. *For a compact operator* T *on a normed linear space* $(X, \|\cdot\|)$, $\ker(I-T)$ *is finite dimensional.*

Proof. Write $B \equiv \{x \in \ker(I-T) : \|x\| \le 1\}$ which is the closed unit ball in $(\ker(I-T), \|\cdot\|_{\ker(I-T)})$. Since $\ker(I-T) = \{x \in X : Tx = x\}$, then $T(B) = B$. Since T is compact and $\ker(I-T)$ is closed in $(X, \|\cdot\|)$ then $T(B)$ is closed in $(X, \|\cdot\|)$ and so is compact. Then B is compact and so by Riesz Theorem 2.1.9, $\ker(I-T)$ is finite dimensional. □

18.14 **Corollary**. *Given a compact operator* T *on a complex normed linear space* $(X, \|\cdot\|)$, *for an eigenvalue* $\lambda \neq 0$, *the eigenspace corresponding to* λ *is finite dimensional.*

Proof. The eigenspace $\{x \in X : (\lambda I-T)(x) = 0\} = \ker(I - T/\lambda)$. □

18.15 **The invariant subspace property of compact operators**

It was mentioned in Remarks 14.27 that for certain classes of continuous linear operators on a Banach space there is a positive solution to the Invariant Subspace Problem.

In fact N. Aronszajn and K. Smith, *Ann. Math.* **60** (1954), 345–350, proved that every compact operator on an infinite dimensional complex Banach space has a nontrivial closed invariant subspace. Their result follows with a simpler proof from a more general result proved by V.I. Lomonosov, *Funk. Anal. i Prilozen* **7** (1973), 55–56.

We present a proof based on a modification of Lomonosov's argument.

18.15.1 **Lomonosov's Theorem.**
Every nonzero compact operator T *on a complex Banach space* $(X, \|\cdot\|)$ *has a nontrivial closed invariant subspace.*

Proof. We write $\Gamma \equiv \{S \in \mathcal{B}(X) : ST = TS\}$ which is a closed subalgebra of $\mathcal{B}(X)$. For each $y \in X$, $y \neq 0$ we write

$$\Gamma(y) \equiv \{Sy : S \in \Gamma\}.$$

Since Γ is closed under composition, $S(\Gamma(y)) \subseteq \Gamma(y)$ for all $S \in \Gamma$ so if $\overline{\Gamma(y)} \neq X$ for some $y \in X$ then $\overline{\Gamma(y)}$ is a suitable nontrivial closed invariant subspace for T.
Suppose that $\overline{\Gamma(y)} = X$ for all $y \in X$. Choose $x_0 \in X$ such that $Tx_0 \neq 0$. Then $x_0 \neq 0$ and by the continuity of T there exists an open ball B centred at x_0 such that

$$\|x\| \geq \tfrac{1}{2}\|x_0\| \quad \text{and} \quad \|Tx\| \geq \tfrac{1}{2}\|Tx_0\| \quad \text{for all } x \in B.$$

Given $y \in \overline{T(B)}$, since $\overline{\Gamma(y)} = X$ then $x_0 \in \overline{\Gamma(y)}$. Since B is an open neighbourhood of x_0 there exists an open neighbourhood W of y and some $S \in \Gamma$ such that $S(W) \subseteq B$.
Since T is a compact operator, $\overline{T(B)}$ is compact.
Now $0 \notin \overline{T(B)}$. But since $\overline{T(B)}$ is compact there exists a finite open cover $W_1, W_2, \ldots, W_n$ of $\overline{T(B)}$ such that $S_k(W_k) \subseteq B$ for some $S_k \in \Gamma$ and $k \in \{1, 2, \ldots, n\}$.
Now $Tx_0 \in \overline{T(B)}$ so Tx_0 lies in some W_{k_1} and $x_1 \equiv S_{k_1}Tx_0 \in B$.
Then $Tx_1 = TS_{k_1}Tx_0 \in \overline{T(B)}$ so $TS_{k_1}Tx_0$ lies in some W_{k_2} and $x_2 \equiv S_{k_2}TS_{k_1}Tx_0 \in B$.
Continuing, we obtain a sequence $\{x_n\}$ where

$$x_n \equiv S_{k_n}T \ldots S_{k_1}Tx_0 \in B.$$

Putting $k \equiv \max\{\|S_{k_1}\|, \|S_{k_2}\|, \ldots, \|S_{k_n}\|\} > 0$ we have

$$\tfrac{1}{2}\cdot\|x_0\| \leq \|x_n\| \leq k^n \|T^n\| \|x_0\| \qquad \text{for all } n \in \mathbb{N}.$$

This gives us $\nu(T) = \lim_{n\to\infty} \|T^n\|^{1/n} = \frac{1}{k} > 0$.
Since T is a compact operator it has an eigenvalue $\lambda \neq 0$. But then $\ker(\lambda I - T)$ is a nontrivial closed invariant subspace of X under T.

The proof actually establishes more than the theorem statement.

18.15.2 **Corollary**. *Given a nonzero compact operator* T *on a complex Banach space* $(X, \|\cdot\|)$ *the nontrivial closed invariant subspace under* T *is also invariant under every continuous linear operator* S *which commutes with* T.

Proof. If $\Gamma(y) \neq X$ for some $y \in X$, $y \neq 0$, then $\Gamma(y)$ is invariant under S.
If $\Gamma(y) = X$ for all $y \in X$, $y \neq 0$ then T has an eigenvalue $\lambda \neq 0$. So, if $x \in \ker(\lambda I - T)$,
$$T(Sx) = S(Tx) = S(\lambda x) = \lambda S(x)$$
then $Sx \in \ker(\lambda I - T)$ which implies that $\ker(\lambda I - T)$ is invariant under S. □

18.16 **Remarks.**
(i) *A quasinilpotent operator* T on a complex Banach space X is a continuous linear operator whose spectrum $\sigma(T) = \{0\}$. C.J. Read *J. London Math. Soc.* (2), **56** (1997), 595–606, has given an example of a quasinilpotent operator which does not have an invariant subspace.
(ii) The theory of compact operators on a normed linear space is structurally very rich. We have only given sufficient of the theory to develop the spectral properties we need for the Spectral Theorem in Section 19. The possible further development of the theory can be glimpsed in the proof of Theorem 18.5. The illuminating Riesz–Schauder Theory of compact operators encompasses the study of the ranges and kernels of iterations of the mapping $I - T$ where T is a compact operator. We leave some of these developments to be explored in Exercise 18.17.6. A fuller account of this theory can be found in A.L. Brown and A. Page, *Elements of Functional Analysis*, van Nostrand Reinhold, 1970, pp.248–255. □

18.17 **EXERCISES**

1. Prove that for a continuous finite rank operator T on a Banach space $(X, \|\cdot\|)$, the spectrum $\sigma(T)$ is a finite set consisting only of eigenvalues.

2. Consider a compact operator T on a Banach space $(X, \|\cdot\|)$.
   (i) Prove that, for any $\lambda \neq 0$, $(T - \lambda I)(X)$ is a closed linear subspace of $(X, \|\cdot\|)$.
   (ii) Prove that for any regular operator S on $(X, \|\cdot\|)$, $(S - T)(X)$ is closed in $(X, \|\cdot\|)$ and has finite codimension.

3. The right shift operator S on $(\ell_2, \|\cdot\|_2)$ is defined by
$$S(\{\lambda_1, \lambda_2, \ldots, \lambda_n, \ldots\}) = \{0, \lambda_1, \lambda_2, \ldots, \lambda_n, \ldots\}.$$
   (i) Prove that
      (a) $\lambda I - S$ is one-to-one for all $\lambda \in \mathbb{C}$,
      (b) $\sigma(S) = \{\lambda \in \mathbb{C} : |\lambda| \leq 1\}$.
   (ii) The operator T on $(\ell_2, \|\cdot\|_2)$ is defined by
$$T(\{\lambda_1, \lambda_2, \ldots, \lambda_n, \ldots\}) = \left\{\frac{\lambda_1}{2}, \frac{\lambda_2}{3}, \ldots, \frac{\lambda_n}{n+1}, \ldots\right\}.$$
   Prove that
      (a) TS is a compact operator,
      (b) $P\sigma(TS) = \varnothing$.

4. The operator T on $(\ell_2, \|\cdot\|_2)$ is defined by

$$T\{\lambda_1, \lambda_2, \lambda_3, \ldots, \lambda_n, \ldots\} = \{0, \lambda_1, \frac{\lambda_2}{2}, \frac{\lambda_3}{3}, \ldots, \frac{\lambda_n}{n}, \ldots\}.$$

Prove that

(i) T is compact,

(ii) $\sigma(T) = R\sigma(T) = \{0\}$.

5. Consider a sequence $\{T_n\}$ of compact operators on a Banach space $(X, \|\cdot\|)$ and an operator T on $(X, \|\cdot\|)$ such that $T_n \to T$ as $n \to \infty$ in $(\mathfrak{B}(X), \|\cdot\|)$. Prove that if $\lambda_n \in \sigma(T_n)$ for all $n \in \mathbb{N}$ and $\lambda_n \to \lambda$ as $n \to \infty$ then $\lambda \in \sigma(T)$.

6. Consider a compact operator T on a normed linear space $(X, \|\cdot\|)$.
In Theorem 18.5 it was actually proved, writing $\mathcal{R}_n \equiv (I-T)^n(X)$ for each $n \in \mathbb{N}$, that there exists a $\mu \in \mathbb{N}$ such that

$$\mathcal{R}_n \underset{\neq}{\supset} \mathcal{R}_{n+1} \text{ for all } n \in \{1, 2, \ldots, \mu-1\} \text{ and}$$

$$\mathcal{R}_n = \mathcal{R}_{n+1} \text{ for all } n \geq \mu.$$

(i) Writing $\mathcal{N}_n \equiv \ker(I-T)^n$ for each $n \in \mathbb{N}$, prove that there exists a $\nu \in \mathbb{N}$ such that

$$\mathcal{N}_n \underset{\neq}{\subset} \mathcal{N}_{n+1} \text{ for all } n \in \{1, 2, \ldots, \nu-1\} \text{ and}$$

$$\mathcal{N}_n = \mathcal{N}_{n+1} \text{ for all } n \geq \nu.$$

(ii) Prove that

(a) $X = \mathcal{R}_u \oplus \mathcal{N}_\mu$,

(b) $\mathcal{R}_\mu$ and $\mathcal{N}_\mu$ are invariant subspaces of T, and deduce that

(c) $\nu = \mu$.

(iii) Prove that, as for a linear operator on a finite dimensional linear space,

$$\operatorname{codim}\,(I-T)(X) = \dim \ker(I-T).$$

7. Consider an incomplete normed linear space $(X, \|\cdot\|)$ and a compact linear operator T on $(X, \|\cdot\|)$ with a unique continuous linear extension $\widetilde{T}$ on $(\widetilde{X}, \|\cdot\|)$.
Writing $\mathcal{R}_n \equiv (I-T)^n(X)$ and

$$\widetilde{\mathcal{R}}_n \equiv (I-\widetilde{T})^n(\widetilde{X}) \text{ for each } n \in \mathbb{N}$$

prove that if μ is the least natural number such that

$$\mathcal{R}_n = \mathcal{R}_{n+1} \text{ for all } n \geq \mu$$

then it is the least natural number such that

$$\widetilde{\mathcal{R}}_n = \widetilde{\mathcal{R}}_{n+1} \text{ for all } n \geq \mu.$$

§19. THE SPECTRAL THEOREM FOR COMPACT NORMAL OPERATORS ON HILBERT SPACE.

We have seen in Theorem 17.4.3 that for a normal operator on a complex Hilbert space, the eigenspaces associated with distinct eigenvalues are orthogonal. It is this property which enables us to construct a spectral theorem for such operators on a finite dimensional inner product space.

19.1 The Finite Dimensional Spectral Theorem.
For a normal operator T *on a complex finite dimensional inner product space* H_n *there exists an orthonormal basis for* H_n *consisting of eigenvectors of* T; *that is,* T *can be represented as a diagonal matrix with respect to some orthonormal basis for* H_n.

We have seen in Section18 that compact operators have a simple spectral structure. We show in this Section that we can establish a spectral theorem for compact normal operators on Hilbert space which includes the finite dimensional case.

Consider a compact normal operator T on a complex Hilbert space H. From Theorem 18.8 the set of eigenvalues of T is countable. Consider the eigenvalues in sequence $\{\lambda_n\}$ where $|\lambda_n| \geq |\lambda_{n+1}|$ for all $n \in \mathbb{N}$. By Corollary 18.14, for each $n \in \mathbb{N}$, $N_n \equiv \ker(\lambda_n I-T)$ is finite dimensional and is the eigenspace of λ_n. Since we are in an inner product space, for each $n \in \mathbb{N}$ we can choose an orthonormal basis for N_n, say $\{u_{n1}, u_{n2}, \dots, u_{nk_n}\}$. Since T is normal, by Theorem 17.4.3 we have for $n \neq m$ that N_n is orthogonal to N_m. So the sequence of eigenvectors

$$\{u_{11}, u_{12}, \dots, u_{1k_1}, u_{21}, u_{22}, \dots, u_{2k_2}, \dots, u_{n1}, u_{n2}, \dots, u_{nk_n}, \dots\}$$

is an orthonormal set in X. We denote by F the closed linear span of this set.

We now relate the range and kernel of T to the subspace F and its orthogonal complement $F^\perp$.

19.2 Theorem. *For a compact normal operator* T *on a complex Hilbert space* H, $T(F) \subseteq F$ *and* $F^\perp = \ker T$.

Proof. Since F is the closed linear span of eigenvectors it is clear that $T(F) \subseteq F$.
Since T is normal, it follows from Theorem 17.4.3 that the eigenvectors of T are eigenvectors of T^* so $T^*(F) \subseteq F$. Therefore, for $x \in F$ and $y \in F^\perp$ we have

$$(x, Ty) = (T^*x, y) = 0 \text{ so } T(F^\perp) \subseteq F^\perp.$$

Similarly, $$(x, T^*y) = (Tx, y) = 0 \text{ so } T^*(F^\perp) \subseteq F^\perp.$$

So F reduces both T and T^* and $T|_{F^\perp}$ is a normal operator on Hilbert space $F^\perp$.
Suppose that $T|_{F^\perp} \neq 0$. Then by Corollary 17.3.4 there exists a $\lambda \in \sigma(T|_{F^\perp})$ such that $|\lambda| = \|T|_{F^\perp}\|$. But $T|_{F^\perp}$ is also compact on $F^\perp$ so by Theorem 18.8, λ is an eigenvalue

of $T|_{F^\perp}$, but then λ is an eigenvalue of T. So $\lambda = \lambda_n$ for some $n \in \mathbb{N}$. But $N_n \subseteq F$ and this contradicts $F \cap F^\perp = \{0\}$. Therefore $T|_{F^\perp} = 0$ which implies that $F^\perp \subseteq \ker T$.
If $F^\perp \neq \ker T$ then by Theorem 2.2.19 and Corollary 14.8 there exists a $z \neq 0$ and $z \in \ker T$ and $z \in F$. But this contradicts T being one-to-one on F. So we conclude that $F^\perp = \ker T$. □

We are now in a position to present the spectral theorem for compact normal operators on Hilbert space.

19.3 The Spectral Theorem.
For the compact normal operator T *on the complex Hilbert space* H, *for each* $x \in H$

$$Tx = \sum_{n=1}^{\infty} \sum_{k=k_1}^{k_n} \lambda_n (x, u_{nk}) u_{nk}.$$

Proof. For $x \in H$ we can write $x = y + z$ where $y \in F$ and $z \in F^\perp$. Then from Theorem 19.1 we have that $Tx = Ty$. Since F is the closed linear span of the orthonormal set

$$\{u_{11}, u_{12}, \ldots, u_{1k_1}, u_{21}, u_{22}, \ldots, u_{2k_2}, \ldots, u_{n1}, u_{n2}, \ldots, u_{nk_n}, \ldots\}$$

then by Theorem 3.10 we have that

$$\begin{aligned} Tx &= \sum_{n=1}^{\infty} \sum_{k=k_1}^{k_n} (Tx, u_{nk}) u_{nk} \\ &= \sum_{n=1}^{\infty} \sum_{k=k_1}^{k_n} (x, T^* u_{nk}) u_{nk} \\ &= \sum_{n=1}^{\infty} \sum_{k=k_1}^{k_n} \lambda_n (x, u_{nk}) u_{nk} \quad \text{by Theorem 17.4.3(i).} \end{aligned}$$ □

From this theorem we can derive an extension of Corollary 17.3.4.

19.4 Corollary. *Given a compact normal operator* T *on a complex Hilbert space* H, *for each* $n \in \mathbb{N}$

$$|\lambda_n| = \nu(T|_{F_{n-1}^\perp})$$

where $F_n = \mathrm{sp}\,\{u_{11}, u_{12}, \ldots, u_{1k_1}, u_{21}, u_{22}, \ldots, u_{2k_2}, \ldots, u_{n1}, u_{n2}, \ldots, u_{nk_n}\}$.

Proof. For $x \in F_{n-1}^\perp$, $Tx = \sum_{p=n}^{\infty} \sum_{k=k_1}^{k_p} \lambda_p (x, u_{pk})\, u_{pk}$.

So
$$(Tx, x) = \sum_{p=n}^{\infty} \sum_{k=k_1}^{k_p} \lambda_p |(x, u_{pk})|^2.$$

Since $\{u_{11}, u_{12}, \ldots, u_{1k_1}, u_{21}, u_{22}, \ldots, u_{2k_2}, \ldots, u_{n1}, u_{n2}, \ldots, u_{nk_n}, \ldots\}$ is an orthonormal set, from Bessel's inequality 3.13

$$\sum_{n=1}^{\infty} \sum_{k=k_1}^{k_p} \lambda_p |(x, u_{nk})|^2 \leq \|x\|^2.$$

Also since $|\lambda_m| \le |\lambda_n|$ for all $m \ge n$,

$$
\begin{aligned}
|(Tx, x)| &\le \sum_{p=n}^{\infty} \sum_{k=k_1}^{k_p} |\lambda_p| \, |(x, u_{pk})|^2 \\
&\le |\lambda_n| \sum_{p=n}^{\infty} \sum_{k=k_1}^{k_p} |(x, u_{pk})|^2 \\
&\le |\lambda_n| \, \|x\|^2.
\end{aligned}
$$

But clearly, $(Tu_{nk}, u_{nk}) = \lambda_n$ for all $k \in \{k_1, k_2, \ldots, k_n\}$ and the result follows from Corollary 17.3.4. □

We now show how the Spectral Theorem 19.3 can be used to give an expression for inverses related to a compact normal operator on a Hilbert space.

19.5 **Theorem**. *Given the compact normal operator* T *on the complex Hilbert space* H, *for any* $\lambda \notin \sigma(T)$ *and all* $x \in H$

$$(\lambda I - T)^{-1} x = \frac{1}{\lambda} x + \frac{1}{\lambda} \sum_{n=1}^{\infty} \sum_{k=k_1}^{k_n} \frac{\lambda_n}{\lambda - \lambda_n} (x, u_{nk}) u_{nk}.$$

Proof. For $x \in H$ write $y = (\lambda I - T)^{-1} x$. Then from the Spectral Theorem 19.3,

$$x = (\lambda I - T) y = \lambda y - \sum_{n=1}^{\infty} \sum_{k=k_1}^{k_n} \lambda_n (y, u_{nk}) u_{nk}.$$

So
$$y - \frac{1}{\lambda} x = \sum_{n=1}^{\infty} \sum_{k=k_1}^{k_n} \frac{\lambda_n}{\lambda} (y, u_{nk}) u_{nk}.$$

But for $m \in \mathbb{N}$ and $\ell \in \{k_1, k_2, \ldots, k_m\}$,

$$
\begin{aligned}
(y, u_{m\ell}) - \frac{1}{\lambda} (x, u_{m\ell}) &= (y - \frac{1}{\lambda} x, u_{m\ell}) \\
&= \sum_{n=1}^{\infty} \sum_{k=k_1}^{k_n} \frac{\lambda_n}{\lambda} (y, u_{nk})(u_{nk}, u_{m\ell}) \\
&= \frac{\lambda_m}{\lambda} (y, u_{m\ell})
\end{aligned}
$$

and so
$$(y, u_{m\ell}) = \frac{1}{\lambda - \lambda_m} (x, u_{m\ell}).$$

Consequently,
$$y = \frac{1}{\lambda} x + \frac{1}{\lambda} \sum_{n=1}^{\infty} \sum_{k=k_1}^{k_n} \frac{\lambda_n}{\lambda - \lambda_n} (x, u_{nk}) u_{nk}.$$ □

19.6 **Example**. *The solution of Fredholm integral equations.*

In Section 15 we considered the Fredholm operator K defined on $\mathfrak{C}[a, b]$ by

$$(Kf)(x) = \int_a^b k(x,t) \, f(t) \, dt$$

with kernel k a continuous complex function on a square region

$$S \equiv \{(x,t) : a \le x \le b, \ a \le t \le b\}.$$

We showed that K is a compact operator on $(\mathfrak{C}[a,b], \|\cdot\|_\infty)$.

It is not difficult to show that K is a compact operator on the incomplete inner product space $(\mathcal{C}[a,b], \|\cdot\|_2)$, (see Exercise 15.23.9(ii)(a)). Further the unique continuous linear extension $\tilde{K}$ on the Hilbert space completion $(\mathcal{L}_2[a,b], \|\cdot\|_2)$ is also a compact operator, (see Exercise 15.23.9(ii)(b)). Consider $x_0 \in [a, b]$ and $f \in \mathcal{L}_2[a,b]$.

Now
$$|(\tilde{K}f)(x) - (\tilde{K}f)(x_0)| \leq \int_a^b |k(x,t) - k(x_0,t)|\,|f(t)|\,dt$$
$$\leq \left(\int_a^b |k(x,t) - k(x_0,t)|^2\,dt\right)^{1/2} \|f\|_2.$$

Since k is continuous then $\tilde{K}f \in \mathcal{C}[a,b]$. Therefore, if λ is an eigenvalue of $\tilde{K}$ with eigenvector f so that $\tilde{K}f = \lambda f$, then we can conclude that $f \in \mathcal{C}[a,b]$.

Now $\tilde{K}^*$ the adjoint of $\tilde{K}$ on $\mathcal{L}_2[a, b]$ is defined by
$$(\tilde{K}^* f)(x) = \left(\int_a^b \overline{k(t,x)}\, f(t)\,dt\right.$$
since $\int_a^b \left(\int_a^b k(x,t)\, f(t)\,dt\right) \overline{g(x)}\,dx = \int_a^b \left(\int_a^b \overline{k(t,x)\, g(x)}\,dx\right) f(t)\,dt$ for all $f, g \in \mathcal{C}[a,b]$.

If the kernel k has the property that
$$|k(x,t)| = |k(t,x)| \quad \text{for all } (x, t) \in S$$
then
$$\|\tilde{K}f\| = \left(\int_a^b |k(x,t)|^2\,|f(t)|^2\,dt\right)^{1/2} = \left(\int_a^b |k(t,x)|^2\,|f(t)|^2\,dt\right)^{1/2}$$
$$= \|\tilde{K}^* f\| \quad \text{for all } f \in \mathcal{L}_2[a,b]$$
and so we would conclude from Theorem 13.11.3 that $\tilde{K}$ is a normal operator.

Applying the Spectral Theorem 19.3 to the compact normal operator $\tilde{K}$ on $(\mathcal{L}_2[a,b], \|\cdot\|_2)$ we have the following decompositon for the generating operator K on $(\mathcal{C}[a,b], \|\cdot\|_2)$,
$$K(f) = \sum_{n=1}^{\infty} \sum_{k=k_1}^{k_n} \lambda_n (f, f_{nk}) f_{nk}$$
where $\{f_{n1}, \ldots, f_{nk_n}\}$ is an orthonormal basis for the eigenspace associated with the eigenvalue λ_n for each $n \in \mathbb{N}$.

But further we can apply Theorem 19.5 to solve the *Fredholm integral equation*
$$g(x) = \lambda f(x) - \int_a^b k(x,t)\, f(t)\,dt$$
where $g \in \mathcal{C}[a, b]$ and $|k(x,t)| = |k(t,x)|$ for all $(x,t) \in S$. This equation has the form
$$g = (\lambda I - K)(g)$$
and when $\lambda \notin \sigma(K) = \sigma(\tilde{K})$, (see Exercise 4.12.5) then it has a unique solution $f \in \mathcal{L}_2[a,b]$ where
$$f = (\lambda I - \tilde{K})^{-1} g = \frac{1}{\lambda} g + \frac{1}{\lambda} \sum_{n=1}^{\infty} \sum_{k=k_1}^{k_n} \frac{\lambda_n}{\lambda - \lambda_n} \left(\int_a^b g(t)\, \overline{f_{nk}(t)}\,dt\right) f_{nk}.$$
□

19.7 EXERCISES

1. Consider a Hilbert space H with orthonormal sequence $\{e_1, e_2, \ldots, e_n, \ldots\}$ and a sequence of nonzero complex numbers $\{\lambda_1, \lambda_2, \ldots, \lambda_n, \ldots\}$ where

$$|\lambda_1| \geq |\lambda_2| \geq \ldots \geq |\lambda_n| \geq \ldots$$

and $\lambda_n \to 0$ as $n \to \infty$.

(i) Prove that for any $x \in H$, $\Sigma\lambda_n (x, e_n)e_n$ defines an element of H.

(ii) Prove that the operator T on H defined by

$$Tx = \Sigma\lambda_n(x, e_n)e_n$$

is linear and continuous and that $\|T\| = |\lambda_1|$.

(iii) Prove that T is a compact operator on H.
(Hint: Use Theorem 15.19.)

(iv) Prove that the adjoint T^* has the form

$$T^*x = \Sigma\bar{\lambda}_n (x, e_n)e_n$$

and deduce that T is normal.

(v) For $\lambda \neq \lambda_n, 0$ for all $n \in \mathbf{N}$, prove that $\lambda I-T$ is invertible and derive an expression for $(\lambda I-T)^{-1}x$ in terms of the orthonormal sequence $\{e_1, e_2, \ldots, e_n, \ldots\}$ and the sequence of complex numbers $\{\lambda_1, \lambda_2, \ldots, \lambda_n, \ldots\}$.

2. Consider a compact normal operator T on a Hilbert space H. For each eigenvalue λ denote by P_λ the orthogonal projection onto $\ker(T-\lambda I)$. Prove that the projection operators are orthogonal, that is,

$$P_{\lambda_j} P_{\lambda_k} = 0 \quad \text{for } j, k \in \mathbf{N}, j \neq k,$$

and

$$T = \sum_{n=1}^{\infty} \lambda_n P_{\lambda_n}$$

where $\{\lambda_1, \lambda_2, \ldots, \lambda_n, \ldots\}$ are the eigenvalues of T.

3. Prove that for every positive compact operator T on a complex Hilbert space H there exists a unique positive operator S on H such that $S^2 = T$ and S is also compact.

4. Given a compact normal operator T on a complex Hilbert space H, prove that

(i) $\overline{W(T)} = \overline{co}\, P\sigma(T)$,

(ii) T is self-adjoint if all its eigenvalues are real,

(iii) T is positive if all its eigenvalues are non-negative.

5. Consider a compact normal operator T on a complex Hilbert space H. Prove that if T has only a finite number of eigenvalues and is one-to-one then H is finite dimensional.

6. Consider the real function k defined on the square region

$$S \equiv \{(x,t) : 0 \le x \le 1, 0 \le t \le 1\} \text{ by}$$

$$\left.\begin{aligned} k(x,t) &= (1-x)t \quad 0 \le x \le t \\ &= (1-t)x \quad 0 \le t \le x \end{aligned}\right\}$$

and the operator K on $(\mathcal{C}[a,b], \|\cdot\|_2)$ defined by

$$(Kf)(x) = \int_0^1 k(x,t)\, f(t)\, dt.$$

(i) Prove that K is a compact operator and that its extension $\tilde{K}$ on $(\mathcal{L}_2[a,b], \|\cdot\|_2)$ is self-adjoint.

(ii) Show that the eigenvalues of T are $\frac{1}{(n\pi)^2}$ with corresponding eigenvectors $\sin n\pi$ for $n \in \mathbb{N}$.

(iii) Given a continuous function g on [0,1] solve the Fredholm integral equation

$$g(x) = f(x) - \int_0^1 k(x,t)\, f(t)\, dt.$$

§20. THE SPECTRAL THEOREM FOR COMPACT OPERATORS ON HILBERT SPACE

To extend the Spectral Theorem to all compact operators we need to develop other technical properties of self-adjoint and positive operators on Hilbert space. We study them via their mapping into a more amenable space.

20.1 Functional Calculus

Given a self-adjoint operator T on a complex Hilbert space H we have from Theorem 16.2 and Corollary 17.3.2 that the spectrum $\sigma(T)$ is a compact subset of the real numbers. The set $\mathfrak{C}(\sigma(T))$ of complex valued continuous functions on $\sigma(T)$ is a Banach space with respect to the supremum norm $\| f \|_\infty = \sup\{| f(\lambda) | : \lambda \in \sigma(T)\}$.
But also $\mathfrak{C}(\sigma(T))$ with pointwise definition of multiplication and pointwise definition of involution by taking conjugates, is a unital commutative B* algebra. On $\sigma(T)$, a polynomial p has the form $p(\lambda) = a_0 + a_1\lambda + \ldots + a_n\lambda^n$ where $a_1, a_2, \ldots, a_n \in \mathbb{C}$
we can define a polynomial operator p in $\mathfrak{B}(H)$ by

$$p(T) = a_0 I + a_1 T + \ldots + a_n T^n.$$

We denote by $\mathfrak{P}(\sigma(T))$ the linear subspace of polynomials on $\sigma(T)$. The formal association of polynomials in $\mathfrak{C}(\sigma(T))$ with polynomial operators in $\mathfrak{B}(H)$ has considerable structure.

20.1.1 **Theorem**. *Given a self–adjoint operator* T *on a complex Hilbert space* H, *the mapping* $p(\lambda) \mapsto p(T)$ *from* $\mathfrak{P}(\sigma(T))$ *into* $\mathfrak{B}(H)$ *is an algebra * homomorphism; that is,*

$$\begin{aligned}(p_1 + p_2)(T) &= p_1(T) + p_2(T)\\ p(\lambda T) &= \lambda p(T)\\ p_1p_2(T) &= p_1(T) \circ p_2(T)\\ \bar{p}(T) &= p(T)^*.\end{aligned}$$

Further, $\| p \|_\infty = \| p(T) \|$ *for all* $p \in \mathfrak{P}(\sigma(T))$
so the mapping is an isometric isomorphism.

Proof. The * property is the only algebraic property that needs checking.
Since T is self-adjoint, $\lambda \in \sigma(T)$ is real, so we have for $p(\lambda) = a_0 + a_1\lambda + \ldots + a_n\lambda^n$, that

$$\bar{p}(\lambda) = \bar{a}_0 + \bar{a}_1\lambda + \ldots + \bar{a}_n\lambda^n.$$

Now $$p(T)^* = (a_0 I + a_1 T + \ldots + a_n T^n)^*$$
which by Theorem 13.7 gives

$$p(T)^* = \bar{a}_0 I + \bar{a}_1 T + \ldots + \bar{a}_n T^n$$

and $$p(T)^* = \bar{p}(T).$$
Clearly p(T) is a normal operator, so by Corollary 17.3.4,

$$\begin{aligned}\| p(T) \| &= v(p(T)) \equiv \sup\{| \lambda | : \lambda \in \sigma(p(T))\}\\ &= \sup\{| p(\lambda) | : \lambda \in \sigma(T)\} \text{ by the Spectral Mapping Theorem 16.9}\\ &= \| p \|_\infty.\end{aligned}$$

□

By the Stone–Weierstrass Theorem (AMS §9), $\mathcal{P}(\sigma(T))$ is dense in $(\mathcal{C}(\sigma(T)), \|\cdot\|_\infty)$. By Theorem 4.8 there exists a unique continuous linear extension from $\mathcal{C}(\sigma(T))$ into $\mathcal{B}(H)$ which is an isometric isomorphism. Further, the continuity of the algebraic and * operations on $\mathcal{C}(\sigma(T))$ guarantee that the isometric isomorphism is also an algebra * isomorphism.

20.1.2 **Definition.** Given a self-adjoint operator T on a complex Hilbert space H, the mapping $f \mapsto f(T)$ of $\mathcal{C}(\sigma(T))$ into $\mathcal{B}(H)$ discussed in Theorem 20.1.1 is called a *functional calculus* for T.

20.2 Square roots of positive operators

We have seen in Section 13.10 that the set of self-adjoint operators in the set of continuous linear operators on a complex Hilbert space has a role analogous to the set of real numbers in the set of complex numbers and the set of positive operators has a role analogous to the set of positive real numbers. Just as there always exists a unique positive square root of a positive real number so we have similar behaviour exhibited for positive operators.

20.2.1 **Definition.** Given T a positive operator on a complex Hilbert space H, we say that the self-adjoint operator S on H is a *square root* of T if $S^2 = T$ and when S is a positive operator we call S a *positive square root* of T.

To show that every positive operator has a unique positive square root we need the following elementary properties from the functional calculus.

20.2.2 **Lemma.** *Consider a self-adjoint operator* T *on a complex Hilbert space* H *and* $f \in \mathcal{C}(\sigma(T))$.

(i) *If* $f(\lambda)$ *is real for all* $\lambda \in \sigma(T)$ *then* $f(T)$ *is a self-adjoint operator.*

(ii) *If* $f(\lambda)$ *is non-negative real for all* $\lambda \in \sigma(T)$ *then* $f(T)$ *is a positive operator.*

Proof.

(i) Given $\varepsilon > 0$, by the Stone–Weierstrass Theorem there exists a $p \in \mathcal{P}(\sigma(T))$ such that

$$\| f-p \| < \varepsilon.$$

If $f(\lambda)$ is real for all $\lambda \in \sigma(T)$ we can choose p such that $p(\lambda)$ is real for all $\lambda \in \sigma(T)$. Then $p(T)$ is self-adjoint and

$$\| f(T)-p(T) \| < \varepsilon.$$

But by Remark 13.10.12(ii), the set of self-adjoint operators in $\mathcal{B}(H)$ is closed, so $f(T)$ is self-adjoint.

(ii) If $f(\lambda) \geq 0$ for all $\lambda \in \sigma(T)$ then there exists a real valued function $g \in \mathcal{C}(\sigma(T))$ where $g(\lambda) \geq 0$ for all $\lambda \in \sigma(T)$ and $g^2 = f$. Now by (i), $g(T)$ is self-adjoint. Since $f(T) = (g(T))^2$ we have

$$(f(T)x, x) = \| g(T)x \|^2 \quad \text{for all } x \in H$$

and so $f(T)$ is a positive operator. □

20.2.3 **Theorem**. *Every positive operator* T *on a complex Hilbert space* H *has a unique positive square root* S.

Proof. By Corollary 17.3.3, $\sigma(T)$ consists of non–negative real numbers, so the real valued function f on $\sigma(T)$ defined by

$$f(\lambda) = \lambda^{1/2}$$

is continuous on $\sigma(T)$ and $(f(\lambda))^2 = \lambda$. So the continuous linear operator $S = f(T)$ has the property that $S^2 = T$.
From Lemma 20.2.2(ii) we conclude that S is a positive operator.
We now show that S is unique.
Given $\varepsilon > 0$ there exists a $p \in \mathfrak{C}(\sigma(T))$ such that $\| f-p \|_\infty < \varepsilon$. Then $\| S-p(T) \| < \varepsilon$.

Suppose that V is a positive operator such that $V^2 = T$. Then by the Spectral Mapping Theorem 16.8, $\sigma(T) = \{\mu^2 : \mu \in \sigma(V)\}$.
For s: $\sigma(V) \to \sigma(T)$ defined by $s(\mu) = \mu^2$ we have $q \in \mathfrak{C}(\sigma(V))$ where $q = p \circ s$ and since $\sigma(V)$ consists of non-negative real numbers $f \circ s = id$ on $\sigma(V)$.

But
$$\begin{aligned} \| V - p(T) \| &= \| V - p \circ s\,(V) \| \\ &= \| id - q \|_\infty \\ &\leq \| f - p \|_\infty \| s \|_\infty < \varepsilon \| x \|_\infty . \end{aligned}$$

We conclude that V = S and that S is unique. □

20.2.4 **Notation**. Given a positive operator T on a complex Hilbert space H we denote by $T^{1/2}$ the unique positive square root of T.

Theorem 20.2.3 has many interesting consequences.

20.2.5 **Corollary**. *For a positive operator* T *on a complex Hilbert space* H, *if* $0 \in W(T)$ *then* $0 \in P\sigma(T)$.

Proof. $\| T^{1/2}x \| = (T^{1/2}x, T^{1/2}x) = (Tx, x)$.
So if $(Tx, x) = 0$ for some $x \in H$ then $T^{1/2}x = 0$. Then $Tx = T^{1/2}T^{1/2}x = 0$. □

20.2.6 **Corollary**. *Given a continuous linear operator* T *on a complex Hilbert space* H, *for the positive operator* T*T,

$$\| (T^*T)^{1/2}x \| = \| Tx \| \text{ for all } x \in H.$$

Furthermore, $(T^*T)^{1/2}$ *is the only positive operator on* H *with this property.*

Proof. $\| (T^*T)^{1/2}x \|^2 = ((T^*T)^{1/2}x, (T^*T)^{1/2}x) = (T^*Tx, x) = \| Tx \|^2$ for all $x \in H$.
Suppose that S is a positive operator on H such that

$$\| Sx \| = \| Tx \| \qquad \text{for all } x \in H.$$

Then $(S^2x, x) = \| Sx \|^2 = \| Tx \|^2 = (T^*Tx, x)$ for all $x \in H$.
So by Lemma 13.10.6(i), $S^2 = T^*T$. □

For a compact positive operator, using the Spectral Theorem 19.3, we have an explicit form for its positive square root.

20.2.7 **Theorem.** *For a compact positive operator* T *on a complex Hilbert space* H, $T^{1/2}$ *is a compact operator.*

Proof. Using Theorem 19.3 and its notation we have that, for each $x \in H$,

$$Tx = \sum_{n=1}^{\infty} \sum_{k=k_1}^{k_n} \lambda_n (x, u_{nk}) u_{nk}.$$

From Corollary 17.3.3 it follows that the eigenvalues $\{\lambda_1, \lambda_2, \ldots, \lambda_n, \ldots\}$ are positive real numbers.

We define the operator S on H by

$$Sx = \sum_{n=1}^{\infty} \sum_{k=k_1}^{k_n} \sqrt{\lambda_n} (x, u_{nk}) u_{nk}.$$

Clearly S is linear.

Now $\sum_{n=1}^{\infty} \sum_{k=k_1}^{k_n} |\sqrt{\lambda_n} (x, u_{nk})|^2 \leq \sqrt{\lambda_1} \sum_{n=1}^{\infty} \sum_{k=k_1}^{k_n} |(x, u_{nk})|^2 \leq \sqrt{\lambda_1} \| x \|^2.$

Then $$\| Sx \| \leq \sqrt{\lambda_1} \| x \|$$

and so S is continuous. Again clearly S is positive and $S^2 = T$. So by Theorem 20.2.3, $S = T^{1/2}$. For each $m \in \mathbb{N}$, we define the finite rank operator S_m by

$$S_m(x) = \sum_{n=1}^{m} \sum_{k=k_1}^{k_n} \sqrt{\lambda_n} (x, u_{nk}) u_{nk}.$$

Then $$(T^{1/2} - S_m)(x) = \sum_{n=m+1}^{\infty} \sum_{k=k_1}^{k_n} \sqrt{\lambda_n} (x, u_{nk}) u_{nk}$$

and $$\| T^{1/2} - S_m \| \leq \sqrt{\lambda_{m+1}} \to 0 \text{ as } m \to \infty.$$

So by Theorem 15.19(iii), $T^{1/2}$ is also compact. □

20.2.8 **Corollary.** *For a compact operator* T *on a complex Hilbert space* H, *the linear operator* $(T^*T)^{1/2}$ *is also compact.*

20.3 Partial Isometries

In Section 13.10 we saw that any continuous linear operator on a Hilbert space can be expressed as the sum of self-adjoint operators. There is another class of continuous linear operators on a Hilbert space which is particularly useful for the decomposition of any continuous linear operator.

20.3.1 **Definition.** A continuous linear operator V on a Hilbert space H is called a *partial isometry* if there exists a closed subspace M of H such that

$$\| Vx \| = \| x \| \quad \text{for all } x \in M$$

and

$$Vx = 0 \quad \text{for all } x \in M^{\perp}.$$

We call M the *initial space* of V and $N = V(M)$ the *final space* of V. From Theorem 14.9, $M \oplus M^{\perp} = H$ so $N = V(H)$.

Clearly, every isometric isomorphism and every orthogonal projection is a partial isometry.

Partial isometries have the following elementary properties.

20.3.2 **Theorem.** *Consider a partial isometry* V *on a complex Hilbert space* H *with initial and final spaces* M *and* N. *Denoting by* P_1 *and* P_2 *the orthogonal projections of* H *onto* M *and* N, *then*

(i) $V^*V = P_1$

(ii) $VV^* = P_2$ *and*

(iii) V^* *is a partial isometry with initial and final spaces* N *and* M.

Proof.

(i) Now $V = VP_1$ so for $x \in H$,

$$\begin{aligned}(V^*Vx, x) &= (Vx, Vx) = (VP_1x, VP_1x)\\ &= (P_1x, P_1x) \quad \text{since } \| Vx \| = \| x \| \text{ for all } x \in M\\ &= (P_1x, x).\end{aligned}$$

By Lemma 13.10.6(i) we have that $V^*V = P_1$.

(ii) Again $V = P_2V$ so $V^* = V^*P_2$.

For $x \in H$, $P_2x = Vy$ for some $y \in H$.

Then $VV^*x = VV^*P_2x = VV^*Vy = VP_1y = Vy = P_2x$ and so $VV^* = P_2$.

(iii) For $x \in N$ we have $\| V^*x \|^2 = (VV^*x, x) = (P_2x, x) = \| x \|^2$

and for $x \in N^{\perp}$ we have $P_2x = 0$ so $V^*x = V^*P_2x = 0$.

Therefore, V^* is a partial isometry with initial space N. □

For the decomposition we need the following elementary properties.

20.3.3 **Lemma.** *For a continuous linear operator* T *on a complex Hilbert space* H,

(i) $\ker T^*T = \ker T$ *and*

(ii) $\overline{T^*T(H)} = \overline{T^*(H)}$.

Proof.

(i) This follows from $\| Tx \|^2 = (T^*Tx, x)$ for all $x \in H$, and Corollary 13.10.6.

(ii) From Theorem 14.11(ii), $\overline{T^*T(H)} = (\ker T^*T)^{\perp} = (\ker T)^{\perp}$ from (i)

$= \overline{T^*(H)}$ from Theorem 14.11(ii) □

20.3.4 The Polar Decomposition Theorem.
Consider a continuous linear operator T *on a complex Hilbert space* H *and write* $M \equiv \overline{T^*(H)}$ *and* $N \equiv \overline{T(H)}$. *There exists a partial isometry* V *with initial space* M *and final space* N *and a positive operator* S *on* H *such that*

$$T = VS \quad \textit{and} \quad S = V^*T$$

and such V *and* S *are uniquely determined.*

Proof. Consider $S \equiv (T^*T)^{1/2}$.
By Corollary 20.2.6, $\|Sx\| = \|Tx\|$ for all $x \in H$.
By Lemma 20.3.3(ii), $\overline{S(H)} = \overline{S^2(H)} = \overline{T^*T(H)} = \overline{T^*(H)}$.
So there exists an isometric isomorphism $Sx \mapsto Tx$ mapping from $S(H)$ a dense subspace of M onto $T(H)$ a dense subspace of N.
Now from Theorem 4.8 this mapping has a unique isometric isomorphic extension W from M onto N such that $WSx = Tx$ for all $x \in H$.
The continuous linear operator V on H defined by

$$\begin{aligned} Vy &= Wy \quad \text{for } y \in M \\ &= 0 \quad \text{for } y \in M^\perp \end{aligned}$$

has the property that $\|VSx\| = \|Tx\| = \|Sx\|$ for all $x \in H$
so $\|Vy\| = \|y\|$ for all $y \in M$.
Then V is a partial isometry with initial space M and final space N.
Further, $Tx = VSx$ for all $x \in H$.

Now by Theorem 20.3.2(i), V^*V is the orthogonal projection of H onto M and $M = \overline{S(H)}$, so we have $Sx = V^*VSx = V^*Tx$ for all $x \in H$.

To prove uniqueness, suppose that there exists a partial isometry V_1 and a positive operator S_1 such that $T = V_1S_1$ and $S_1 = V_1^*T$.
Then by Theorem 20.3.2(ii), $V_1V_1^*$ is the orthogonal projection onto N and $N = \overline{T(H)}$, so we have $S_1^2 = S_1^*S_1 = T^*V_1V_1^*T = T^*T$.
But the positive square root of T^*T is unique so $S_1 = S$. For each $x \in H$,

$$VSx = Tx = V_1Sx.$$

That is, $Vy = V_1y$ for all $y \in S(H)$.
But as V and V_1 are continuous on M so

$$Vy = V_1y \quad \text{for all } y \in M = \overline{S(H)}.$$

Also from Theorem 20.3.2(i), $V_1^*V_1$ is the orthogonal projection onto M and by Lemma 20.3.3(i) and Theorem 14.14, $\ker V_1 = \ker V_1^*V_1 = M^\perp$
so $Vx = V_1x$ for all $x \in M^\perp$. □

20.4 **The Spectral Theorem.**

We now apply the Polar Decomposition Theorem 20.3.4 to extend the Spectral Theorem 19.3 for compact normal operators to compact operators in general.

Consider a compact operator T on a complex Hilbert space H, then by Corollary 20.10, $S \equiv (T^*T)^{1/2}$ is also a compact positive operator. So we have a sequence $\{\lambda_1, \lambda_2, \ldots, \lambda_n, \ldots\}$ of positive eigenvalues of S where $\lambda_n \geq \lambda_{n+1}$ for all $n \in \mathbb{N}$ and an orthonormal sequence

$$\{u_{11}, u_{12}, \ldots, u_{1k_1}, u_{21}, u_{22}, \ldots, u_{2k_2}, \ldots, u_{n1}, u_{n2}, \ldots, u_{nk_n}, \ldots\}$$

where for each $n \in \mathbb{N}$, $\{u_{n1}, u_{n2}, \ldots, u_{nk_n}, \ldots\}$ is an orthonormal basis for N_n the eigenspace associated with λ_n.

20.4.1 **Theorem.** *Given a compact operator* T *on a complex Hilbert space* H, *there exists an orthonormal sequence*

$$\{v_{11}, v_{12}, \ldots, v_{1k_1}, v_{21}, v_{22}, \ldots, v_{2k_2}, \ldots, v_{n1}, v_{n2}, \ldots, v_{nk_n}, \ldots\}$$

such that

$$Tx = \sum_{n=1}^{\infty} \sum_{k=k_1}^{k_n} \lambda_n (x, u_{nk}) v_{nk}.$$

Proof. By Theorem 19.3, $Sx = \sum_{n=1}^{\infty} \sum_{k=k_1}^{k_n} \lambda_n (x, u_{nk}) u_{nk}$.

But by the Polar Decomposition Theorem 20.3.4

$$Tx = VSx = \sum_{n=1}^{\infty} \sum_{k=k_1}^{k_n} \lambda_n (x, u_{nk}) v_{nk}$$

where $Vu_{nk} = v_{nk}$ for all $n \in \mathbb{N}$ and all $k \in \{k_1, k_2, \ldots, k_n\}$.

Now V is a partial isometry with initial space $\overline{S(H)}$. But also by Theorem 20.3.2(i), V^*V is an orthogonal projection onto $\overline{S(H)}$. Now the orthogonal sequence

$$\{u_{11}, u_{12}, \ldots, u_{1k_1}, u_{21}, u_{22}, \ldots, u_{2k_2}, \ldots, u_{n1}, u_{n2}, \ldots, u_{nk_n} \ldots\}$$

lies in $\overline{S(H)}$, so $V^*Vu_{nk} = u_{nk}$ for all $n \in \mathbb{N}$ and all $k \in \{k_1, k_2, \ldots, k_n\}$.

Therefore, $(v_{nk_m}, v_{nk_p}) = (Vu_{nk_m}, Vu_{nk_p}) = (V^*Vu_{nk_m}, u_{nk_p}) = (u_{nk_m}, u_{nk_p})$ so $\{v_{11}, v_{12}, \ldots, v_{1k_1}, v_{21}, v_{22}, \ldots, v_{2k_2}, \ldots, v_{n1}, v_{n2}, \ldots, v_{nk_n}, \ldots\}$ is an orthogonal sequence and we have our result. □

20.5 **Remark.** The Spectral Theorem extends to normal operators on Hilbert space, but to develop the general case would require more background than we allow for here. A gentle account of this development can be found in G. Bachman and L. Narici, *Functional Analysis*, Academic Press, 1966, or a more sophisticated account can be found in W. Rudin, *Functional Analysis*, McGraw Hill, 2nd edn. 1991. □

20.5 EXERCISES

1. Consider a self-adjoint operator T on a complex Hilbert space H.
   (i) Prove that if $\sigma(T) \subseteq [0, \infty)$ then T is a positive operator on H.
   (ii) Generalise this result in (i) to prove that $\overline{W(T)} = \text{co}\,\sigma(T)$.
   (Hint: Use the functional calculus.)

2. Consider a continuous linear operator T on a complex Hilbert space H.
   Prove that
   (i) if S is a self–adjoint operator on H and
   $$\| Sx \| = \| Tx \| \quad \text{for all } x \in H$$
   then $S = (T^*T)^{1/2}$,
   (ii) if T is a positive operator then $\| T \| = \| T^{1/2} \|^2$.

3. Consider a positive operator T on a complex Hilbert space H.
   Prove that
   (i) $| (Tx, y) | \leq (Tx, x)^{1/2} (Ty, y)^{1/2}$ for all $x, y \in H$,
   (ii) $\| Tx \| \leq \| T \|^{1/2} (Tx, x)^{1/2}$ for all $x \in H$
   and deduce that $(Tx, x) = 0$ if and only if $Tx = 0$.

4. Consider a self–adjoint operator T on a complex Hilbert space H. Prove that there exist positive operators T^+ and T^- so that $T = T^+ - T^-$ and $T^+T^- = T^-T^+ = 0$.
   (Hint: Consider $T^+ = \frac{1}{2}(\sqrt{T^2}+T)$ and $T^- = \frac{1}{2}(\sqrt{T^2}-T)$.)

5. (i) Consider a positive operator T on a complex Hilbert space H. Determine $\sigma(T^{1/2})$ in terms of elements of $\sigma(T)$.
   (ii) Consider positive operators T_1 and T_2 on H where $T_1T_2 = T_2T_1$.
   (a) Prove that $(T_1T_2)^{1/2} = T_1^{1/2}T_2^{1/2}$.
   (b) Determine $\sigma((T_1T_2)^{1/2})$ in terms of elements of $\sigma(T_1)$ and $\sigma(T_2)$.

6. Consider a continuous linear operator T on a complex Hilbert space H. Prove that
   (i) T is a partial isometry if and only if T^*T is a projection,
   (ii) T is a partial isometry if and only if $T = TT^*T$.

7. Consider $\{\mu_n\}$ a decreasing sequence of positive real numbers which is finite or convergent to 0 and $\{e_n\}$ and $\{u_n\}$ orthonormal sequences in a Hilbert space H.
   (i) Prove that the operator T defined on H by
   $$Tx = \sum \mu_n (x, e_n) u_n$$
   is a compact operator on H.
   (ii) Prove that $T^*y = \sum \mu_n (y, u_n) e_n$.
   (iii) Prove that $(T^*T)^{1/2} x = \sum \mu_n (x, e_n) e_n$.

8. (i) Given a positive operator T on a complex Hilbert space H, prove that if T is invertible on H then so is $T^{1/2}$.

(ii) Given a continuous linear operator T on a complex Hilbert space H, prove that if T is invertible then it has a unique polar decomposition T = US where U is a unitary operator and S is a positive operator on H.

APPENDIX

A.1 Zorn's Lemma

We confine ourselves to developing sufficient set theory to explain the statement of Zorn's Lemma and to enable us to apply it.

A.1.1 **Definitions.** Given a nonempty set X, a *partial order relation* $\precsim$ on X is a relation defined by the properties:

(i) $x \precsim x$ for every $x \in X$ (reflexivity)

(ii) $x \precsim y$ and $y \precsim x \Rightarrow x = y$ (antisymmetry)

(iii) $x \precsim y$ and $y \precsim z \Rightarrow x \precsim z$ (transitivity).

The set X together with the partial order relation $\precsim$, sometimes denoted $(X, \precsim)$, is called a *partially ordered set.*

It is clear that any subset of a partially ordered set is a partially ordered set in its own right.

Elements $x, y, \in X$ are said to be *comparable* with respect to the partial order relation $\precsim$ if $x \precsim y$ or $y \precsim x$. There may be pairs of elements in X which are not comparable.

If every pair of elements is comparable with respect to the partial order relation $\precsim$ then the relation is said to be a *total order relation* and $(X, \precsim)$ is called a *totally ordered set,* (or a *chain*).

Every subset of a totally ordered set is a totally ordered set in its own right. A partially ordered set has totally ordered subsets.

A.1.2 **Examples.**

(i) Consider $\mathbb{R}$ the set of real numbers.

The relation "less than or equal to" denoted by $\leq$ is a total order relation on $\mathbb{R}$.

(ii) Consider X a nonempty set and $\mathcal{S}$ the family of all subsets of X.

The relation "set inclusion" denoted by $\subseteq$ is a partial order relation on $\mathcal{S}$.

Not all subsets are comparable; for example, for $A, B \subseteq X$, $A, B, \neq \emptyset$ and $A \cap B = \emptyset$ we have $A \not\subseteq B$ and $B \not\subseteq A$.

An example of a totally ordered subset is an increasing sequence of sets; that is, $\{A_n\}$ where $A_1 \subseteq A_2 \subseteq \ldots \subseteq A_n \subseteq \ldots$.

(iii) Consider $\mathbb{N}$ the set of natural numbers.

The relation "divides" denoted by $|$ is a partial order relation on $\mathbb{N}$.

Again not all elements are comparable; for example, for 3 and 5, $3 \nmid 5$ and $5 \nmid 3$. The set $\{2^n : n \in \mathbb{N}\}$ is a totally ordered subset.

(iv) Consider $\mathcal{C}[0, 1]$ the linear space of continuous real mappings on [0, 1].

The relation "is dominated by" denoted by $\leq$ defined by

$$f \leq g \text{ when } f(x) \leq g(x) \quad \text{for all } x \in [0, 1]$$

is a partial order relation on $\mathcal{C}[0, 1]$.

Not all elements of $\mathcal{C}[0, 1]$ are comparable; for example, f and g where

$$f(x) = x^2, \quad g(x) = 1 - x^2.$$

The set $\{f_n : f_n(x) = 1 - x^n, n \in \mathbb{N}\}$ is a totally ordered subset. □

A.1.3 **Definitions.** Given a partially ordered set set $(X, \precsim)$ we say that an element $x_0 \in X$ is *maximal* if

$$x_0 \precsim x \quad \text{for } x \in X \Rightarrow x = x_0.$$

So a maximal element is not 'less than or equal to' any other element of X. This does not mean that it is the 'greatest' element in X, for it is not necessarily comparable with every element of X. A maximal element does not necessarily always exist, and if it does it is not necessarily a unique element of X.
Similarly, we say that an element $y_0 \in X$ is *minimal* if

$$x \precsim y_0 \quad \text{for } x \in X \Rightarrow x = y_0.$$

A.1.4 **Examples.**
(i) In $(\mathcal{B}, \subseteq)$, $\mathcal{B}$ has a single maximal element X.
(ii) In $(\mathbb{N}, |)$, $\mathbb{N}$ does not have a maximal element.
In the subset $A \equiv \{1, 2, \ldots, 10\}$ we see that 6, 7, 8, 9, 10 are maximal elements of A.

□

A.1.5 **Definitions.** Given a partially ordered set $(X, \precsim)$, for subset $A \subseteq X$ we say that $x_0 \in X$ is *an upper bound* for A if

$$a \precsim x \quad \text{for all } a \in A.$$

Similarly, we say that $y_0 \in X$ is *a lower bound* for A if

$$y_0 \precsim a \quad \text{for all } a \in A.$$

An upper or lower bound for A is not necessarily a member of A but is comparable with every member of A.
We say that $x_0 \in X$ is a *least upper bound* or a *supremum* of A if x_0 is an upper bound for A and

$$x_0 \precsim x \quad \text{for all upper bounds x of A,}$$

and $y_0 \in X$ is a *greatest lower bound* or an *infimum* of A if y_0 is a lower bound for A and

$$y \precsim y_0 \quad \text{for all lower bounds y of A.}$$

A subset A of a partially ordered set $(X, \precsim)$ need not have an upper bound and if it does it does not necessarily have a supremum. However, if A has a supremum then it is a unique element of X. A similar statement could be made for lower bounds and infimum. In view of this we denote the supremum of the set A by sup A and the infimum by inf A.

A.1.6 **Examples.**
(i) In $(\mathcal{B}, \subseteq)$, consider a subfamily $\mathcal{Q} \subseteq \mathcal{B}$.
Now X is an upper bound for $\mathcal{Q}$ and $\varnothing$ is a lower bound for $\mathcal{Q}$.
sup $\mathcal{Q}$ is the union of all sets in $\mathcal{Q}$ and inf $\mathcal{Q}$ is the intersection of all sets in $\mathcal{Q}$.

(ii) In $(\mathbb{N}, |)$ consider a subset $A \equiv \{4, 6, 8\}$.

An upper bound for A is any $n \in \mathbb{N}$ divisible by 4, 6 and 8; for example, 24, 48, ...

sup A is the lowest common multiple of the set A, so sup A = 24.

A lower bound for A is any $n \in \mathbb{N}$ which divides into 4, 6 and 8; for example, 1, 2.

inf A is the greatest common divisor of the set A so inf A = 2.

(iii) In $(\mathbb{R}, \leq)$, consider the subset $\mathbb{Q}$ of rational numbers. Now $\mathbb{Q}$ does not have an upper bound.

Consider the subset $A \equiv \{x \in \mathbb{Q} : x^2 \leq 2\}$.

An upper bound for A is 2, but sup A does not exist in $\mathbb{Q}$.

(iv) In $(\mathcal{C}[0, 1], \leq)$ consider any finite set $A \equiv \{f_k : k \in \{1, \dots, n\}\}$.

Now
$$\sup A = \vee\{f_k : k \in \{1, \dots, n\}\} \text{ and}$$
$$\inf A = \wedge\{f_k : k \in \{1, \dots, n\}\};$$

For the set $B \equiv \{f_n : f_n(x) = 1 - x^n, n \in \mathbb{N}\}$ we see that B has f_0 as an upper bound where
$$f_0(x) = 1 \quad \text{for } x \in [0, 1],$$
and in fact $f_0 = \sup B$. □

We are now in a position to state Zorn's Lemma which asserts the existence of certain elements in certain partially ordered sets.

A.1.7 **Zorn's Lemma.**

In a nonempty partially ordered set where every totally ordered subset has an upper bound, there exists at least one maximal element.

It can be shown that Zorn's Lemma is equivalent to the Axiom of Choice which is a set theory axiom independent of and additional to those more generally accepted axioms of set theory. Among mathematical logicians there is considerable discussion about the Axiom of Choice. Among analysts, the constructive analysts are very critical of its use. However, classical functional analysis does rely at significant points on the Axiom of Choice in the form of Zorn's Lemma.

A.2 Numerical equivalence

A.2.1 **Definitions.** Given a family of all subsets of some universal set we say that a subset X is *numerically equivalent* to a subset Y if there exists a one-to-one mapping of X onto Y.

Clearly, numerical equivalence is an equivalence relation on the family of all subsets which partitions subsets into equivalence classes. We call the equivalence classes *cardinal numbers*.

For a universal set which contains the set of real numbers, given $n \in \mathbb{N}$, the equivalence class which contains the subset $\{1, 2, \dots, n\}$ is called the cardinal number n. Any set which belongs to this class is said to be *finite* and have n elements. A set is said to be *infinite* if it is not finite and is said to be *countably infinite* if it is numerically equivalent to the set of natural numbers $\mathbb{N}$. A set is said to be *countable* if it is either finite or countably infinite and is said to be *uncountable* if it is not countable.

The following theorem is important in establishing numerical equivalence of sets.

A.2.2 The Schroeder–Bernstein Theorem.

For nonempty sets X *and* Y *if there exists a one-to-one mapping* f *of* X *into* Y *and a one-to-one mapping* g *of* Y *into* X, *then there exists a one-to-one mapping* h *of* X *onto* Y.

Proof. We divide X into three subsets as follows.
For $x \in X$, if $x \in g(Y)$ call $g^{-1}(x) \in Y$ the *first ancestor* of x,
if $g^{-1}(x) \in f(X)$ call $f^{-1}(g^{-1}(x)) \in X$ the *second ancestor* of x,
if $f^{-1}(g^{-1}(x)) \in g(Y)$ call $g^{-1}\big(f^{-1}(g^{-1}(x))\big) \in Y$ the *third ancestor* of x.

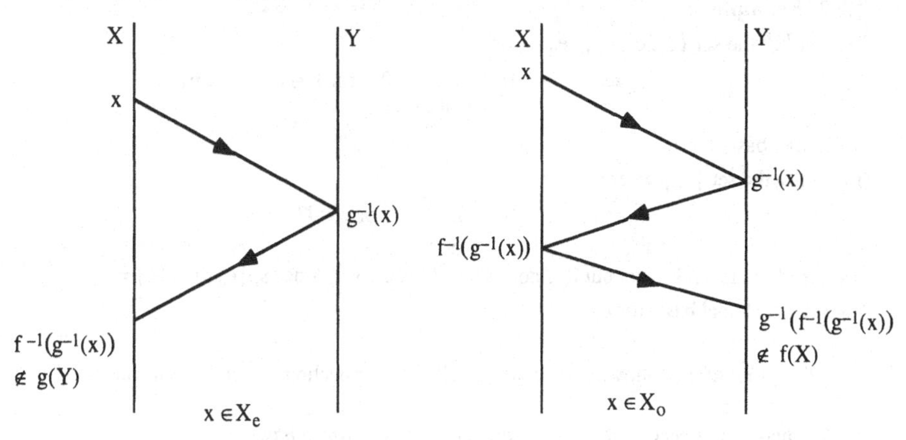

Figure 20. Tracing ancestors.

Continuing this process we see that there are three possibilities:

(i) x has an infinite number of ancestors, we define the set
$$X_i \equiv \{x \in X : x \text{ has an infinite number of ancestors}\}.$$

(ii) x has an even number of ancestors; that is, the last ancestor of x is a member of X; we define the set
$$X_e \equiv \{x \in X : x \text{ has an even number of ancestors}\}.$$

(iii) x has an odd number of ancestors; that is, the last ancestor of x is a member of Y; we define the set
$$X_o \equiv \{x \in X : x \text{ has an odd number of ancestors}\}.$$

Now the sets X_i, X_e and X_o are mutually disjoint and their union is X. We form a division of Y into sets Y_i, Y_e and Y_o in the same way.

We have that f maps X_i onto Y_i and X_e onto Y_o and g^{-1} maps X_o onto Y_e, so we define the mapping $h: X \to Y$ by
$$\left.\begin{aligned} h(x) &= f(x) \quad x \in X_i \cup X_e \\ &= g^{-1}(x) \quad x \in X_o \end{aligned}\right\}$$
and h is a one-to-one mapping of X onto Y. □

A.3 Hamel basis

Important theory for infinite dimensional linear spaces is constructed on the assumption of Zorn's Lemma.

A.3.1 **Definition.** A *Hamel basis* for a linear space X is a linearly independent set which spans X.

It is clear that a nonempty subset A of X is a Hamel basis for X if and only if each element $x \in X$ can be expressed uniquely as a linear combination of elements of A.

A.3.2 **Examples**

(i) In $\mathbb{R}^n$ the set $\{e_1, e_2, \dots, e_n\}$ where

$$e_k \equiv \{0, \dots, 0, \underset{k\text{th place}}{1}, 0, \dots, 0\} \quad \text{for } k \in \{1, \dots, n\}$$

is a Hamel basis for $\mathbb{R}^n$.

(ii) In c_0 the set $\{e_n\}$ where

$$e_n \equiv \{0, \dots, 0, \underset{n\text{th place}}{1}, 0, \dots\} \quad \text{for } n \in \mathbb{N}$$

is a linearly independent set but it is not a Hamel basis for c_0 since $sp\{e_n : n \in \mathbb{N}\} = E_0$. The set is a Hamel basis for E_0. □

The proof of the following existence result is an application of Zorn's Lemma.

A.3.3 **Theorem.** *Every nontrivial linear space X has a Hamel basis.*

Proof. Consider the family $\mathfrak{F}$ of all linearly independent sets in X, partially ordered by set inclusion.
For any $x \in X$, $x \neq 0$, the set $\{x\}$ is a linearly independent set so $\mathfrak{F}$ is nonempty.
Consider a totally ordered subfamily in $\mathfrak{F}$.
It is clear that the union of sets in this subfamily is a linearly independent set and is an upper bound for the totally ordered subfamily.
By Zorn's Lemma $\mathfrak{F}$ has a maximal linearly independent set A.
Now for any $x \notin A$ it follows that $A \cup \{x\}$ is a linearly independent set in X.
So there exist scalars $\{\lambda, \lambda_1, \dots, \lambda_n\}$ and a subset $\{e_1, \dots, e_n\}$ of A such that
$\lambda x + \sum_{k=1}^{n} \lambda_k e_k = 0$. Now $\lambda \neq 0$ since $\{e_1, \dots, e_n\}$ is a linearly independent set, so

$$x = -\frac{1}{\lambda} \sum_{k=1}^{n} \lambda_k e_k.$$

We conclude that A spans X and therefore A is a Hamel basis for X. □

We have a significant cardinality result for Hamel bases of linear spaces.

We need the following cardinality property, which is itself dependent on Zorn's Lemma.

A.3.4 **Proposition**. *Given an infinite set* A, *the set* $\bigcup \{N_a : a \in A\}$ *where* N_a *is countable for each* $a \in A$, *has the same cardinality as* A.

A.3.5 **Theorem**. *In any given nontrivial linear space* X, *any two Hamel bases are numerically equivalent.*

Proof. The case when X is finite dimensional we take as a well known linear algebra result. Consider the case when X is infinite dimensional.

Suppose that there exist two Hamel bases, $B_1 \equiv \{e_\alpha\}$ and $B_2 \equiv \{f_\beta\}$ in X. Each $e_\alpha \in B_1$ is a linear combination of elements from B_2.

For each $e_\alpha \in B$, write $B_2(e_\alpha)$ as the finite subset of B_2 which produces e_α by linear combination.

Every member of B_2 is a member of at least one of the sets $B_2(e_\alpha)$; for otherwise, if there exists an $f_\beta \in B_2$ such that $f_\beta \notin B_2(e_\alpha)$ for all α, then f_β is a linear combination of elements from B_1, say $\{e_{\alpha_k} : k \in \{1, \ldots, n\}\}$ and each e_{α_k} is a linear combination of elements from B_2, so f_β is a linear combination of elements from B_2 and then B_2 is not linearly independent.

Therefore, $B_2 \subseteq \bigcup_\alpha B_2(e_\alpha)$.

So by Proposition A.3.4. there exists a one-to-one mapping of B_2 into B_1. By symmetry of argument and the Schroeder–Bernstein Theorem A.2.2 we have our result. □

So we are led to the following definition.

A.3.6 **Definition**. The *Hamel dimension* of a linear space X is the cardinal number of a Hamel basis for X.

This definition is consistent with the definition of dimension for finite dimensional linear spaces and is simply an extension of this notion for infinite dimensional linear spaces.

HISTORICAL NOTES

§1. Introduction

A persistent theme of mathematics in the nineteenth century was the *pursuit of rigour* in establishing foundations, in definition of terms and in the presentation of proofs. In analysis, the great proponents of rigour were Augustin-Louis Cauchy early in the century and Karl Weierstrass towards the end. The axiomatising of the real number system by Julius Dekekind and Georg Cantor had a determining effect and became the foundation on which analysis depends.

The outstanding development of mathematics in the twentieth century has been the study of *structure* and the promotion of the *axiomatic method* in exploring the implications of that structure. It is worthwhile reflecting on the principal stages which led up to this development to appreciate the unifying effect of the study of structure and the simplifying power that an axiomatic analysis of structure has brought to a variety of applications.

The significance of structure first became apparent from an algebraic point of view. Until the middle of the nineteenth century mathematicians had been dealing with well determined mathematical objects: numbers, points, curves, functions, ... But it came to be realised that the algebraic manipulation of these different objects had a remarkable similarity. The essence of these manipulations did not lie in the nature of the objects but in the rules for handling them which were often the same for different types of objects. The precise formulation of this perception had to wait for the development of set-theoretic concepts and language and it was only towards the end of the century that an abstract group defined on an arbitrary underlying set was to become an area of serious study.

By the turn of the century no similar development had occurred in analysis. Extensions of the ideas of limits and continuity which had been formulated were always relative to special objects such as functions or curves. The possibility of defining such notions on an arbitrary set by a generalised notion of distance was first put forward by Maurice Fréchet in his doctoral thesis of 1906. On the simple structure of a metric space it is possible to extend most of the arguments concerning neighbourhoods, limits and continuity which are familiar in Euclidean space.

Until the middle of the nineteenth century Euclid's geometry was considered a beautiful logical and orderly description of the natural spatial world. In the first half of the nineteenth century the advances in projective geometry and the geometry of many dimensions had started to change this view. The solution of the problem of the Parallel Axiom of Euclid radically changed the outlook of mathematicians. The axioms were no longer self-evident truths but independent hypotheses although they may have been suggested by some real world situation. The importance of Euclid's achievement is that it is a prototype of the *axiomatic method*; its usefulness is that it is a convenient model of our spatial world locally. David Hilbert's refinement of Euclid's axioms at the turn of the century, extracting from them any remaining intuitive ideas, had considerable influence in promoting the axiomatic method. He stressed that the axioms concerned undefined objects called, point, line and plane and the relations between these objects and that they were independent of physical reality. All of Euclid's geometry was then deduced consistently from Hilbert's axioms by the axiomatic method.

The outcome of the pursuit of rigour in the nineteenth century has been the analysis of structures by the axiomatic method in the twentieth century. Jean Dieudonné declares: "there can be no rigorous proof except in the context of an axiomatic theory in which objects and basic relations have been specified and the axioms by which they are connected have been exhaustively listed".†

§2. **The development of function spaces**

It was Weierstrass who defined the concept of uniform convergence in 1841 and had made its importance generally appreciated. Weierstrass' Approximation Theorem of 1885, showing that polynomials are uniformly dense in the continuous functions, (AMS §9), gave further impetus to the idea of a *function space*, where the functions are treated as points in a space of functions. The Ascoli–Arzelà Theorem of 1883 which characterises compactness of a set of continuous functions, (AMS §9), was a landmark in establishing the focus on function spaces.

Many of the problems in mathematical physics studied by mathematicians in the nineteenth century resolved themselves into finding the solution of an integral equation. For example, Fourier studying heat flow in 1811 considered the equation

$$f(x) = \int_0^\infty \cos(xt)\; y(t)\; dt.$$

Abel's investigation of the tautochrome problem in 1822 led to the equation

$$f(x) = \int_0^x \frac{y(t)}{\sqrt{x-t}}\; dt, \quad f(0) = 0.$$

In both cases the aim is to find the function y given the function f.

Early in the century individual problems were considered in isolation and solved by particular methods. But towards the end of the century mathematicians had begun the quest for a general theory for the solution of *integral equations.*

In 1895 Jean-Marie Le Roux and in 1896 Vita Volterra published the first existence and uniqueness theorems for general classes of integral equations. Volterra studied equations of the form

$$y(x) = f(x) + \lambda \int_0^x K(x,t)\; y(t)\; dt.$$

† Jean Dieudonné, *Mathematics – the Music of Reason*, translated by H.G. and J.C. Dales, Springer Verlag, Berlin 1992, p 237.

The function K(x,t) is the kernel of the equation and is symmetric if K(x, t) = K(t, x). Given functions f and K the aim is to find the function y. Abel's equation is related to this form.

A decisive contribution to the theory of integral equations was made by Ivar Fredholm in 1900 and published in full in 1903. He studied a class of integral equations more general than those of Volterra and of the form

$$y(x) = f(x) + \lambda \int_a^b K(x,t)\, y(t)\, dt.$$

Fourier's heat flow equation is related to this form.

The papers of Volterra and particularly Fredholm devising a general theory gave impetus to the consideration of integral operators acting on a function space. Hilbert was immediately attracted to Fredholm's work. He published six papers on integral equations and by the fourth of these papers, published in 1906, he had begun *spectral analysis* of compact operators on function spaces. He had laid the foundation for what came to be called spectral theory on Hilbert space.

For the Fredholm integral equation with f and symmetric K continuous he showed that the eigenvalues for the homogeneous equation

$$y(x) = \lambda \int_a^b K(x,t)\, y(t)\, dt$$

are real and can be ordered. He then defined eigenfunctions corresponding to eigenvalues and showed that they form an orthogonal family which give solutions to the homogeneous equation. He defined Fourier coefficients for a function with respect to normalised eigenfunctions and showed that any function f of the form

$$f(x) = \int_a^b K(x,t)\, g(t)\, dt$$

for some continuous function g, can be expressed as a Fourier series in the eigenfunctions of the homogeneous equation. He then reworked his theory based on a complete orthogonal system of functions and showed that what is essential for the spectral theory of an integral operator is that it is a compact operator not that it has a representation in terms of integrals.

It was Hilbert's student Erhard Schmidt who in his doctoral thesis of 1905 introduced a simplified approach to Hilbert's theory of integral equations. In his paper of 1908, he developed a geometric approach to Hilbert space theory and set the pattern for the modern theory of abstract Hilbert space. Hilbert, in his work on integral equations, had considered a function as represented by its Fourier coefficients in an expansion with respect to an orthonormal sequence of functions. These coefficients are sequences $\{x_n\}$ where $\sum x_n^2 < \infty$. Schmidt considered these sequences as co-ordinates of a point in an infinite dimensional space, *ℓ_2-space*.

In his analysis of ℓ_2-space which included complex sequences, Schmidt introduced the notation for an inner product and norm. He used the Cauchy–Schwarz and Bessel inequalities and proved that ℓ_2-space is complete, a vital property for the development of the general theory. He had in his thesis used the Gram–Schmidt orthogonalisation process and he now developed the notion of orthogonal projection.

Of crucial importance was the theory of *Lebesgue integration* which was developed by Henri Lebesgue in his doctoral thesis presented in 1902. Building on the work of Emile Borel on measure, Lebesgue constructed a new general and powerful theory of integration. This theory was much more satisfactory than Riemann's and had the advantage that under mild conditions the limit of an integral is the integral of the limit. It was then possible to consider function spaces which are complete metric spaces.

In 1907, Frederic Riesz making use of Lebesgue's new theory of integration and simultaneously with Ernst Fischer, proved the *Riesz–Fischer Theorem* which implied that the space $\mathcal{L}_2[a,b]$ of all Lebesgue square integrable functions on [a,b] with mean square norm is complete and that it is isometrically isomorphic to ℓ_2-space. Following the publication of this theorem both Schmidt and Fréchet remarked that the space $\mathcal{L}_2[a,b]$ has a geometry analogous to ℓ_2-space.

In the same year Fréchet and Riesz independently obtained the representation theorem for continuous linear functionals on $\mathcal{L}_2[a,b]$ which was later generalised by Riesz for abstract Hilbert space. In 1909, Riesz obtained a representation theorem for continuous linear functionals on $\mathcal{C}[a,b]$ with the supremum norm by Stieltjes integrals.

In 1910 Riesz generalised the study of $\mathcal{L}_2[a,b]$-space by considering the $\mathcal{L}_p[a,b]$ spaces where $1 < p < \infty$ consisting of functions whose pth power is Lebesgue integrable on [a,b] with the p-norm. He used Hölder and Minkowski inequalities and established the duality between spaces $\mathcal{L}_p[a,b]$ and $\mathcal{L}_q[a,b]$ where $\frac{1}{p} + \frac{1}{q} = 1$. He extended the Riesz–Fischer Theorem for these spaces and proved that these spaces are complete. He obtained the representation theorem for continuous linear functionals for these spaces. This was a major step in generalising from Hilbert function space to more general function spaces.

Riesz's study of integral equations in $\mathcal{L}_p[a,b]$-space where $1 < p < \infty$ laid the foundation for operator theory on abstract normed linear spaces. He considered integral equations which generate linear operators on $\mathcal{L}_p[a,b]$-space and he defined continuity and the norm of continuous linear operators on such spaces. He also introduced the notion of the conjugate of an operator:

For a continuous linear operator T on $\mathcal{L}_p[a,b]$ where $1 < p < \infty$, given $g \in \mathcal{L}_q[a,b]$ where $\frac{1}{p} + \frac{1}{q} = 1$,

$$\int_a^b T(f(x))\, g(t)\, dx$$

defines a continuous linear functional on $\mathcal{L}_p[a,b]$. By the Riesz Representation Theorem there exists a function $h \in \mathcal{L}_q[a,b]$ where $\frac{1}{p} + \frac{1}{q} = 1$, unique in the Lebesgue sense, such that

$$\int_a^b T(f(x))\, g(x)\, dx = \int_a^b f(x)\, h(x)\, dx.$$

The operator T' conjugate to T is defined on $\mathcal{L}_q[a,b]$ by

$$T'(g) = h$$

and he showed that T' is also continuous linear and $\| T' \| = \| T \|$. He considered inverses of operators and related them to inverses of their conjugates.

The work of Riesz on compact operators, which he published in 1918, was extraordinarily powerful. He gave a definition for compactness of an operator which greatly improved on that of Hilbert and he developed it for more general Banach spaces. Although his theory was developed for the function space $\mathcal{C}[a,b]$ he regarded it to be of general applicability. He used the term *norm* for the supremum norm in $\mathcal{C}[a,b]$. He essentially derived the spectral decomposition for compact operators on Banach spaces. His work was revised and extended in 1930 by Juliusz Schauder and for this reason is called the *Riesz–Schauder Theory* of compact operators.

Mathematicians in the eighteenth and nineteenth century were also concerned with problems in the *calculus of variations*. Here the aim is to find a function y which satisfies certain constraints and minimises or maximises an integral of the form

$$J(y) = \int_a^b F(x,y,y')\, dt .$$

For example, in 1696 John Bernoulli proposed the brachistochrone problem, that is, to find the curve joining two points down which a particle will slide in minimum time. This led to finding the function y which minimises the integral

$$J(y) = \int_a^b \sqrt{\frac{1+[y'(x)]^2}{a-y(x)}}\, dx.$$

Volterra interpreted the general calculus of variations problem as one of minimising the functional J on the function space $\mathcal{C}[a,b]$.

Jacques Hadamard in 1903 took up the study of functionals in relation to the calculus of variations. The term *functional* is due to him. He began the study of linear functionals on a function space. The term *functional analysis* was first used by Paul Lévy in his book published in 1922 when the main interest was on functionals acting on a function space.

§3. The axiomatic analysis of normed linear spaces

In his doctoral thesis of 1920, the Polish mathematician Stefan Banach defined an *abstract complete normed linear space* with axiomatic structure as is now commonly in use. Independently and at the same time Hans Hahn published the same set of axioms. We have seen that the norm had been defined for particular function spaces and in 1921 Eduard Helly explored a similar structure for sequence spaces. Banach had considered normed linear spaces over the real numbers but in 1923 Norbert Wiener pointed out that the theory could be extended with wider applications using normed linear spaces over the complex numbers.

After the first World War, mathematics in Poland experienced a particularly influential period of growth. The story of Banach's recruitment into the mathematical academic community is of interest because he came from a poor background and had to begin earning his own living at the age of fifteen. In 1916, Professor Hugo Steinhaus walking in Cracow Park overheard two students discussing the new theory of Lebesgue measure and integration; they were Stefan Banach and Otto Nikodỳm. Steinhaus formed a mathematical club which began to meet in his rooms. Steinhaus claimed that Banach was his "greatest mathematical discovery".

Considering the achievement of Fréchet in defining abstract metric spaces and the similarities in treatment of the different function spaces it seems odd that it took so long to introduce an axiomatic analysis of normed linear spaces. Nevertheless, there was a gradual understanding that many diverse problem areas shared this discernable common structure which was amenable to axiomatic formulation and solution as a consequence of the axiomatic method.

The stunning first success of this abstract analysis was Banach's general formulation of his fixed point theorem for contraction mappings which appeared in his doctoral thesis, (AMS §5).

It was Helly who brought a geometric aspect to the study. Exploring geometrical notions in Euclidean space he showed that a closed symmetric convex body with 0 as an interior point, as discussed by Herman Minkowski, could be considered as a closed unit ball under an appropriate norm. This enabled a useful visualisation of problems. The geometry of normed linear spaces is a significant metric generalisation of Euclidean geometry.

Normed linear spaces form a subclass of metric spaces. But there had been an earlier generalisation to a larger class of spaces. In his book of 1914, Felix Hausdorff defined a *general topological space* based on the idea of neighbourhood rather than distance and he was able to develop a rich analysis of limits and continuity. His second book, published in 1927, where he concentrated on metric topology, was particularly influential for the Banach school. The development of general topology led to the significant study of weak topologies on normed linear spaces which was used to great effect in Banach's book of 1932. But this aspect of the analysis of normed linear spaces follows naturally as a particular application of the more general analysis of locally convex spaces.

What makes the study of normed linear spaces particularly fruitful is the linkage between the algebraic and topological structures, the axiomatic prescription of a continuity condition associated with the basic algebraic operations of addition and multiplication by a scalar. It seems that Fréchet was the first to make this observation explicitly. This led naturally to the definition of a *linear topological space* which is both a linear space and a topological space where there is a similar continuity condition associated with the algebraic operations, in particular where neighbourhoods of a point are translations of scalar multiples of neighbourhoods of the origin.

The convexity of the balls in a normed linear space is a significant property and a linear topological space with a neighbourhood base of convex subsets was first given general definition as a *locally convex space* by John von Neumann in 1935. Many more example spaces were available to be investigated by this expansion of the theory.

Progress in the analysis of abstract normed linear spaces depends crucially on the *Hahn–Banach Theorem.* Minkowski working geometrically had established what amounted to the Hahn–Banach theorem for finite dimensional normed linear spaces, that is, to each point in the boundary of the closed unit ball there exists at least one hyperplane of support containing the point. He established the dual property with the space of all continuous linear functionals under the dual norm.

Helly generalised these ideas for certain sequence spaces which are separable normed linear spaces. However, he confronted the problem that his infinite dimensional spaces might not be self-dual. He gave the first example of a non-reflexive Banach space by exhibiting a continuous linear functional which does not attain its norm on the closed unit ball of the space.

In 1927 Hahn, examining the work of Helly, proved the extension form of the theorem for normed linear spaces without any assumption of separability. For a separable space induction can be used, but he introduced the method of transfinite induction for the first time into the study of normed linear spaces. He finally defined the dual of a normed linear space with dual norm and considered the natural embedding of the space into its second dual and also the notion of reflexivity. He had, with this paper, begun the study of duality theory for general normed linear spaces.

In 1929 Banach, unaware of Hahn's work, published the same theorem but in a more general form and more obviously as an extension theorem. Given a linear functional on a subspace dominated by a convex functional, his argument guaranteed the existence of a dominated extension of the linear functional to the whole space. Banach's form of the theorem really belongs in the theory of linear spaces quite apart from any topology. It was crucial in the development of the theory of locally convex spaces.

Although the general proof was devised using transfinite induction, since the 1940s the proof is generally given using Zorn's Lemma. The generalisation of the theorem to normed linear spaces over the complex numbers was given by Bohnenblust and Sobczyk.

The Hahn–Banach Theorem in its more general form implies many fundamental existence properties on a normed linear space X over the real numbers:

A continuous convex function ϕ on an open convex subset A is said to be *subdifferentiable* at $x_0 \in A$ if there exists a continuous linear functional f on X such that

$$f(x-x_0) \leq \phi(x) - \phi(x_0) \qquad \text{for all } x \in A.$$

The Hahn–Banach Theorem implies that ϕ is subdifferentiable at each point of A.

A closed convex set K is said to have a *support point* $x_0 \in K$ if there exists a continuous linear functional f on X such that

$$f(x_0) = \sup f(K).$$

The Hahn–Banach Theorem implies that if K has non-empty interior then every point in the boundary of K is a support point of K.
Two nonempty convex sets K_1 and K_2 are said to be *separated* by a hyperplane if there exists a continuous linear functional f on X and $\alpha \in \mathbb{R}$ such that

$$K_1 \subseteq \{x \in X : f(x) \leq \alpha\} \text{ and } K_2 \subseteq \{x \in X : f(x) \geq \alpha\}.$$

The Hahn–Banach Theorem implies that if int $K_1 \neq \varnothing$ and $K_2 \cap$ int $K_1 = \varnothing$ then K_1 and K_2 can be separated by a closed hyperplane.

The development of set theory at the end of the nineteenth century had a profound influence. In 1899, René Baire proved that in Euclidean space a countable intersection of dense open sets is dense, a result which could be extended for any complete metric space. Baire's work enabled the adoption of non-constructive proofs as a technique in analysis.

In 1922 Hahn proved the Uniform Boundedness Theorem using the method of "the gliding hump", but in 1927 Banach and Steinhaus used *Baire Category Theory* to give a much simpler proof of the theorem. In 1929 Banach using Baire Category Theory again proved both the Open Mapping Theorem and the Closed Graph Theorem and included these in his book of 1932. These results provided convincing evidence of the great power of the axiomatic study of abstract normed linear spaces.

The journal *Fundamenta Mathematicae* specialising in point-set topology was first published in 1920 and soon gained international status. Under the impetus of the analysts of Lvov, the journal *Studia Mathematica* was founded in 1929 specialising in functional analysis.

In 1932 Banach published his book, *Théorie des opérations linéaires*, containing a comprehensive account of the state of the theory of normed linear spaces up to that time. The book had great appeal and became a landmark foundational text. Mathematicians began to see the power of the methods and to apply them. In many areas of research normed linear space terminology began to be used. Soon the theory of Banach spaces became a vital study for intending research mathematicians.

The development of the analysis of normed linear spaces was greatly influenced in the 1930s by the group of mathematicians associated with Stefan Banach at Lvov. The friends became known as the *Scottish Café group* because they habitually gathered at the coffee house of an afternoon and worked into the evening setting and solving problems related to their latest research. Eventually in 1935 a large notebook, 'The Scottish Book' was bought to enter problems with rewards and solutions. By the time of the German invasion of Eastern Poland in 1941 the book had entries for 193 problems. The book was hidden during the war and in 1956 was translated and printed. Some of the problems are still unsolved and some have remarkable tales relating to their solution. Problem 153, posed by Stanislaus Mazur in November 1936 was shown by Alexander Grothendieck in 1955 to be an equivalent to the Approximation Problem and is related to the Basis Problem. These problems were solved in the negative by Per Enflo in 1972 and Mazur, a surviving member of the original Scottish Café group, presented him with the designated prize, a live goose.

With hindsight it seems incredible that the identification of Hilbert function space and Hilbert sequence space did not lead sooner to an axiomatic definition of an inner product space. However, this eventuated only after the axiomatic development of metric spaces by Fréchet and normed linear spaces by Banach and the axiomatic method had achieved some degree of acceptance. A stimulus to the study of Hilbert spaces came in 1923 from the newly emerging quantum theory when it was realised that spectral theory begun by Hilbert could be used as a mathematical instrument to study quantum mechanics.

In 1929, John von Neumann presented an *axiomatic approach to Hilbert space* and operators on that space. His aim was to develop spectral theory for classes of operators on this abstract space. His axioms are for a separable space. Since all separable Hilbert spaces are isometrically isomorphic to ℓ_2-space, this might explain why the abstract generalisation took so long to formulate.

In three seminal papers von Neumann developed his spectral theory of operators intrinsically from the axioms defining Hilbert space. The notions of a densely defined operator and of an operator with closed graph were particularly useful for quantum theory. He elucidated a complete spectral theory for normal operators. He formulated the spectral analysis of an operator and determined spectral properties for different types of operators. In short, von Neumann essentially established spectral theory for operators on Hilbert space as it stands today. Von Neumann began the serious study of the algebra of operators on a separable Hilbert space in his second paper on spectral theory. He introduced the weak topology on the algebra of operators and, in collaboration with Francis Murray in 1935, he began to investigate what have been called von Neumann algebras.

Again it is surprising that it was not until 1941 that the study of *abstract normed algebras* was begun by Izrail Gelfand. As with the structure specified for a normed linear space there is an axiomatic prescription of a continuity condition associated with multiplication. He extended spectral theory to elements of normed algebras and, in collaboration with Mark Naimark, began to study algebras with an involution and B*-algebras. The famous Gelfand–Naimark Theorem of 1943 establishes that every unital B*-algebra is isometrically * isomorphic to a subalgebra of the algebra of operators on some Hilbert space and this paved the way for a new interpretation of Hilbert's spectral theory.

§4. Two later significant advances

(i) *Variational principles*

The Bishop–Phelps Theorem published by Errett Bishop and Robert Phelps in 1961 has been generalised out of all recognition to a variety of forms with remarkably wide applications especially in the theory of optimisation. Bishop and Phelps themselves generalised their result to show that for a closed bounded convex set K in a Banach space the support points of K are dense in the boundary of K and the continuous linear functionals supporting K are dense in the dual. A generalisation was given by Arne Brøndsted and Tyrrell Rockafellar in 1965 concerning the subdifferentiability of proper lower semicontinuous convex functions on a Banach space. The outstanding generalisation called Ekeland's Variational Principle proved by Ivar Ekeland in 1964 concerns the points of minimality of real valued proper lower semicontinuous functions on a complete metric space.

The original proof of the Bishop–Phelps Theorem used Zorn's Lemma but subsequently a proof was given using only the completeness of the space. In fact Francis Sullivan in 1981 showed that the Ekeland property actually characterises completeness of the metric space.

The applications of the Ekeland Variational Principle were surveyed by Ivar Ekeland in 1979. The Principle can provide a proof to various fixed point theorems, it is of considerable use in non-convex optimisation situations and it has application in differentiability theory on Banach spaces.

In 1987, Jonathan Borwein and David Preiss produced what they called a Smooth Variational Principle which gives an improved result when the norm of the Banach space has a differentiable norm. A real valued function ψ on an open subset A of a normed linear space X is said to be *Gâteaux differentiable* at $x \in A$ if

$$\lim_{\lambda \to 0} \frac{\psi(x+\lambda y)-\psi(x)}{\lambda}$$

exists and is continuous and linear in y for all $y \in X$, and is *Fréchet differentiable* at x if also this limit is approached uniformly for all $y \in S(X)$. They showed that if a Banach space X has a Gâteaux (Fréchet) differentiable norm at each point of S(X) then every lower semicontinuous function ψ on X is Gâteaux (Fréchet) subdifferentiable at the points of a dense subset of its domain. If ψ is convex then it is Gâteaux (Fréchet) differentiable at such points.

A real valued function ψ on an open subset A of a normed linear space X is said to be *locally Lipschitz* and $x \in A$ if there exists $K > 0$ and $\delta > 0$ such that

$$|\psi(y) - \psi(z)| \le K \|y-z\| \quad \text{for all } y, z \in B(x; \delta).$$

In 1990, David Preiss proved the technically deep result that a locally Lipschitz function on an open subset of an Asplund space is Fréchet differentiable at the points of a dense subset of its domain.

(ii) *Asplund spaces*

A continuous convex function on an open convex subset of a Euclidean space is Fréchet differentiable at the points of a dense G_δ subset of its domain. In 1933, Mazur showed that a comparable Gâteaux differentiability property holds for every continuous convex function on an open convex subset of a separable Banach space. The examination of such differentiability properties for continuous convex functions was taken up again by Edgar Asplund in 1968 and interest in the topic was renewed. In particular, he showed that Mazur's result could be strengthened to Fréchet differentiability for a Banach space with separable dual. Nowadays, a Banach space which possesses the Euclidean space property for Fréchet differentiability is called an *Asplund space.*

The study of this class of spaces gained significance because of the duality between Asplund spaces and spaces with the Radon–Nikodỳm property. For some time there had been investigation into the possible extension of the classical Radon–Nikodỳm Theorem to vector valued measures. In 1967 Mark Rieffel discovered that the extension was possible for vectors from a Banach space which exhibited a special geometrical property. Following a major contribution by Hugh Maynard, in 1973 a geometrical characterisation of the Radon–Nikodỳm property was achieved by Bill Davis and Robert Phelps and independently by Bob Huff. That geometrical property had been linked to the differentiability property defining an Asplund space by Isaac Namioka and Robert Phelps in 1975 and in the same year Charles Stegall proved that a Banach space is an Asplund space if and only if its dual has the Radon–Nikodỳm property.

The Preiss result that a locally Lipschitz function on an open subset of an Asplund space is densely Fréchet differentiable shows that such spaces are particularly significant for optimisation theory.

Principal references:

Garrett Birkhoff and Erwin Kreysig,
The establishment of functional analysis, Historia Mathematica II, (1984), 258–321.
Jean Dieudonné,
History of functional analysis, North Holland, Amsterdam. Mathematics Studies 49, 1981.
Robert Phelps,
Convex Functions, monotone operators and differentiability. Springer Verlag, Berlin.Lecture Notes in Mathematics 1364, 2nd edn, 1993.

LIST OF SYMBOLS

(page where first introduced)

LIST OF SPACES
(page where first introduced)

INDEX